Enriching the Earth

Also by Vaclav Smil

Enriching the Earth
Fritz Haber, Carl Bosch, and the Transformation of World Food Production

Vaclav Smil

The MIT Press
Cambridge, Massachusetts
London, England

This book was set in Sabon by Achorn Graphic Services, Inc., and was printed and bound in the United States of America.

Printed on recycled paper.

Library of Congress Cataloging-in-Publication Data

Smil, Vaclav.
  Enriching the earth: Fritz Haber, Carl Bosch, and the transformation of world food production / Vaclav Smil.
    p. cm.
  Includes bibliographical references.
  ISBN 0-262-19449-X (hc: alk. paper)
  1. Nitrogen fertilizers. 2. Ammonia as fertilizer. I. Title.
S651.S56   2000
631.8′4—dc21
                                                                    00-026291

Cognitio contemplatioque naturae manca quodam modo atque inchoata sit, si nulla actio rerum consequatur.

Knowledge and the study of nature would somehow be weak and incomplete if it were not followed by practical results.

Marcus Tullius Cicero, *De officiis,* I(153)

Contents

Acknowledgments

As is always the case with interdisciplinary books, my most obvious debt is to hundreds of experts—in this case mostly chemists, chemical engineers, historians, agronomists, and fertilizer, plant, and environmental scientists—whose work I have consulted, used, and cited in this book.

Special thanks to Kristina Winzen of the BASF corporate archive in Ludwigshafen for making available the company's records on the early history of nitrogen fixation; to Amitava Roy (president and CEO) and Jean Riley (senior librarian) of the International Fertilizer Development Center (IFDC) in Muscle Shoals, Alabama, for access to the center's excellent library; to Rick Strait, Director for Fertilizers & Synthesis Gas Chemicals of Kellogg Brown & Root in Houston, for information on advanced ammonia synthesis; to Svend Erik Nielsen, Ammonia Technology Supervisor of Haldor Topsøe in Lyngby, for publications on the company's ammonia synthesis; to Patrick Luciani, Director of the Canadian Donner Foundation in Toronto, whose grant covered the cost of research trips to Ludwigshafen and Muscle Shoals; and to E. T. York (Chairman of the IFDC Board) and Donald R. Waggoner for reviewing the typescript.

Transforming the World

What has been the most important technical invention of the twentieth century? Airplanes, nuclear energy, space flight, television, and computers are the most common answers. Yet none of these inventions has been as fundamentally important as the industrial synthesis of ammonia from its elements. Lives of the world's 6 billion people might be actually better without Microsoft Windows and 600 TV channels, and neither nuclear reactors nor space shuttles are critical determinants of human well-being. But the single most important change affecting the world's population— its expansion from 1.6 billion people in 1900 to today's 6 billion—would not have been possible without the synthesis of ammonia.

In order to understand the significance of the link between the world's population growth and the synthesis of a pungent, colorless gas composed of one nitrogen and three hydrogen atoms (appendix A), it is necessary to appreciate first the *yin-yang* nature of nitrogen's biospheric presence. The element is abundant in the biosphere, making up almost 80% of the atmosphere's volume, yet its usable forms are scarce; and although living organisms need it in only small quantities, its shortage is commonly the most important factor that limits both crop production and human growth. This paradox arises from nitrogen's exceedingly stable atmospheric presence as a nonreactive N_2 molecule—and from the paucity of natural ways of transforming this recalcitrant dinitrogen into reactive compounds.

The bulk of living matter is made up of polymerized sugars or alcohols (cellulose, hemicellulose, and lignin) organized in wet tissues. Carbon, making up nearly half of these compounds, is the principal structural element of life, the supplier of quantity: in the biosphere it is about 100 times more abundant than nitrogen, which is a key provider of quality. Although relatively scarce, nitrogen is present in every living cell; in chlorophyll whose excitation by light energizes photosynthesis (the biosphere's most important conversion of energy); in the nucleotides of nucleic acids (DNA and RNA), which store and process all genetic information; in amino acids, which make up all proteins; and in enzymes which control the chemistry of the living world. Consequently, Arthur Needham did not exaggerate the element's importance when he wrote that "every vital phenomenon is due to some change in a nitrogen compound and indeed in the nitrogen atom of that compound."[1] Nitrogen's abundance in plants cannot be missed: it is the nutrient responsible for the vigorous vegetative growth, for the deep green of the leaves, for their large size and delayed senescence, as well as for the size and protein content of cereal grains, the staples of mankind. Nitrogen deficiency cannot be missed either: pale green or yellowing leaves, slow and stunted plant growth, low yields and depressed protein content of seeds.

Nitrogen's importance for human beings is no less critical. We have to ingest ten complete, preformed essential amino acids in order to synthesize our body proteins needed for tissue growth and maintenance. Stunted mental and physical development are the starkest consequences of protein malnutrition. Agricultural crops supply almost $9/10$ of those essential amino acids in food proteins, directly in staple cereal and legume grains, indirectly in animal foods; the remainder of dietary proteins comes from foods derived from grassland grazing and from aquatic species. But while photosynthesis draws readily on fairly small stores of CO_2 in the atmosphere to get the needed carbon, atmosphere's huge dinitrogen store is highly nonreactive, and availability of reactive nitrogen is almost always the most important factor limiting the yields in intensive agricultures.

N_2 molecules must be split into their two constituent atoms before they can be incorporated into an enormous variety of organic and inorganic compounds. Lightning is the only physical process that can fix substantial amounts of nitrogen, that is, to split the tightly bound molecule so the freed atoms can form reactive compounds. Although we cannot pinpoint the annual flux of this high-energy fixation, its global contribution clearly falls far short of agriculture's nitrogen needs. And there is only one group of living organisms capable of nitrogen fixation: about 100 bacterial genera, most notably *Rhizobium* bacteria associated with the roots of leguminous plants.

Not surprisingly, agricultures have been preoccupied with manipulating the flows of reactive nitrogen. Traditional farming relied on a combination of increasingly intensive recycling of organic wastes and cultivation of leguminous plants, but these inputs were insufficient to sustain high crop yields over large cultivated areas. As the concerns about future nitrogen shortages intensified during the latter half of the nineteenth century, chemists tried to prepare ammonia, the simplest of all nitrogen compounds, from its elements.

Industrial synthesis of ammonia thus does not belong to that fascinating class of serendipitous inventions: the compound just did not appear by accident in somebody's laboratory. Its synthesis from nitrogen and hydrogen was sought for more than 100 years, and by the beginning of the twentieth century it became one of the holy grails of synthetic inorganic chemistry. We can pinpoint—much like with Edison's lightbulb or the Wright brothers' flight—the time of the decisive breakthrough. Archives of Badische Anilin- & Soda-Fabrik (BASF) in Ludwigshafen contain a letter that Fritz Haber, at that time a professor of physical chemistry and electrochemistry at the Technische Hochschule in Karslruhe, sent on July 3, 1909, to the company's directors. In it he described the events of the previous day, when two BASF chemists came to witness the first successful demonstration of the synthesis in his laboratory.

Haber's invention was translated with an unprecedented rapidity into a commercial synthetic process by Carl Bosch. The Haber–Bosch process was the breakthrough that removed the most ubiquitous limit on crop yields, opening the way for the development and adoption of high-yielding cultivars and for the multiplication of global harvests. Today's ammonia synthesis has been greatly improved in many details and it operates with much higher energy efficiencies; but Haber and Bosch would easily recognize all principal features of their invention. Global output of ammonia is now about 130 million tonnes (that is, about 110 million tonnes of fixed nitrogen), and $4/5$ of it go into fertilizers, of which urea is by far most important.

Rich countries could fertilize much less by cutting their excessive food production in general, and by reducing their high intakes of animal foods in particular—but even the most assiduous recycling of all organic wastes and the widest practical planting of legumes could not supply enough nitrogen for land-scarce, poor, and populous nations. All the children to be seen running around or leading docile water buffaloes in China's southern provinces, throughout the Nile Delta, or in the manicured landscapes of Java got their body proteins, via urea their parents spread on bunded fields, from the Haber–Bosch synthesis of ammonia. Without this synthesis about $2/5$ of the world's population would not be around—and the dependence will only increase as the global count moves from 6 to 9 or 10 billion people.

Without synthetic ammonia today's global population would not stand at six billion people, but the process also raised the human intervention in the nitrogen cycle to an entirely new level. Ammonia is now one of the two most important synthetic compounds, and the ammonia industry is comprised of hundreds of mostly large plants that feed subsequent syntheses of various nitrogen fertilizer compounds. An adult contains about 1 kilogram of nitrogen in body tissues, but in many countries annual fertilizer applications now prorate to more than 50 kilograms of nitrogen per capita (the global mean is about 13 kilograms).

While the world's population stores no more than about 6 million tonnes of nitrogen (less than one-billionth of the atmosphere's enormous nitrogen stores), in order to maintain, and to gradually expand, this negligible reservoir, the reactive nitrogen in synthetic fertilizers is now perhaps equal to half of the total fixed by all bacteria in all natural terrestrial ecosystems. Moreover, in every intensively cultivated region, and especially where large-scale agriculture neighbors cities and industry, inputs from human activity are greatly surpassing natural flows of fixed nitrogen. On smaller scales, from local watersheds to individual fields, they dominate natural flows by more than an order of magnitude.

Transformations of these nitrogen inputs intensify the rates of microbial processing and increase the atmospheric emissions and deposition of nitrogen compounds and their leaching to fresh and coastal waters. We are thus fertilizing not just fields but, indirectly, also many natural ecosystems—and interfering in nitrogen's natural flows to a much higher extent than in the case of other biospheric cycles.

The fascinating story of nitrogen thus has many facets, many beginnings, and many consequences: traditional agronomy and modern biochemistry, ancient knowledge of valuable legumes, the creation of new industry at the beginning of the twentieth century, the quest for food self-sufficiency and prosperity, and large-scale environmental changes with consequences ranging, literally, from deep wells to the stratosphere.

All of these realities will be taken up in considerable detail in this book. I will first describe nitrogen's unique and indispensable status in the biosphere, its role in crop production, and the traditional means of supplying the nutrient. Then I will concentrate on various attempts to expand natural nitrogen flows by introduction of mineral and synthetic fertilizers. The core of the book is a detailed narrative of the epochal discovery of ammonia synthesis by Fritz Haber and its commercialization by Carl Bosch and BASF.

Subsequent chapters trace the emergence of the large-scale nitrogen fertilizer industry and its various products and analyze the extent of global, and national, dependence on the Haber–Bosch process and its biospheric consequences. I close by looking back—and looking ahead—at the role of nitrogen in civilization. And, in a sad coda to the story, I attach a short postscript describing the lives of Carl Bosch and Fritz Haber after the discovery of ammonia synthesis.

By trying to make the book as comprehensive, and as interdisciplinary, as possible, I recognize that I made its reading harder, or less interesting, for many people who are only curious about some parts of the whole. They should go ahead and create their own books within the book: agronomists can concentrate on chapters 1–3 and 7–9, ecologists on chapters 1–2 and 8–10, historians of technology on chapters 2–6. My hope is that both kinds of readers—those who skip some chapters, and those who persevere and read the whole book—will get new perspectives on how the synthesis of ammonia from its elements came about, and how its diffusion has transformed the world and enriched the Earth.

Enriching the Earth

1

Nitrogen in Agriculture
Discovering the Basics

The uncovering of the complex relations among the atmosphere, plants, and soil could begin only with the birth of modern chemistry, the process dating to the last three decades of the eighteenth century.[1] Its progress was relatively rapid but uneven. Fundamentals of photosynthesis were among the first puzzles to be solved during the 1770s, thanks largely to the ingenious experiments of Joseph Priestley (1733–1804), an English minister, librarian, and private tutor, and Jan Ingenhousz (1730–1799), a Dutch physician.[2] Stated in modern terms (standard chemical nomenclature was codified only a few decades after these experiments took place), atmospheric CO_2 was identified as the source of carbon for photosynthesis, complex carbohydrates its principal products, and O_2 an essential by-product.

Ingenhousz also determined that solar radiation is the energy source of photosynthesis, discovered that the process takes place only in the leaves and green stalks, and was aware of its reversibility (plant respiration releasing CO_2) during the night.[3] He also believed, incorrectly, that atmospheric nitrogen can be directly assimilated by plants. This was denied by Nicholas Theodore de Saussure (1767–1845), a Swiss chemist and plant physiologist, who showed in 1804 that plants can be grown in clean sand and that water can carry to the roots all nutrients, including nitrogen, required for photosynthesis.[4]

In spite of these experiments the idea that humus (a mixture of complex organic compounds resulting from the decomposition of biomass) is the dominant source of carbon, as well as other elements needed by plants, remained highly influential. The humus theory was favored by writings of Daniel Thaer (1752–1828), a Prussian professor of agriculture,[5] and de Saussure himself continued to believe that humus has a key role in plant growth; it was resolutely defeated only two generations later.

Karl Sprengel (1787–1859) and Justus von Liebig (1803–1873) deserve most of the credit for disproving the humus theory.[6] Liebig demonstrated beyond any doubt that plants can grow solely with inputs of inorganic compounds. This meant that plants accelerate the cycling of many elements by transferring them from inorganic stores in the air, soils, and water and concentrating them in plant tissues. But Liebig was on the losing side of another controversy that also took a long time to settle: the debate on the nature of biomass decomposition, the process that returns carbon, nitrogen, and other nutrients bound in complex organic tissues to simpler inorganic compounds that are once again available for assimilation by plants. During the 1830s he maintained that the decomposition of organic matter into acids and alcohols is nothing but a purely inorganic chemical reaction.[7]

Thanks to more powerful microscopes, Theodor Schwann (1810–1882) and Charles Cagniard-Latour (1777–1859) were able to observe a clear correlation between growing *Saccharomyces* yeasts and alcoholic fermentation of grape juice, but Liebig retorted with a mechanistic explanation: atomic motions of the fermenting yeasts were breaking up molecules of grape sugar.[8] The convincing explanation came in 1857 with Louis Pasteur's (1822–1895) demonstration of the microbial nature of organic decomposition, but the process of biomass breakdown was satisfactorily explained only after Hans Buchner's (1850–1902) accidental discovery of the first enzyme in 1897 opened a new era of biochemistry.[9]

Discovering Nitrogen

Advances in the early understanding of intricate transfers of nitrogen among the atmosphere, soils, waters, and living organisms were no less complicated and controversial than was the elucidation of carbon pathways in photosynthesis, respiration, and decay. First realizations that a peculiar gas makes up most of the atmosphere came during the 1770s, when several pioneering chemists discovered nitrogen while trying to decompose the air and investigate various modes of combustion.[10]

Carl Wilhelm Scheele (1742–1786), a Swedish chemist, did his first experiments to prepare oxygen by heating saltpeter (KNO_3) in 1770 (fig. 1.1). He was also the first scientist who realized, sometime between 1771 and 1772, that "the air consists of two fluids differing from each other, the one which does not manifest in the least the property of attracting phlogiston, whilst the other . . . is peculiarly disposed to such attraction."[11] The latter "fluid" (writing in German and Swedish, he called it *Feuerluft* or *elds luft,* fire air) is obviously oxygen, the former one (*verdorbene Luft* or *skamd luft,* spoiled or vitiated air) is nitrogen.

Figure 1.1
Carl Wilhelm Scheele (1742–1786). Courtesy of the E. F. Smith Collection, Rare Book and Manuscript Library, University of Pennsylvania.

Concurrently (in 1772) and independently, Daniel Rutherford (1749–1819), a botany professor in Edinburgh, experimented with mice breathing in a confined volume of air from which the respired CO_2 ("fixed air") was then removed by caustic potash solution. He discovered a residue of *aer malignus* that killed the rodents and extinguished his flame but, unlike CO_2, did not precipitate lime water and was not absorbed by alkalis.[12] Several years later Priestley found that air bubbling from Bath water was, by volume, $1/20$ "fixed air" (i.e., CO_2) and that the rest was "almost perfectly noxious" air.[13]

The element emerged also during extensive combustion experiments performed by Antoine Laurent Lavoisier (1743–1794), the century's greatest chemist, during the 1770s (fig. 1.2). Nitrogen was the "nonvital air" left behind after the air's "vital air" reacted with various tested chemicals.[14] The founder of modern chemistry, who proposed *oxygène* as the new name for "vital air," did not give a new name to "phlogisticated" or "mephitic air." According to Guyton de Morveau, the principal author (with Lavoisier, Berthollet, and de Fourcroy) of *Méthode de Nomenclature Chimique,* published in 1787, it was a group decision. Lavoisier wrote two years later:

Figure 1.2
Antoine Laurent Lavoisier (1743–1794). Courtesy of the E. F. Smith Collection, Rare Book and Manuscript Library, University of Pennsylvania.

The chemical properties of the noxious portion of the atmospheric air being hitherto but little known, we have been satisfied to derive the name of its base from its known quality of killing such animals as are forced to breathe it, giving it the name of *azote,* from the Greek privative particle α, and ζωή, vita. . . . We cannot deny that this name appears somewhat extraordinary.[15]

But the choice appears natural from the point of the element's properties: the gas is odorless, colorless, nonflammable, nonexplosive, nontoxic—and nonreactive under normal environmental conditions. In 1790, four years before Lavoisier's life was cut short by Marat's guillotine, Jean Antoine Claude Chaptal (1756–1832) named the gas *nitrogène.* This designation refers obviously to the element's presence in *nitre* (potassium nitrate), a name whose triple-consonant root goes, via Greek, all the way back to ancient Egyptian.[16] Of course, the French and the Russians still use the old name (*azote* and *azot*), and Germans cling to the onomatopoeic *Stickstoff.*

Claude-Louis Berthollet (1748–1822) had already shown during the 1780s that "all animal substances contain azote, which is readily evolved by the action of nitric acid in the cold."[17] Subsequent rapid advances in organic analysis and in experimental agronomy left no doubt about nitrogen's key presence in the biosphere and about its indispensable role in crop production. Four researchers made particularly outstanding contributions: Jean-Baptiste Boussingault (1802–1887), Justus von Liebig (1803–1873), John Bennet Lawes (1814–1900), and Joseph Henry Gilbert (1817–1901).

Nitrogen in Crop Production

Boussingault had a career as a mining expert and a professor of chemistry in Bogotá and Lyon before he settled in 1836 on his father-in-law's farm in Alsace, which became the site of his experiments on crop rotations, manuring, and sources of plant nitrogen (fig. 1.3).[18] He concluded that the nutritional value of fertilizers is proportional to their nitrogen content.[19] Later he published impressively accurate analyses of nitrogen content of about eighty different organic materials, including all major crops, their residues, animal wastes, and an assortment of organic wastes ranging from dried animal blood to spoiled salted cod (appendix B).[20]

Boussingault was also a good conceptualizer. One of the most eloquent descriptions of nitrogen's key biospheric role appeared in a lecture he co-authored with Jean-Baptiste André Dumas (1800–1884) for the conclusion of Dumas's 1841 course at the *École de Medicine* at the Sorbonne. Inevitably, the text has factual deficiencies. Most importantly, the authors had no inkling of nitrogen fixation by bacteria, the process of splitting the nonreactive molecule of atmospheric dinitrogen and incorporating the element into ammonia (NH_3), which can be used by plants.

Consequently, the authors greatly overestimated the importance of lighting in forming reactive nitrogen.[21] But the lecture's dramatic narrative still reads well more than 150 years later:

As it is from the mouths of volcanoes, then, whose convulsions often make the crust of our globe to tremble, that the principal food of plants, carbonic acid, is incessantly poured out; so it is from the atmosphere on fire with lightnings, from the bosom of the tempest, that the

Figure 1.3
Jean-Baptiste Boussingault (1802–1887). Courtesy of the E. F. Smith Collection, Rare Book and Manuscript Library, University of Pennsylvania.

second and scarcely less indispensable aliment of plants, nitrate of ammonia, is showered down for their behoof. . . . Scarcely are carbonic acid and nitrate of ammonia formed, than a calmer, though not less energetic force begins to act upon them for new purposes: this force is light. By the agency of light, carbonic acid yields up its carbon, water its hydrogen, nitrate of ammonia its nitrogen. These elements combine, organic matters are formed, and the earth is clothed with verdure.[22]

Liebig's contributions were even more far-ranging: during the first half of the nineteenth century no scientist shaped the debate on the chemical foundations of agriculture as much as he did (fig. 1.4).[23] Trained in Bonn, Erlangen, and Paris, Liebig was, together with his lifelong friend Friedrich Wöhler (1800–1882), one of the founders of organic chemistry in general, and of accurate organic analysis in particular.[24] His

Figure 1.4
Justus von Liebig (1803–1873). From the author's collection.

activities ranged from the editorship of the era's leading pharmacology journal to investigating reports of spontaneous combustion of human bodies. Among his more than 2,000 publications are numerous detailed and elaborate analyses of organic compounds as well as broad conceptualizations and grand generalizations.

In agricultural studies Liebig's name remains most commonly invoked because of his law of the minimum: plant growth is limited by the element, or the compound, that is present in the soil in the least adequate amount.[25] Liebig's most famous summary of the understanding of the nitrogen cycle and nutrient needs in agriculture came with the publication of *Die organische Chemie in ihrer Anwendung auf Agricultur und Physiologie*.[26] Eventually there were ten German, six English, and nineteen American editions, as well as translations in all the other major European languages. The text, which grew from 195 pages of the first edition to 1,150 pages twenty-two years later, is a fascinating mixture of lasting truths, exaggerated conclusions, and plainly erroneous assumptions—but, as it turned out, Liebig's errors were no less influential for our understanding of nitrogen's role in agriculture than his enduring generalizations.

Although organic chemistry was at its very beginning, Liebig's explanations of basic chemical processes in the nutrition of plants today need only slight emendations rather than fundamental rewriting. He was one of the first scientists to offer a coherent image of global biospheric cycles:

Carbonic acid, water and ammonia, contain the elements necessary for the support of animals and vegetables. The same substances are the ultimate products of the chemical processes of decay and putrefaction. All the innumerable products of vitality resume, after death, the original form from which they sprung. And thus death—the complete dissolution of an existing generation—becomes the sources of life for a new one.[27]

He recognized the critical importance of nitrogen in farming by juxtaposing agriculture and forestry:

Agriculture differs essentially from the cultivation of forests, inasmuch as its principal object consists in the production of nitrogen under any form capable of assimilation; whilst the object of forest culture is confined to the production of carbon.[28]

After noting nitrogen's "indifference to all other substances, and apparent reluctance to enter into combination with them," he stressed its importance and its incessant biospheric cycling:

Yet nitrogen is an invariable constituent of plants, and during their life is subject to the control of the vital powers. But when the mysterious principle of life has ceased to exercise its influence, this element resumes its chemical character, and materially assists in promoting the decay of vegetable matter, by escaping from the compounds of which it formed a constituent.[29]

By the late 1830s the liberation of nitrogen from complex organic compounds and its return to simpler inorganic forms were well-accepted facts, but the question Liebig posed on the origin and assimilation of nitrogen—"How and in what form does nature furnish nitrogen to vegetable albumen, and gluten, to fruits and seeds?"—remained to be answered. Liebig thought that "this question is susceptible of a very simple solution."[30]

His analyses of animal manures, which contain (especially after prolonged storage) only small amounts of nitrogen, convinced him that their recycling cannot be the most important source of the element for crops. The element's "indifference" led

Liebig to conclude, correctly, that "we have not the slightest reason for believing that the nitrogen of the atmosphere takes part in the process of assimilation of plants."[31] But in his search for the source of the nutrient he erred by ascribing all of the nitrogen available to plants to ammonia present in precipitation. By what seemed to him to be a process of logical elimination he identified the atmosphere as the only source of agricultural nitrogen that was at least balancing, and often surpassing, the amount of the nutrient removed annually by crop harvests: "Whence, we may ask, comes this increase of nitrogen? The nitrogen in the excrements cannot reproduce itself, and the earth cannot yield it. Plants, and consequently animals, must, therefore, derive their nitrogen from the atmosphere."[32]

In the first edition of *Chemistry* he stressed that this supply is not sufficient to maintain good harvests: "Cultivated plants receive the same quantity of nitrogen from the atmosphere as trees, shrubs, and other wild plants; and this is not sufficient for the purposes of agriculture."[33] Three years later, in the book's third edition, he made a complete reversal by dismissing any need for additional applications of ammonia or nitrates and changing the passage to "and this is quite sufficient for the purposes of agriculture." He stressed the point:

It is of great importance for agriculture, to know with certainty that the supply of ammonia is unnecessary for most of our cultivated plants, and that it may even be superfluous, if only the soil contains a sufficient supply of the mineral food of plants, when the ammonia required for their development will be furnished by the atmosphere.[34]

This belief was based on his exaggerated estimates of ammonia typically present in the atmosphere. At that time there were no analytical methods sensitive enough to register ammonia's low concentrations in the atmosphere, but as the compound could be always detected in rain and snow water, and as it is so commonly the final product of organic decomposition, Liebig argued correctly "that it is invariably present in the atmosphere." In the only quantitative example he provided in his book, he assumed that "if a pound of rain-water contains only 1/4th of a grain of ammonia, than a field of 26,910 square feet must receive annually upwards of 88 lbs. of ammonia, or 71 lbs. of nitrogen."[35]

That assumption, equal in metric units to about 35 mg NH_4/L, was much too liberal. Recent worldwide measurements of the chemical composition of precipitation show values ranging mostly between 0.1 and 1 mg/L, with the U.S. continental mean of about 0.3 mg/L.[36] Given the much less concentrated and much less productive agricultural activities of the first half of the nineteenth century, typical concentrations of ammonia in Europe in the 1840s could not have been above today's levels.[37] Extensive monitoring of nitrogen deposition shows typical rates of just a few kg NH_3-N/ha, values generally two orders of magnitude smaller than Liebig's assumed mean, which implies average annual deposition of almost 130 kg N/ha.[38]

Because of the constant supply of atmospheric ammonia, and because even the poorest sandy soils contain relatively large amounts of the compound, Liebig believed that "the soil cannot be exhausted by the exportation of products containing nitrogen (unless these products contain at the same time a large amount of mineral ingredients)."[39] What was not known at that time is that normally only a very small share of that ammonia is readily available to plants, with the bulk of the compound fixed within the crystalline lattices of soil clays.[40] Liebig also erred by believing that analyzing crop ashes provides a reliable guide for the crop's nutritional requirements and for the rate of the needed mineral applications.

Obviously, in contrast to ammonia, numerous mineral elements required for plant growth cannot be supplied from the atmosphere, and so Liebig argued that their provision is the critical challenge in maintaining long-term agricultural productivity. He became convinced that what soils lacked most were the minerals. By analyzing the ash content of crops he divided them into "potash" (beets, turnips, potatoes, corn), "lime" (legumes and tobacco), and "silica" (wheat, oats, rye, barley) plants, and he believed that if appropriate minerals were added according to the kind of crop, then the harvests could continue indefinitely without fallowing.[41]

This rigid adherence to mineral theory left only a marginal place for nitrogenous fertilizers in Liebig's otherwise correct vision of the future agriculture. He thought that ammonia compounds merely facilitate the absorption of minerals by increasing their solubility, but that "A time will come when fields will be manured with a solution of glass (silicate of potash), with the ashes of burnt straw, and with the salts of phosphoric acid prepared in chemical manufactories, exactly as at present medicines are given for fever and goitre."[42]

In 1843 he told British prime minister Robert Peel that "the most indispensable nourishment taken up from the soil is the phosphate of lime,"[43] and two years later he began marketing in England his line of low-solubility mineral fertilizers. These mixes were prepared from plant ashes, gypsum, calcined bones, potassium silicate, and magnesium sulfate. Even those mixtures that contained some ammonium sulfate had too little nitrogen to become commercially successful, and the enterprise, launched jointly with Sheridan Muspratt (1793–1886), an alkali manufacturer in Liverpool, failed after only about three years.[44]

Liebig's incorrect but aggressively held opinion that plants receive all of their nitrogen from the atmospheric deposition and benefit only from mineral fertilizer had a very beneficial effect: it inspired a series of field trials that still continue as the world's longest-running crop experiment. In 1843 John Bennet Lawes (fig. 1.5, an Eton- and Oxford-educated landowner, began experimenting with unfertilized and variously fertilized crops on Broadbalk field in Rothamsted, a family estate in Hertfordshire he inherited in 1834. In the same year he invited Joseph Henry Gilbert, a young chemist with a doctorate from Liebig's laboratory, to help with the design of experiments and to perform the necessary chemical analyses, a cooperative effort that ended only with Lawes's death.[45]

Figure 1.5
John Bennet Lawes (1814–1900). Courtesy of Rothamsted Experimental Station, Harpenden, Hertfordshire, England.

In order to test the validity of Liebig's mineral theory, Lawes and Gilbert began continuous cultivation of wheat on plots receiving either no fertilizer or the following combinations of nutrients: minerals only (P, K, Na, Mg), minerals with nitrogen, or farmyard manure. A decade later there was no doubt that adding only the minerals recommended by Liebig produced no significant increase in grain yields when compared to plots receiving no fertilizer. In contrast, plots receiving combination of minerals and higher rates of ammonium sulfate had yields roughly twice as high as those treated only with minerals, and their yields were also somewhat higher than those produced with applications of farmyard manure.[46]

Unyieldingly, Liebig, in a confidential letter to his friend, wondered how "such a set of swindlers" could produce research that "is all humbug, most impudent humbug."[47] Publicly, he claimed that the Rothamsted experiments actually support his mineral theory, emphasizing again his view that ammonium fertilizer acts merely as a facilitator for the absorption of minerals by increasing their solubility. That was arguing against the evidence. Rothamsted results showed that clover, even when grown in the absence of any ammonia, removed more minerals than did grain crops while acquiring a much larger concentration of nitrogen, and that the succeeding wheat crop withdrew quantities of minerals as large as the previous crop even when it was not assisted by any ammonia fertilizers.

Rothamsted experiments also proved beyond any doubt that applications of nitrogen fertilizers offered the best path to higher grain crop yields; after nitrogen, phosphorus was the most important nutrient in raising the yields. Lawes concluded that it is "hardly possible to have two opinions on the subject."[48] Moreover, in 1853, ten years after Liebig reversed his position on the sufficiency of atmospheric nitrogen, his claim concerning the magnitude of airborne ammonia available to crops was disproved. Boussingault and George Ville (1824–1899) published careful, long-term measurements proving that the amount of ammonia in rain is much less than was assumed by Liebig; and two years after these French studies Lawes and Gilbert demonstrated that the total amount of atmospheric ammonia absorbed by soils is far below the mass of the nutrient assimilated by crops.[49]

Several key facts concerning nitrogen in agriculture were thus clear by the beginning of the second half of the nineteenth century: the element's ubiquitous presence in plant and animal tissues, its indispensability for vigorous plant growth, its constant cycling between organic and inorganic compounds, and beneficial effects of its application on grain crop yields. But many questions remained unanswered, the foremost one being the ability of leguminous crops to thrive without any additions of nitrogen.

Nitrogen and Legumes

In 1838 Boussingault became the first researcher to demonstrate that legumes can restore nitrogen to the soil.[50] He showed that clover could be grown in sterilized sand that could not be a source of any nitrogen, and that clover and peas grown without any additions of fertilizer in open pots of sterile sand could do something wheat or oats were not capable of: they actually increased the sand's nitrogen content by the time of the harvest. The only obvious explanation of these puzzling phenomena was that legumes had an ability to fix atmospheric nitrogen directly—but he could not explain how they could do it. Nor could anybody else for almost another fifty years—and not for any want of trying.

Boussingault's experiments during the 1850s—done with lupins, beans, and oats grown in large closed glass containers, some without an air flow, others receiving a current of washed air enriched with CO_2—showed either no changes or small losses of nitrogen, leading him to conclude that the plants did not assimilate any free nitrogen from the atmosphere.[51] Similarly, an extensive series of experiments done at Rothamsted in 1857 and 1858 failed to show any substantial difference among non-leguminous and leguminous species grown in glass bells in sterilized soil and supplied with purified air and water: both kinds of plants did poorly, and neither could assimilate the atmospheric nitrogen.[52]

But under field conditions legumes clearly did acquire more nitrogen than indicated by the totals derived from seeds, and from nitrogen compounds present in soils and water. A clover plot at Rothamsted showed particularly high nitrogen yields: in 1855 more than 400 kg N/ha were harvested from this plot, and the average for the years 1856–1863 was almost 280 kg N/ha.[53] There had to be another source of nitrogen aside from the measured inputs. Lawes and Gilbert believed that large stores of reactive nitrogen in fertile soils might be the best explanation of the puzzle, as both of them, together with Boussingault, continued to discount the possibility that plants could use atmospheric nitrogen.

They also did not connect the unexplained capacity of leguminous plants with the obvious nodules on their roots (fig. 1.6). During the 1870s and 1880s the nodules were considered to be absorptive roots, storage organs, or products of plant association with fungi.[54] The first hint of a solution came in 1885 when Marcelin Berthelot (1827–1907) noted that nitrogen gas can be absorbed and fixed in an unknown way in uncultivated clay soil in pots—but not when the soil was sterilized: it appeared that something in the soil, rather than something in the plant, could be carrying out nitrogen fixation. But then why would legumes be so much better at it?

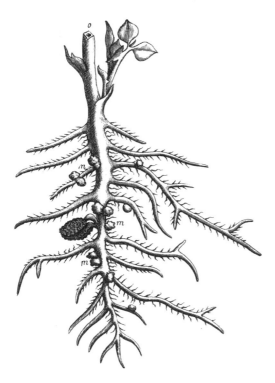

Figure 1.6
An early depiction of root nodules of *Vicia faba* by Malpighi (1679).

The breakthrough came a year later when in September 1886 Hermann Hellriegel (1831–1895) reported to the 59th Conference of German Natural Scientists and Physicians held in Berlin on the experiments he had done together with Hermann Wilfarth (1853–1904) during several months preceding the meeting.[55] These experiments arose from trials done with potted crops over the previous three years. Comparisons of the average weight of seeds and crop residues (cereal straw and legume vines) harvested at the end of the growing period showed that cereal yields responded in an almost perfectly linear manner to additions of nitrogen (as calcium nitrate, $Ca(NO_3)_2$). In contrast, the response of legumes was unpredictable, and unrelated either to investigated species and soil types or to environmental conditions and agronomic practices.

Hellriegel and Wilfarth advanced three hypotheses: that the erratic responses in legumes were caused by an unknown uncontrolled property residing in the soil; that this activity should be transferable; and, if it is of microbial origin, that it should be inactivated by heat. To test their hypotheses, they added either a suspension of raw or sterilized soil to pea plants growing in a sterile medium free of combined nitrogen. Addition of raw soil inoculum increased yields dramatically.

A more elaborate set of experiments followed immediately, comparing the responses of five species of nonleguminous and six kinds of leguminous crops both to additions of calcium nitrate and to soil suspensions prepared by decanting the supernatant from a vigorous mixing of two kinds of fresh topsoil, loamy marl, and glacial sand soil with five times its mass of distilled water. While cereals and oil crops showed a good response to nitrate and none to soil suspensions, legumes responded erratically to nitrate but strongly to added suspensions, with peas, clover, and beans growing better in the presence of loamy soil inoculum.

Potted plants from these persuasive experiments were displayed at the Berlin meeting. There could be no doubt about the microbial origins of nitrogen fixation, and Hellriegel and Wilfarth correctly concluded that nodules, so obvious on roots of vigorously growing legumes, were not just the storage organs of nitrogen but the sites where the fixation took place.

Additional experiments during 1887 confirmed the original conclusions, and in 1888 Hellriegel and Wilfarth published a detailed account of their research whose basic conclusions remain unchanged more than a century later:

The behavior of the *Leguminosae* is fundamentally different from the *Gramineae* in regard to the absorption of their nitrogenous nutrition. The *Gramineae* are solely dependent on the assimilable nitrogenous compounds present in the soil, and their development always stands in direct relation to the nitrogenous supply available in the soil. To the *Leguminosae* a second source is available besides the nitrogen of the soil, from which they are able in a highly efficient manner to cover their needs entirely or in part, in case the first source is insufficient. The free elemental nitrogen of the air is this second source. The *Leguminosae* do not themselves posses the ability to assimilate the free nitrogen of the air, but the active participation of living microorganisms in the soil is absolutely necessary. . . . [I]t is necessary that certain kinds of the latter enter into a symbiotic relation with the former.[56]

Although there was some inevitable initial scepticism, the findings of symbiotic nitrogen fixation in legumes received rapid experimental verification in a number of European and American research centers. The first species of *Rhizobium* bacteria, one of only three genera responsible for symbiotic fixation in legumes, was isolated from pea nodules a year later by Martinus Beijerinck (1851–1931), a Dutch microbiologist.[57] Other rhizobia, and many species of nonsymbiotic, free-living nitrogen fixers, were reported within five years of Hellriegel's discovery. And in 1889 Albert Frank (1839–1900), who initially questioned Hellriegel's report, discovered the first case of nitrogen fixation in cyanobacteria, at that time classified as blue-green algae.[58]

Completing the Nitrogen Cycle

Discovery of bacterial nitrogen fixation was yet another impressive illustration of the critical dependence of the element's biospheric pathways on microorganisms. Less than a decade before Hellriegel's lecture, a convincing experiment demonstrated the bacterial origins of nitrification, conversion of ammonia to nitrate. This reaction is of great importance for all crops because nitrates are much more soluble than ammonia is, and plant roots can absorb them more readily from solutions in soil.

But it was only in 1877 when Théophile Schloesing (1824–1919) designed a convincing experiment proving the bacterial origins of nitrification.[59] All of the ammonia in sewage trickling through a tube filled with sand and chalk was converted to nitrate in a few days. Addition of chloroform to the liquid terminated the process by killing all microbes, but recharging of the tube with a small lump of fresh soil restarted the conversion. But identifying the bacteria responsible for the oxidation proved elusive as long as the researchers used organic media.

Only in 1889, after the Russian microbiologist Sergei Nikolaevich Winogradsky (1856–1953) used an inorganic substrate (ammonium chloride, NH_4Cl), were the two nitrifiers—*Nitrosomonas* and *Nitrobacter*—finally isolated in a pure culture (fig. 1.7).[60] Substrate-specific bacteria of the first genus derive their energy by oxidizing ammonia to nitrite, and the other specialized microbes complete the process by converting nitrite to nitrate. Obviously aerobic while metabolizing, both nitrifiers can survive long periods in anaerobic environments.

Return of the element from nitrate to its huge atmospheric reservoir as a simple dinitrogen molecule was explained shortly before Hellriegel's presentation. In 1885 Ulysse Gayon (1845–1929) and his assistants finally isolated pure cultures of two bacteria that could reduce nitrates, and they named the process denitrification.[61] This stepwise process first reduces nitrates to nitrites; NO and N_2O may be the subsequent intermediate products, and N_2 is its final product. Unlike nitrification, which is dominated by just two bacterial genera, denitrification is carried out by a variety of microorganisms able to use the oxygen in nitrates for their respiration as they feed on many organic substrates. *Pseudomonas, Bacillus,* and *Alcaligenes* are the most common denitrifying genera.[62]

Gayon's fundamental discovery literally closed the biospheric nitrogen cycle. Since that time we have not discovered any other major flows in nitrogen's agricultural cycle—but our research has provided enormous amounts of detailed information concerning every important reservoir of the nutrient and its intricate transfers among

Figure 1.7
Sergei Nikolaevich Winogradsky (1856–1953). From the author's collection.

the atmosphere, plants, soils, and waters.[63] But major uncertainties remain: while our qualitative understanding is rather satisfactory, we are still unable to quantify reliably all but one of the element's large biospheric reservoirs (we know accurately its total atmospheric presence made up overwhelmingly of N_2), and we have to use rather wide ranges when estimating the element's biospheric flows. Figure 1.8 shows all important reservoirs and flows of the biospheric nitrogen cycle centered on agricultural crops.

The atmosphere is nitrogen's largest biospheric reservoir, containing some 3.9×10^{15} t of the gas. Stable N_2 molecules dominate, forming 78% of atmospheric volume. Highly variable concentrations of nitric oxide (NO) and nitrogen dioxide (NO_2), often designated jointly as NO_x, nitrous oxide (N_2O), nitrates (NO_3^-),

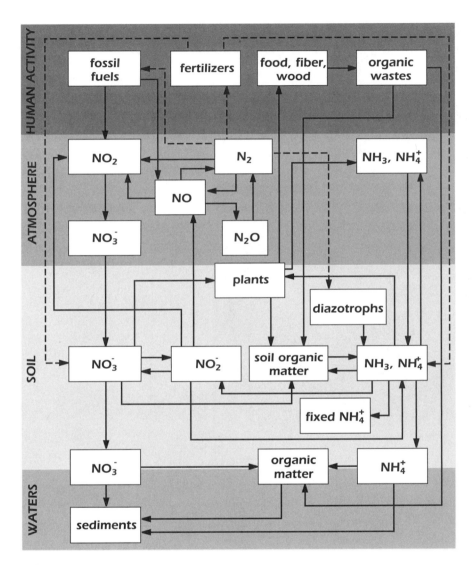

Figure 1.8
Nitrogen cycle centered on crops.

ammonia (NH_3), and ammonium (NH_4^+) are measured merely in parts per billion (ppb) or parts per trillion (ppt). Nitrogen content of soils varies widely, ranging from less than 1,000 kg N/ha in the poorest glacial soils to well over 10,000 kg N/ha in excellent chernozems. But the organic nitrogen, stored in long- and short-lived humus, almost always dominates.[64] The hydrosphere stores very little nitrogen: nitrate concentrations in uncontaminated streams (ammonia is much less soluble) were very low (below 0.1 mg NO_3-N/L), and even today clean rivers generally carry less than 1 mg/L.[65] Because most of the plant tissues are composed of nitrogen-poor polymers, nitrogen stored in crops adds up to a minuscule reservoir of the nutrient compared to the element's store in soils.[66]

Fixation, nitrification, and denitrification are the basic flows of the cycle. Biofixation moves nitrogen from the atmosphere's enormous N_2 stores to NH_3; it is performed by symbiotic and free-living diazotrophs (nitrogen-fixing bacteria), which can sever dinitrogen's strong molecular bond at atmospheric pressure and at ambient temperature thanks to nitrogenase, a specialized enzyme no other organisms carry.[67] *Rhizobium* is by far the most important symbiotic diazotroph forming nodules on leguminous roots. Symbiosis between *Azolla pinnata*, a small, floating, freshwater fern common in tropical Asia, and *Anabaena azollae*, an N-fixing cyanobacterium living in a cavity of the fern's leaflet, is important in Asian wet fields.[68] A recent discovery of endophytic diazotrophs (*Acetobacter, Herbaspirillum*) living inside sugarcane roots, stems, and leaves explains high yields of unfertilized crops after years of consecutive cultivation: these endophytes fix at least 50 kg N/ha.[69]

Among free-living fixers, only cyanobacteria can add substantial amounts of nitrogen to cultivated soils: *Anabaena, Nostoc,* and *Calothrix* are by far the most important genera, enriching soils by as much as 20–30 kg N/ha a year.[70] They fix nitrogen in special heterocysts formed at regular intervals along their filaments. *Oscillatoria* and *Plectonema,* common anaerobic fixers, need no heterocysts. Fixation by other free-living bacteria adds much less nitrogen: reported values for such common species as *Azotobacter* and *Clostridium* are typically less than 0.5 kg N/ha, and the shortage of available organic soil carbon is often the most important factor limiting their growth.[71] Lightning severs nitrogen's triple bond thanks to the high pressures and high temperature of electrical discharges, and the element then forms NO and NO_2. After atmospheric oxidation most of this fixed nitrogen is deposited as NO_3.

Nitrifying bacteria present in soils and waters transform NH_3 to NO_3^-, a more soluble compound plants prefer to assimilate. Assimilated nitrogen is embedded mostly in amino acids that form plant proteins. Heterotrophs (animals and people) are incapable to synthesize de novo a number of amino acids needed to form their body proteins, and they must ingest them in feed and food.[72] Food proteins are required to supply ten of these essential amino acids in children and nine in adults.[73] Naturally, relative protein requirements (g/kg of body mass) are highest during infancy when rapid growth requires plenty of essential amino acids to synthesize new tissues; at that time breakdown rates of proteins will be also high.[74] Ingested nitrogen also becomes a part of nucleic acids and neurotransmitters.[75]

After plants and heterotrophs die, enzymatic decomposition (ammonification) moves nitrogen incorporated in their tissues to NH_3, which, much like the newly fixed nitrogen, is oxidized by bacterial nitrifiers. Denitrification returns the element from NO_3^- via NO_2^- to the atmospheric N_2. There are considerable leakages, detours, and backtrackings along this main cyclical route. Volatilization of ammonia from soils, plant tissues, and animal and human urine and feces (representing a large part of the ingested nitrogen) links plants, soils, and waters with the atmosphere. After a rather short atmospheric residence time ammonia is deposited, in dry form or in precipitation, on plants, soils, and waters.[76]

Both nitrification and denitrification release NO_x (mostly NO) and N_2O from soils. Nitrogen in NO_x returns to the ground in atmospheric deposition, mostly after oxidation to NO_3.[77] In contrast, N_2O is basically inert in the troposphere, but it is a powerful greenhouse gas, absorbing the outgoing long-wave radiation.[78] The high solubility of nitrates means that the compounds leak readily into ground and surface waters, and both organic and inorganic nitrogen in soils can be moved to waters by soil erosion.

Armed with this understanding we can now focus on the ways in which traditional agricultures managed their nitrogen flows. As we can make some fairly reliable estimates for older practices (for which we have few, or no, contemporary quantitative details), and as there is some surprisingly detailed information about more recent ways of preindustrial agronomic management, it is possible to go beyond qualitative descriptions and to make revealing quantitative reconstructions of nitrogen supplies, removals, and limits in traditional agricultures.

2

Traditional Sources of Nitrogen
Preindustrial Agricultures

The two notable achievements of nineteenth century science whose history was traced in the preceding chapter—understanding of nutrient needs in agriculture, and the discovery of principal nitrogen pathways in the element's grand biospheric cycle—validated a number of traditional agronomic practices and showed the direction cropping has to take in order to increase future harvests. Higher crop yields were impossible without higher nitrogen inputs, and agricultural advances could be seen as a continuous quest for higher nutrient supply.

Shifting farmers, the earliest agriculturalists, took no direct steps in order to replenish nutrient supplies: they alternated short periods of crop cultivation with long spans of forest or woodland regrowth, which conserved and restored soil nitrogen. Sedentary agricultures were engaged in progressively intensifying recycling of various kinds of organic wastes ranging from crop residues and animal manures to canal mud and human excrements.

But their most remarkable way of providing additional nitrogen was a nearly universal and highly effective practice of cultivating leguminous species, be it in rotations with cereal or root staples or in the form of intercropping. Interestingly, as we have already seen, the efficacy of this ancient practice was just about the last major puzzle to be explained by the scientific quest for an understanding of the nitrogen cycle.

More than 3/4 of all nitrogen available for crop growth in the most productive traditional agricultures came from managed inputs, and their output was sufficient to feed, albeit on largely vegetarian diets, more than 10 people per hectare of arable land in places where continuous cultivation was possible. Densities of about 5 people/ha were achievable as averages even for large regions or countries in less hospitable climates. These impressive achievements marked the limits of traditional farming: to go beyond them, intensifying agricultures had to find new sources of nitrogen.

Recycling of Organic Matter

Shifting farmers—the earliest practitioners of the least intensive form of agriculture—did not engage in any active recycling of organic matter. Their cropping sequences began with the removal of natural phytomass, a task accomplished by a combination of slashing and burning in forests, or with setting of grassland fires. Depending on the fire's intensity, a large share, or nearly all, of the nitrogen in the original vegetation was lost as NO_x during the combustion, as was a great deal of nitrogen from the surface soil layer. But mineral nutrients liberated by the fire helped to produce a few acceptable harvests.[1]

Subsequent lengthy regeneration—lasting for up to 25–30 years, and getting shorter where higher population densities reduced the availability of land—helped to raise, if not fully restore, nitrogen level in soils protected by renewed plant cover from erosion and from rapid decomposition of organic matter. Although shifting cultivation could support population densities between 0.1 and 0.6 people/ha—an order of magnitude higher than even the relatively most affluent hunting and gathering societies—sedentary cropping could eventually feed ten times as many people from the same area of continuously farmed land.[2]

In contrast to shifting farming, no permanent traditional agriculture could do without the recycling of organic matter. Decomposition and mineralization proceed faster in soils exposed by plowing and concentrations of organic matter, the principal reservoir of naturally accumulated nitrogen, begin falling rapidly with continued cropping (fig. 2.1).[3] Recycling of organic wastes is thus essential to replenish these nitrogen stores. During the second century B.C. Marcus Cato advised farmers to "spread pigeon dung on meadow, garden and field crops. Save carefully goat, sheep, cattle, and all other dung. . . . You may make compost of straw, lupines, chaff, bean stalks, husks, and ilex and oak leaves."[4]

Sung Ying-hsing, writing in 1637 during the last years of the Ming dynasty, listed the fertilizers suitable for rice and even provided an interesting quantitative estimate of the nutritional benefit of unusual soybean recycling:

human and animal excretions, dry cakes and pressed seeds (. . . of these sesame and turnip seed cakes are best . . .), grass and tree leaves. . . . When the price of beans is low, soy beans can be cast into the field, each bean enriching an area of about three inches square; the cost is later twice repaid by the grain yield.[5]

Recycling was particularly important in areas with adequate precipitation or in regions where irrigation either removed or at least alleviated the recurrent threat of

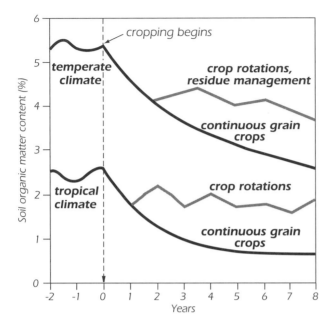

Figure 2.1
Declines of soil organic matter with crop cultivation.

water shortages, which would have been otherwise the most important yield-limiting factor—but its impact was always restricted by competing uses for crop residue as well as by commonly inefficient means of recycling both plant and animal and human wastes.

The return of crop residues was the most obvious option to slow down the extraction of soil nitrogen. In every Old World society in which cereals were the dominant staple, their straws were the largest source of recyclable nitrogen. Although the nitrogen content is fairly low—cereal straws have between 0.4 and 0.9% N, with 0.6% N a common average, and leguminous straws have 0.8–1.4% N (appendix B)[6]— their output was relatively large. In contrast to modern short-stalked varieties the traditional cultivars yielded much more straw than grain (straw/grain ratios for staple grains ranged mostly between 2 and 3).[7]

The simplest way of recycling crop residues is to burn cereal stubble, crop stalks, or vines in the field, a practice still common in all traditional agricultures.[8] As with the burning of original vegetation by shifting farmers, this combustion of crop residues in harvested fields returns all mineral nutrients to the soil in the remaining ash, but it releases from about half (in smoldering, inefficient fires) to more than $9/10$ (in rapid burns) of all nitrogen as NO_x to the atmosphere. Plowing in stubble and any other residues not needed for other uses is always a much more efficient way of nitrogen recycling—but it may actually result in temporary decline of soil nitrogen available to plants.

This is because the C/N ratios of crop residues are commonly above 50, and as high as 150 (only leguminous residues having C/N ratios below 40), much higher than those of fresh leafy phytomass or animal manures.[9] Biomass with C/N ratios below 20 will decompose fairly rapidly, making its nitrogen available for plant growth. In contrast, the decomposers trying to mineralize high C/N residues will require nitrogen from the soil to synthesize their own tissues, temporarily immobilizing the nutrient during the early stages of decay, and hence reducing the short-term productivity of the soil.[10] Most of the immobilized nitrogen will become available weeks or months later after the death of microorganisms that decomposed the residues.

The amount of nitrogen directly recycled in crop residues was relatively small. Recycling only low stubble and roots from a medieval European wheat or barley crop yielding often only 500 kg/ha would have returned as little as 2 kg N/ha, or just 10% of the crop's nitrogen needs. Burning all straw from a good harvest of traditional Southeast Asian rice would have returned, with low-temperature smoldering fire preserving about $2/5$ of all nitrogen, about 10 kg N/ha, no more than $1/5$ of the nutrient assimilated by the crop.[11]

But direct recycling of crop residues was frequently very limited because it was either impossible or impractical to plow most of this waste phytomass back into soils: impractical because the long-stemmed residues would have to be chopped up first before being incorporated into soils,[12] impossible because the residues had a number of competing uses. Cereal and leguminous straws, corn and sorghum stover, stalks and vines of oil and tuber crops were an important source of household fuel— and they still are, particularly in deforested and densely populated alluvial regions of Asia as well as in arid countries of Africa (fig. 2.2).[13] Their use as building material (mostly as straw mixed with clay and for roof thatching) was also very common.

Crop residues have also provided indispensable feed (sometimes from stubble grazing, mostly as stall-fed roughage) and bedding for domestic animals, especially for ruminants whose nutrition requires a minimum share of roughage in order to maintain normal rumen activity.[14] Cereal straws and a few other residues (cotton waste, banana leaves) are also an excellent substrate for cultivation of white button and straw mushrooms,[15] and in some societies residues have been used for making paper.

Figure 2.2
Chinese rice straw dried for feed, fuel, and animal bedding. Photo by V. Smil.

Farmyard Manures

Nitrogen in residues used for feed and bedding was generally recycled as a part of animal manures. The difficulties of quantifying actual rates of nitrogen recycled in manures are due to the high variability of waste output, great fluctuations in its nitrogen content, and often very large nitrogen losses both before and after the application. Manure production and its nitrogen content depend on breed, sex, age, health, feeding, and water intake of animals, as well as on the quality and quantity of bedding materials.

Waste output per head can be estimated either as the share of the nutrient consumed by livestock or as a share of animal's live weight. However, variations of body mass—mostly due to the adequacy and quality of feed, prevalence of diseases, and intensity of field work—make it difficult to estimate typical averages.[16] Proportions of ingested nitrogen excreted in dung and urine range from 75 to 85% for dairy cattle and 65 to 75% for pigs.

Typical daily rates of waste generation in traditional agriculture equaled almost 4% of an animal's live weight for sheep and goats, 5% for pigs, horses, and camels, 6% for poultry, and as much as 9% for dairy cattle. Choosing the most likely means of nitrogen content of animal wastes is even more difficult. Feces rarely have more than 0.75% N, and urine's nitrogen content ranges between 0.6 and 1.4% N. The element's presence in fresh organic wastes is below 1% for all but poultry excreta, but it is highly variable (appendix D).

Measured nitrogen contents for dairy, pig, and poultry manures have two- to threefold ranges, and even greater differences have been found for beef manures.[17] Given the poorer diets of animals in traditional agricultures, it would be sensible to assume that the nitrogen content of their wastes was near the lower range of modern rates, or about 0.5% N for cattle and pig wastes and 1.2% N for poultry excrements. Any wastes produced by animals grazing on permanent pastures will be unavailable for recycling to cropland. Animals grazing on cropland pastures, harvested fields, or cover crops in tree plantations would deposit their wastes directly on the cultivated land. Only manure produced in confinement could be economically gathered and distributed to nearby fields. Shares of animals raised in confinement ranged from 100% for many European draft animals to a tiny fraction of Asia's sheep.

Cleaning the stalls and sties, moving the wastes to temporary storages for fermentation or time-consuming composting of mixed wastes, and eventual transportation of liquid or solid manures to fields and their spreading on the cropland required a great deal of unappealing and heavy labor.[18] Because of the low nitrogen content of most manures and composts (between 0.4 and 0.6% N), massive applications were required if the recycled wastes were to provide a significant share of a crop's nitrogen needs. Most of the labor was actually spent on moving water: it makes up about 65–80% of manures.

Manuring is attested by some of the oldest written sources of antiquity. Odysseus returning to Ithaca found his dog Argus "lying neglected on the heaps of mule and cow dung that lay in front of the stable doors till the men should come and draw it away to manure the great close."[19] Manuring was essential in Roman farming, but because of the limited amount of good pastures and generally low quality of feed, cow manure, the single largest type of animal waste, was considered unimportant, ranking below ass, sheep, and goat wastes. Poultry manure was deservedly seen to be the most valuable organic waste.[20] At the very beginning of the common era Lucius Junius Columella, whose estate was near Gades (Cadiz) in southern Spain, advised readers to apply between 18 and 24 manure loads (*vehes*) per *iugerum*, equivalent to about 14–18 t/ha.[21]

Manuring eventually reached the highest intensity in Western Europe and in parts of East Asia. In China annual field applications of composted manures commonly surpassed 10 t/ha, with small farms in the southwest averaging almost 30 t/ha.[22] Typical rates in prerevolutionary France were around 20 t/ha; two centuries later the recommended applications in Iowa's corn fields were almost identical.[23]

Volatilization and leaching cause considerable losses both during often lengthy storage—European manures were recycled only after several years of storage in mostly unprotected manure heaps[24]—and after field spreading, especially when the manures were applied during the fall and winter and on bare, sloping land.[25] These losses reduced the original nitrogen content by at least half, and they would often reach two-thirds, or even more than four-fifths, of the initially voided nitrogen.[26] Nitrogen from recycled manures actually available to crops could thus amount to no more than the nutrient in atmospheric deposition (just a few kg N/ha)—or it could be the single largest nitrogen input in traditional cropping, providing well over 50 kg N/ha a year.

Nitrogen content of human wastes also varies considerably with the quality of average diets, but appreciable amounts of the nutrient were returned to soils in fermented urine and excrements in many traditional societies that did not proscribe handling of these wastes: after all, adults excrete 75–90% of the nitrogen's daily intake in urine. Slightly conservative age-weighted assumptions would be an average daily output of 150 g of feces and 1.2 L of urine. With respective dry matter contents of 25% and 4%, and dry matter's nitrogen shares at 5% and 15%, typical annual output would be about 3.3 kg N/capita.[27]

In traditional Chinese farming 70–80% of all human waste were recycled; huge volumes of night soil were brought from cities and towns, creating a large waste-handling and transportation industry (fig. 2.3).[28] A similar system had evolved after 1650 in Japan: starting around 1649 authorities in Edo (present-day Tokyo) ordered all toilets directly discharging into rivers and moats to be torn down, and by 1656 farmers began hauling off the waste as fertilizer.[29] Western observers commonly commented on the ubiquity and intensity of Chinese and Japanese night-soil recycling.[30] But, as with manure nitrogen, there were considerable volatilization losses

Figure 2.3
Collection of night soil in Chengdu, the capital of Sichuan, in the early 1980s. Only the installation of sewers and modern water-treatment plants has done away with this ancient practice. Photo by V. Smil.

during storage, collection, and application. Only in densely settled communities practicing assiduous recycling could the contributions of human wastes surpass 20 kg N/ha.

A huge variety of other recycled organic wastes contained materials with both very high and very low nitrogen content. Cakes remaining after pressing oil from various seeds could have up to 7% N, while canal, pond, and river mud traditionally spread on fields throughout South China had no more than 0.1–0.3% of the nutrient (appendix B). Market and kitchen wastes, made up mostly of plant remains, were not particularly high in nitrogen, but as they were often available in large quantities they became important ingredients of composts: Zola's exuberant description of their recycling is perhaps the most memorable vignette of organic farming.[31]

Cultivation of Legumes

Of course, recycling of crop residues represented no net additions of nitrogen to the agroecosystems in question, and recycling of animal and human wastes brought net nitrogen supply only to the small extent to which the nitrogen in feed and food came from sources outside the cultivated fields (that is, largely from grazing on permanent pastures, canal banks, or roadsides, from hunting and gathering, and from fishing and aquaculture).

The only means to provide an additional supply of nitrogen in traditional agricultures was to boost the rate of natural biofixation beyond the rate resulting from the presence of leguminous weeds and free-living diazotrophs, that is, through the cultivation of leguminous plants. That otherwise highly disparate cultures on four continents have independently discovered varieties of this effective approach is one of the best examples of evolutionary convergence.

Nor did these discoveries come only after millennia of agricultural experience: benefits of leguminous grains were known since the very beginning of cropping. Archaeological finds place peas and lentils with barley and wheat in the earliest settled agricultures of the Middle East (as early as 10,000 B.C.); by 7000–6000 B.C. both species, together with different vetches, were very common in sites ranging from Jericho to southern Turkey.[32]

Because pulses ripen in the Near East earlier than wild cereals—in late winter and early spring, at the time when hunting societies faced the worst meat shortages—and because their patchy distribution might have eased possible territorial conflicts, legumes might have been actually the earliest domesticated plants.[33] Textual evidence documents beans, broad beans, lentils, chickpeas, and vetch in Egypt's Old Kingdom, and biblical references include Jacob's proverbial pottage of lentils.[34]

The pattern of leguminous cultivation is remarkable for the fascinating regional differences as individual agricultures combined a variety of legumes with cereal (or tuber) staples. In traditional Chinese farming soybeans, beans, peas, and peanuts were rotated with millets, wheat, and rice (fig. 2.4).[35] Indian legume cultivation has

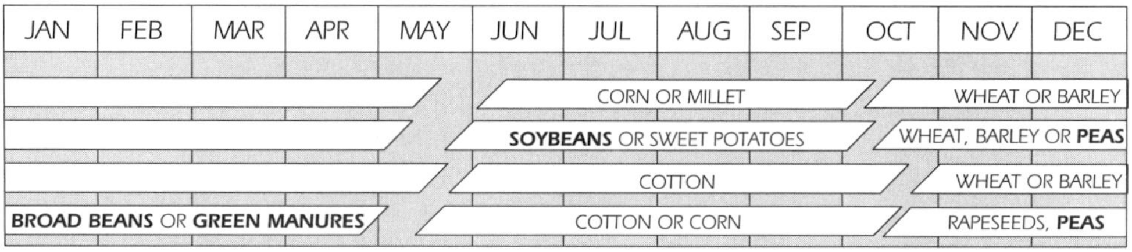

JAN	FEB	MAR	APR	MAY	JUN	JUL	AUG	SEP	OCT	NOV	DEC
						CORN OR MILLET				WHEAT OR BARLEY	
					SOYBEANS OR SWEET POTATOES				WHEAT, BARLEY OR **PEAS**		
						COTTON				WHEAT OR BARLEY	
BROAD BEANS OR **GREEN MANURES**					COTTON OR CORN				RAPESEEDS, **PEAS**		

Figure 2.4
An example of a traditional crop rotation from South China including the cultivation of legumes.

been dominated by lentils, but it also included peas, chickpeas, and pigeon peas in both wheat- and rice-growing areas of the subcontinent.[36]

Peasants of Southeast Asia still cultivate a number of species not widely planted elsewhere, including wing beans, mung beans, and rice beans; sub-Saharan Africa cultivated peanuts, cowpeas, and bambara groundnuts with its staple root crops; and beans and corn were grown by all settled New World cultures.[37] Peas and beans were the European favorites. Intercropping of legumes with cereals or tubers was common in many regions of Africa, Asia, and in both Americas. Rotations of cereals and legumes were perhaps the most remarkable example of optimization in traditional farming as they combined nitrogen enrichment of soils with production of complementary proteins.

Use of legumes as green manure has also a long tradition. This practice incorporates plants of various legume species (mostly cover crops) into soil by hoeing or plowing, or, less commonly, by harvesting the plants and burying them in other fields. The incorporation comes sometime after just 40, but usually after 60–120, days after their planting. In a mild climate several months of their winter growth added enough nitrogen to the soil to produce a good summer cereal crop.

The first written record of Chinese use of leguminous green manures dates to the fifth century B.C., but the practice was almost certainly more ancient.[38] Theophrastus (370–285 B.C.) noted that "the bean best reinvigorates the ground," and it also rots easily, "wherefore the people of Macedonia and Thessaly turn over the ground when it is in flower."[39] Romans adopted and extended this Greek practice. All great sources of Roman agronomic wisdom—Cato, Varro, Pliny, and Columella—extol legumes. Columella correctly recognized that of those legumes "that find favor alfalfa is outstanding for several reasons."[40]

Various species of two clover genera (*Trifolium* and *Melilotus*) and several vetches (*Vicia*) were the most common green manures in Europe. Field peas (*Pisum*), chickpeas (*Cicer*), and lupines (*Lupinus*) were also used. *Astragalus sinicus* (Chinese vetch or *genge*) was by far the leading choice in China and Japan; *Crotalaria juncea* (sunn-hemp), *Sesbania* (*aculeata* and *rostrata*), *Phaseolus trilobus*, *P. mungo* (mung bean), *Vigna* (cowpea), and *Cajanus* (pigeon pea) were favorite choices in the Indian subcontinent and in Southeast Asia, accumulating anywhere between 40 and 160 kg N/ha.[41]

Since the late 1950s monsoonal Asia has added an intriguing form of green manuring by cultivating and distributing floating *Azolla* ferns, which harbor N-fixing cyanobacterium *Anabaena*.[42] Fern stocks are preserved in nurseries for distribution to flooded fields where they could double their mass in just a few days, producing sometimes more than 1 t/ha of phytomass a day. After the plantlets die and settle to the paddy bottom, their mineralized nitrogen (typically anywhere between 10 and 50 kg N/ha) helps to raise rice yields in paddies.

Amounts of nitrogen provided by traditional legumes is almost as difficult to estimate as the contributions of animal manures. Because the nodulation of different legumes requires the presence of particular rhizobial strains in the soil, biofixation rates vary enormously with both abundance and persistence of specific bacteria. Fixation rates also vary with the host's photosynthetic output and with environmental stresses.

Major factors reducing, or even inhibiting, nitrogen fixation include high soil temperature, both too little and too much moisture, low soil O_2 levels, high concentrations of inorganic forms of nitrogen (especially of nitrates), shortages of potassium and of essential micronutrients—particularly molybdenum, vanadium, and iron present in the redox center of nitrogenase—and low soil pH.[43] As a result, we are still unable to offer reliable, representative values of average annual fixation rates even for the most important leguminous cultivars (appendix C).[44]

Conservative values applicable to traditional farming would be at least 100 kg N/ha a year contributed by clover and 150 kg N/ha for alfalfa grown on cropland pastures—and at least as much as that when planted, as was common both in Europe and in Asia, as green manure winter crops.[45] In contrast, fixation rates in common beans in Latin America, the crop's region of origin, could be less than 10 kg N/ha.[46] Conservative rates for common pulses (lentils, peas, chickpeas) would be below 30 kg N/ha, while values for soybeans could be easily twice as high—but only when virtually all crop residues were also recycled (appendix C).[47]

Nitrogen Balances in Traditional Farming

The effects of traditional management of nitrogen supply can be reconstructed from a variety of historic sources. Naturally, the oldest extant information is overwhelmingly qualitative, but ancient Chinese records contain a sprinkling of useful numbers, and more recent sources—ranging from Stamford Raffles's (1781–1826) meticulous accounts of early-nineteenth-century farming in Java to John L. Buck's (1890–1965) comprehensive surveys of cropping in China during the 1920s and 1930s[48]—have a variety of numbers on inputs and crop yields, providing the base for a revealing reconstruction of nitrogen balances.

I have used this information to prepare nitrogen balances for a number of pre-industrial agroecosystems, ranging from ancient single-cropping with extensive fallowing in the Huang He valley to intensive cultivation of China's best subtropical farmlands, which had supported higher population densities than had any other purely organic agriculture.

As long as the managed nitrogen inputs were limited only to recycling of relatively small shares of crop residues or to returns of some straws, stalks, and animal manures, anthropogenic nitrogen inputs in early agricultures were almost always equal to just around 10–15% of the total nutrient supply, and they amounted to less than ¼ of all nitrogen actually assimilated by plants. Such early agricultures could not support more than 2 people/ha of farmland even in regions of fertile soils and adequate water supply.

For example, straw recycling in ancient Egypt returned no more than 20% of all nitrogen required by barley or wheat, and the average staple grain yields could support just 1.5–2 people/ha.[49] Low-intensity recycling of crop residues, manures, and other organic wastes would have provided at best around ⅓ of all nitrogen required by North China's millet crops, and the country's carrying capacity remained below 2 people/ha until the Tang dynasty (A.D. 618–906).[50]

Residue and manure recycling and limited planting of legume crops in Europe of late antiquity and the medieval era managed to supply no more than ⅓ or ⅖ of all nitrogen needed by the continent's staple grain crops.[51] These shares rose substantially only during the eighteenth and nineteenth century when managed inputs accounted for ½–¾ of all nitrogen removed by common three-crop rotations. The main reason for this increase was the diffusion of more intensive farming with its markedly more frequent cultivation of leguminous crops.

The Low Countries aside, perennial legumes began to be widely cultivated as field (as opposed to pasture) crops only around 1760, but by 1880 nearly 20% of all cultivated land in northwestern Europe was sown to legumes.[52] Clover (many species of *Trifolium*) and alfalfa (*Medicago*) dominated, but vetches (*Vicia*) and lupins (*Lupinus*) were also grown. A detailed examination of six centuries of farming in the county of Norfolk provides one of the best illustrations of this great transition.[53]

At around 13% of the total, the county's area sown to legumes remained fairly constant between 1250 and 1740, but by 1836 the rate reached 26.9%, and the composition of leguminous plantings had shifted away from traditional peas, beans, and vetches to clover. While no clover was grown in the county before the middle of the seventeenth century, its plantings were a key component of what came to be known as the Norfolk cycle of rotation (wheat-turnips-barley-clover), and they accounted for about 90% of the area sown to legumes by the 1830s. Doubling of the land under legumes and the shift to clover resulted in at least tripling the rate of symbiotic nitrogen fixation.

Although the modernization of European farming was a multifaceted affair, Chorley concluded that

there was one big change of overriding importance: the generalization of leguminous crops and the consequent increase in the nitrogen supply. Is it not fanciful to suggest that this neglected innovation was of comparable significance to steam power in the economic development of Europe in the period of industrialization?[54]

Given the unmistakeable increases in average yields of staple grains and potatoes, Chorley's comparison is not at all exaggerated: industrialization would have been impossible without population growth, and higher nitrogen supply allowed not only for higher population density per unit of arable land, but also for the slow, but steady, improvement of average diets.

European staple grain yields during the early Middle Ages could not support population densities higher than the pre-dynastic Egyptian average: harvests were often no more than twice the seeded amount. The best available long-term reconstructions of average yields show thirteenth-century English wheat returning between three and four times its seed weight in the harvest, or just above 500 kg/ha; the national mean for 1500 was doubled by the middle of the nineteenth century, largely as a result of the widespread adoption of crop rotation, including legumes, and intensive manuring.[55]

By 1850 the best English and Dutch wheat yields were three times as high as the average medieval harvests, but this increase did not result in a commensurate rise in the average carrying capacity because these higher yields were supporting diets containing much higher shares of animal foodstuffs. This can be illustrated by a particularly detailed nitrogen balance that was reconstructed for a large Dutch grain-and-dairy farm around the year 1800.[56] Of the farm's 55 ha, about 25 ha were in field crops (the most important ones being wheat, barley, beans, flax, and oats), roughly the same amount of grazing land under clover and a grass mixture, with the rest in fallow. About three-quarters of all nitrogen supply (about 100 kg N/ha) came from managed inputs, and about 60% of these inputs were recycled.

Because of its extensive cultivation of legumes (beans, peas, clover), biofixation was the largest nitrogen input (almost 50 kg N/ha), followed by the recycling of manure and animal droppings on pasture (a total of nearly 40 kg N/ha); recycled crop residues added annually about 20 kg N/ha. Such a farm could feed no more than 3–4 people/ha, albeit on diets relatively rich in dairy products and meat.

Perhaps the highest nitrogen inputs in traditional agriculture were reached during the late nineteenth and early twentieth centuries in parts of South China where year-round cropping is possible, particularly in Sichuan's Red Basin and in the lowlands of the southern provinces, especially in Guangdong's Zhujiang (Pearl River) delta.

There the combination of nitrogen inputs—intensively recycled human and animal wastes, regular cultivation of green manures and food and feed legumes, and biofixation by cyanobacteria in paddy fields—provided annually well in excess of 100 kg N/ha of arable land. My detailed reconstructions of nitrogen flows on small Sichuanese and Hunanese farms show that their highest managed inputs amounted typically to between 120 and 150 kg N/ha.[57]

Yet another traditional Chinese agroecosystem managed even higher nitrogen inputs, but as it involved a major component of aquaculture, its performance is not directly comparable with crop-based schemes. The dike-and-pond region in the Zhujiang delta in South China's Guangdong province evolved over many centuries into the world's most productive traditional food production system: carp polyculture in ponds and continuous cropping of sugarcane, rice, vegetables, mulberries, and fruit on dikes were nourished by annual inputs of 50–270 t/ha of organic wastes.

Pig, duck, and water buffalo manure and human excrements were the main ingredients, augmented by aquatic plants and pond mud.[58] Production of 1 t of fish required inputs of 7.5 t of duck manure or 45 t of pig manure, while dike crops received from 35 t/ha (bananas) to 180 t/ha (vegetables) of organic wastes as well as dressings of pond mud.

Organic recycling at rates around 100 t/ha would imply, even with a highly conservative assumption of just 0.4% of average nitrogen content, inputs of more than 400 kg N/ha, able to support average yields of 37 t/ha of fish and plant biomass. Calculations assuming typical yields and half of all area in ponds produce carrying capacities of between 17 and 25 people/ha of land.[59] In reality, one cannot call agriculture here a subsistence form of farming, as most of the harvest, be it fish or cane sugar, has been always destined for sale outside of this exceptionally productive region.

By far the highest recorded applications of manure in history were the foundation of the nineteenth-century practice in Paris of producing salad and vegetable crops in *marais,* suburban plots (about one-quarter of them covered by glass) receiving annually about 1 million tonnes of stable manure from the horses powering the city's transportation.[60] Peak applications in greenhouses reached 1,060 t/ha, and the mean for all *marais* was about 675 t/ha.

Even with 50% losses the mean application would have added annually more than 1,500 kg N/ha; the resulting food production supplied enough protein to feed more than 50 people/ha. But because of the low energy density of vegetables, the carrying capacity in terms of food energy was no higher than 15 people/ha. These are, of course, only theoretical estimates, as no population would survive on a salad-and-vegetables diet.

But such specialized agroecosystems as the Zhujiang delta's dike-and-pond system or the Parisian *marais* were spatially restricted oddities whose incredibly high rates of nitrogen recycling could not be replicated on large scales. Reconstructions of traditional nitrogen balances show that the best among the common continuous cropping systems managed inputs of nitrogen accounting for about four-fifths of the total supply of the nutrient, and that up to two-thirds of nitrogen assimilated by crops was recycled.

With the efficiency of nitrogen utilization ranging between 50 and 60%, managed inputs in excess of 120 kg N/ha could produce 200–250 kg of protein per hectare, enough to feed as many as 12–15 people from a hectare of arable land—providing they were subsisting on an overwhelmingly vegetarian, and often rather monotonous but adequate, diet enriched only occasionally by some animal foods.[61]

But the carrying capacity of 12–15 people/ha could not be a large-scale mean. The mean was lowered even in the best farming areas by the necessity of growing nonfood (above all fiber) crops, and on regional and national scales it was much reduced by climatic imperatives, above all by the limits on multicropping and by inadequate water supply. Consequently, the early-twentieth-century mean for South China was about 7 people/ha, and Buck's surveys indicate that the national average, depressed by northern dryland farming, was about 5.5 people/ha in the early 1930s.[62] Still, this rate was higher than the contemporary mean for Java, Indonesia's most densely populated island, it was at least 40% above the Indian mean, and it came very close to the Egyptian average.[63]

Limits to Recycling and Legume Cultivation

Recycling rates were obviously limited by the availability of suitable biomass. As already noted, even relatively nitrogen-poor crop residues were often unavailable for recycling. Wastes with the highest nitrogen content—poultry manures and oilseed cakes—were produced in limited quantities.[64] Densely populated regions where all farmland had to be devoted to food crops could support only small numbers of large domestic animals and hence could generate only limited quantities of manure; moreover, because of relatively poor feeding such animals produced low-quality dung.[65]

Because of their relatively large shares of farmland under pasture, traditional European agroecosystems could support many more domestic animals, but then only a part of their waste was produced in confinement and was available for recycling to cropland.[66] High labor demands for gathering and distributing (and often also composting) large volumes of low-quality wastes precluded intensive applications on low-value crops or on more distant fields.

Intensive use of nightsoil often carried a high health cost. Even during the late 1980s Chinese researchers found that the prevalence of ascariasis, anchylostomiasis, and trichuriasis among vegetable farmers using untreated nightsoil was, respectively, 94%, 65%, and 93%.[67] And while a few parasite eggs survive on tomatoes or pepper, they are more common on harvested leafy vegetables (spinach, cabbage, mustard). Parasites are also common in fish reared in ponds receiving untreated sewage. Anaerobic fermentation of wastes eliminates most of these health threats—but it is difficult to eliminate oxygen in traditional settings.[68]

Low yields have been the most obvious disadvantage of pulses. When good traditional yields of European staple grains were around 1 t/ha, those of peas and beans were barely half that rate; during the last decades of the nineteenth century the best European wheat yields were between 1.5 and 2 t/ha, but peas and beans yielded less than 1 t/ha.[69] Similar gaps were evident in China, where even the national mean of rice yields was around 2.5 t/ha after 1900, while soybean harvests averaged about 900 kg/ha and were less than 700 kg/ha in northern provinces.[70]

Typical yields of many traditional legume varieties had remained so low that there was no incentive to expand their cultivation in areas of inadequate or marginal food supply experiencing higher population growth. This discrepancy still matters: even a poor cereal crop in the arid Sahelian zone of Africa will yield 1–1.5 t/ha, while traditional legumes may bring less than 500 kg/ha.[71]

Low yields of edible legumes also translated into unfavorable energy returns.[72] In China unirrigated northern wheat returned between 25 and 30 units of food energy in unmilled grain for every unit of food energy needed for field work and crop processing. Higher-yielding rice also required more labor, and its gross energy returns in the early decades of the twentieth century were between 20 and 25 (gross energy returns for southern corn were up to 40, but cornmeal was never a favored food). In contrast, gross energy returns for pulses were rarely higher than 15 and commonly just around 10.

Nutritional considerations have been no less important than low yields and unfavorable energy returns in limiting the expansion of pulse cultivation. Complications in cooking and digesting mature legume seeds have been known since antiquity, as well as the toxicity of some species.[73] But only modern chemical analyses uncovered the frequency of antinutritive factors present in pulses: enzyme inhibitors and lectins (toxins agglutinating red blood cells) are present in relatively high amounts in nearly all pulses, and goitrogens, cyanogens (glycosides releasing HCN), estrogens, antivitamin factors (most notably those causing rachitis and vitamin E antagonists), and toxic amino acids are prominent in some species.[74]

Many traditional cultures learned to live with the poorer digestibility of legumes, and they also discovered how to destroy, or at least substantially eliminate, many antinutritive factors by prolonged cooking and by specific processing. But the resulting inconvenience and the remaining risks accompanying insufficiently modified foodstuffs certainly do not improve the mass appeal of legumes. Consequently, it is hardly surprising that the foods not requiring such elaborate preparation yet providing equal, or better, nutritional value will be preferable. As soon as modernizing societies get richer and increased purchasing power enables people to buy more convenient and less risky foodstuffs, they follow one of the most notable universal nutritional shifts and begin reducing their consumption of legumes. Declining consumption of traditional pulses, mainly beans and peas, has been the worldwide norm (fig. 2.5).[75]

Increased pressure to produce more food on a limited amount of farmland had been restricting the share of land devoted to green manures. Accurate surveys of

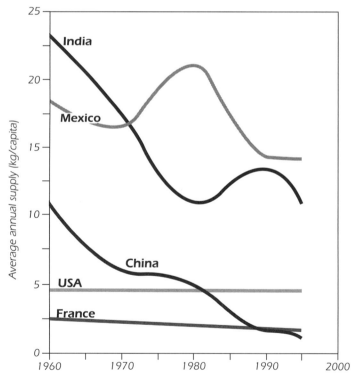

Figure 2.5
Declining consumption of pulses, 1960–1995.

traditional Chinese farming show that green manures had never occupied more than 10% of the planted area.[76] Modern studies of crop-yield response to green manuring show that gains after plowing in 50–60 day-old phytomass can be as low as 30–80 kg/ha per tonne of fresh green matter; the average Indian increase in rice yields was about 240 kg/ha after applications of 5.6 t/ha (dry matter) of green manure.[77]

Such relatively low gains make it impossible to precede every grain crop by 50–60 days of green manure growth: in double-cropping regions those 100–120 days could be used instead to grow a pulse crop, or a nonleguminous crop whose overall yield would be still higher than the phytomass added with the help of the green manuring. The only way to break through this nitrogen barrier was to apply nitrogen that originated outside of agroecosystems. Not surprisingly, the first inputs of this kind—Peruvian guano and Chilean nitrates—were in high demand by farmers eager to boost their yields, providing they were able to pay relatively high prices for these nitrogen-rich compounds.

3

New Sources of the Nutrient
Searching for Fixed Nitrogen

Once the need for nitrogen fertilization was clearly understood, European chemists and agronomists began searching for new sources of the nutrient in order to extend the limited supply provided by traditional recycling of organic wastes and planting of leguminous crops. Imports of guano (solidified bird excrement accumulated on a few arid tropical and subtropical islands) and of South American sodium nitrate ($NaNO_3$) introduced two materials with nitrogen content up to about thirty times higher than that of common barnyard manures. But in just a few decades after many European countries began these imports it became clear that the rising demand had already denuded the richest guano deposits, and that the much more abundant nitrates do not offer the long-term security of a reliable supply.

Rapid exhaustion of rather rare high-quality guano deposits and, as we know now, overly conservative estimates of ultimately available $NaNO_3$ resources led to the search for substitutes. Three options became available during the second half of the nineteenth century: recovery of by-product ammonia from the coking of coal; high-temperature synthesis of cyanamide ($CaCN_2$) from CaC_2 and N_2; and fixation of atmospheric N_2 by electrical discharge, an idea more than a century old given new possibilities by the availability of relatively cheap electricity from large hydrostations.

Only the first of these three techniques eventually contributed relatively large amounts of nitrogen, and none of them solved the problem of long-term provision of fixed nitrogen: they represented welcome additions of fixed nitrogen but not lasting solutions of the nutrient's reliable supply. Not surprisingly, concerns about the prospects of global food production had grown more acute by the turn of the twentieth century. While the global population had grown from 1 to 1.6 billion during the nineteenth century, this rate of growth had been more than matched by the expansion of cultivated land. This combination could not be repeated during the twentieth century: population growth was certain to accelerate in the new century, but it was obvious that increased harvests would have to come largely from intensified cultivation rather than from expansion of cultivated lands.

Guano

Two conditions are needed for the formation of large deposits of guano containing high concentrations of nitrogen: huge colonies of nesting sea birds feeding on abundant fish stocks, and a virtually rainless climate. Of course, bat guanos can be also preserved in caves in both arid and rainy climates, but such deposits are rarely large enough to justify commercial exploitation. The Chincha Islands off the coast of Peru offered the best combination of the two desirable conditions: a nearly rainless climate that allowed the retention of soluble nitrates in the accumulated excreta, and large nesting sites of fish-eating cormorants, pelicans, gannets, and other birds benefiting from high ocean productivity due to the upwelling of the cool, nutrient-rich waters of the Peru Current (fig. 3.1).[1]

Numerous analyses performed during the nineteenth century on guanos from different locations showed a fairly wide range of moisture content (11–25%) and even more variable nitrogen content ranging from about 15% (exceptionally up to about 20%) for the best guanos from arid locations to less than 1% for heavily leached materials from wetter tropical and subtropical sites (appendix E).[2] The element was bound in urates and phosphates, as well as in rather volatile ammonium carbonate, $(NH_4)_2CO_3$.

Because of the carbonate's presence, the nitrogen content of guano deposits often tended to increase linearly with the depth of accumulated layers. All guanos also had relatively high (typically 4–5%) phosphorus content, usually with a smaller, but significant, portion in soluble compounds and the bulk of the nutrient in insoluble phosphates. Guanos were sometimes treated with H_2SO_4 in order to prevent excessive nitrogen volatilization and to increase the solubility of phosphorus: additions of smaller amounts of the acid, diluted with water and often mixed with sand or peat, merely fixed some of the volatile nitrogen in a sulfate, whereas larger additions produced a superphosphate.

Guano was used to fertilize crops along the western coast of South America long before the Inca empire, and the Incas extended the practice to highland crops.[3] Spanish conquerors were impressed by the fertilizer's effects but made no effort to introduce it to Europe. The first small lump of guano was brought to the continent from

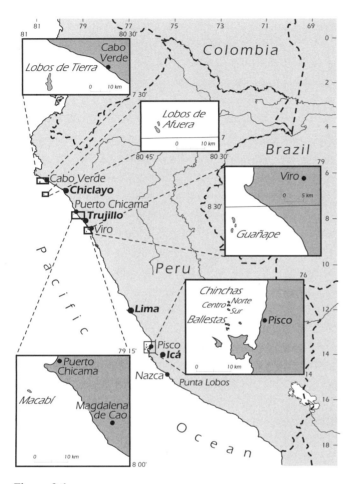

Figure 3.1
Guano islands offshore Peru.

the Chincha Islands in 1804 by Alexander von Humboldt (1769–1859), a famous German naturalist who spent several years at the beginning of the nineteenth century in explorations of South America. The first shipment to the United States was ordered by the editor of *The American Farmer,* who imported two casks of Peruvian guano to Baltimore in 1824.[4]

Antoine-François de Fourcroy (1755–1809), an experienced French chemist and a co-author of the Lavoisier-inspired *Méthode de Nomenclature Chimique,* was one of the researchers who analyzed Humboldt's sample and confirmed its high nitrogen content. The governor of Maryland tested the fertilizer on his farm and found it the most effective manure he had ever applied to corn. But it was only thanks to Liebig's promotion of mineral fertilizers that the European and American commercial imports started. Lima merchants shipped the first trial cargo of twenty casks of Chincha guano to England in 1840. Enthusiastic acceptance of the product by European farmers led the Peruvian government to nationalize its guano resources in 1842. Regular commercial shipments to New York began in 1843, and the following years saw runaway growth of guano exports: by the early 1850s the United Kingdom imported over 200,000 t a year, and U.S. imports totaled about 760,000 t during the 1850s.

The guano mania of the 1850s bore an amazing similarity to the situation in the world crude oil market of the 1970s: rising prices in an oligopolistic market, fears of resource exhaustion, attempts at price controls, the U.S. government getting involved in schemes of armed intervention, American entrepreneurs—taking the advantage of the 1856 Guano Islands Act—rushing to explore and exploit new deposits on tiny islands and reefs in the Caribbean and in the Pacific.[5]

Peruvian guano exports rose rapidly and steadily during the 1850s and 1860s, reaching an annual peak of close to 600,000 t in the late 1860s. The material was first sold by weight, but later the prices took into account guano's variable nitrogen content. The guano deposits of Chincha Islands, which originally contained almost 12.5 Mt of nitrogen-rich (up to 14–15% N) bird droppings, were the first to be exhausted: by 1872, when Peru stopped further exports and reserved limited extraction for domestic use, only about 150,000 t of Chincha guano remained in place.[6]

As Chincha guanos were disappearing, extraction for export shifted to much smaller and poorer Peruvian deposits: after 1869 to the Guañape Islands off Santa (7% N), after 1870 to the Macabí Islands off Malabrigo, and beginning in 1871 to Lobos de Tierra and Lobos de Afuera, two islands much closer to the equator (6.5° and 7° S, respectively) whose guanos were considerably leached. Finally, the extraction shifted to the mainland deposits along the coast of Punta de Lobos south of Pino (15°30′ S) where the material contained at most 5–6% N (fig. 3.1).

Ichaboe Island off the coast of Southwest Africa was the source of the first non-Peruvian guano: by 1843, just three years after their discovery, these deposits (amounting to some 300,000 t) were stripped bare. Later non-Peruvian exports included shipments from Caribbean islands (Sombrero in the Leeward Islands, Navassa west of Haiti, tiny islands off Yucatán and Venezuela), as well as from Bolivia, Brazil, Australia, and the island of St. Helena in the South Atlantic. None of these deposits had as much nitrogen as the Chincha guanos. So-called rock or crust guanos contained hardly any nitrogen, and if not powdery they had to be ground first before being treated with acid.

Combination of limited resources and aggressive extraction made the large-scale exports of high-quality guano only a short-lived phenomenon. Between 1840 and 1870 Peruvian guano was the world's most important commercial nitrogen fertilizer: in 1860 it contained at least five times as much nitrogen as the exported Chilean nitrates, and in 1870 the difference was still more than twofold.[7] Shipments of high-quality guano began declining during the early 1870s, and they were less than half of their peak production by the mid-1880s.

In 1890 the ninth edition of the *Encyclopedia Britannica* still devoted more than a dozen full and closely printed columns to guano; twenty years later the eleventh edition noted that "this material, though once a name to conjure with, has now not much more than an academic interest, owing to the rapid exhaustion of the supplies."[8] But the once burgeoning Pacific guano trade was not over yet: high-nitrogen deposits were gone, but exploitation of inferior guanos, valued largely for their phosphorus, continued. However, the superior competition from newly opened phosphate rock mines in North America and Europe precluded a reprisal of a new guano frenzy.

During the first two decades of the twentieth century the total guano production from the Peruvian upwelling ecosystem was fairly steady at around 500,000 t a year (production from the Benguela upwelling along South Africa's western coast was two orders of magnitude smaller).[9] But because its nitrogen content was at best only a few percent, the Pacific guano production was equivalent to no more than 1.5% of all fixed nitrogen supplied by commercial fertilizers by the end of the first decade of the twentieth century (appendix F).

Sodium Nitrate

Huge deposits of sodium nitrate ($NaNO_3$) were discovered in 1821 by Mariano Eduardo de Rivero, a Spanish (Peruvian) naturalist and chemist, in the coastal province of Tarapacá. Subsequent exploration showed that these deposits exist in widely dispersed nitrate beds (*salitreras*) extending for more than 700 km from about 18°30′ S in Pampa de Tarapacá to 27° S in Pampa de Tactal (fig. 3.2). This whole territory is an extremely arid plateau (rain once in 3–4, or even 8–10, years) situated between the western slopes of the high Cordillera de Los Andes and the Pacific Coastal Range at altitudes of between 1 and 4 km above the sea level, and at distances of about

Figure 3.2
Major areas of nitrate deposits (marked in black) in northern Chile.

25 to 150 km from the Pacific.[10] The crude mineral, *caliche*, is a conglomerate of insoluble material cemented by soluble oxidized salts, including nitrates, sulfates, and chlorides of sodium, calcium, potassium, and magnesium as well as borates and iodates.

Earlier explanations of these huge nitrate accumulations ranged from decay of aquatic plants in inland arms of the Pacific to nitrification and leaching of seabird guano at the margins of saline lakes. In reality, the deposits were laid down not because of any exceptional source of mineral compounds but because the extraordinary aridity and the paucity of nitrogen-using microorganisms and plants created an environment particularly favorable to accumulation of various salts during the late Tertiary and Quaternary eras and to their subsequent preservation.[11]

These ancient evaporites are typically made up of several distinct layers: thin, powdery *chuca* (10–30 cm of silt, sand, and small rocks) on the surface, moderately to firmly cemented *costra* (0.5–2 m of either hard or brittle material), and then the *caliche*, 1–3 m of firmly cemented nitrate underlaid by *coba*, an uncemented regolith (fig. 3.3).[12] Veins and layers of very pure nitrate are common in some areas, rare in others. Nitrogen content of these deposits varied widely: nitrate accounted for as little as 6.5% of the extracted mass in the poorest *caliche*, and as much as 70% in some *salitreras* of Tarapacá. Typical share of $NaNO_3$ in the exploited deposits was initially 40–50%, but by the beginning of the twentieth century it fell to below

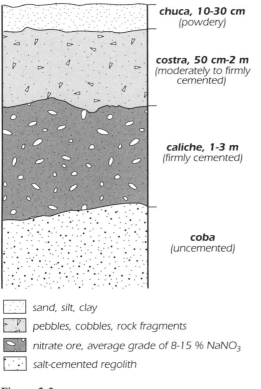

chuca, 10-30 cm
(powdery)

costra, 50 cm-2 m
(moderately to firmly
cemented)

caliche, 1-3 m
(firmly cemented)

coba
(uncemented)

sand, silt, clay

pebbles, cobbles, rock fragments

nitrate ore, average grade of 8-15 % $NaNO_3$

salt-cemented regolith

Figure 3.3
Profile of a typical Chilean *salitrera*.

25%.[13] Pure $NaNO_3$ contains 16% N, roughly 30 times as much as common barnyard manure.

Small-scale extraction was started by a Frenchman, Hector Bacque, in 1826, and exports began in 1830 with shipments of about 8,300 t. Annual exports averaged about 30,000 t in the early 1850s and over 300,000 t by the late 1870s (appendix G).[14] In 1879 the seizure of the Chilean Nitrate Company operating under an agreement at Antofagasta in the Bolivian province of Atacama brought Chile's declaration of war against Bolivia and Peru. The conflict lasted until 1883, and the Chilean victory eventually led to Bolivia's ceding of Antofagasta and Peru's surrender of Tacna and Tarapacá.[15]

The war was very costly to Chile, but the conquest was highly rewarding because the country seized control of all nitrate deposits: customs taxes on exports of Tarapacá and Antofagasta nitrates became the largest source of revenue for the Chilean government. As Hancock concluded a decade later, "it was a desert worth fighting for; it was yielding $10,000,000 a year."[16]

Extraction consisted of removing the *costra* (manually or by blasting) in order to expose the salts whose uneven distribution led to working the deposits in isolated pits or in short trenches rather than in extensive contiguous areas. This kind of mining resulted in very wasteful extraction, with large shares of *caliche* remaining in place. Raw *caliche* was hauled to refineries where it was hand-sorted and coarsely crushed by rollers before being mixed with hot water in order to separate the nitrates. After the mineral was concentrated in tanks and the powdered nitrate spread out to dry, it was packed in large sacks for export through the ports of Iquique and Antofagasta.[17] The final product was composed of 94–96% $NaNO_3$, 1–1.5% NaCl, and 1.5–2.25% water, and it contained about 15% N.

Annual exports began rising rapidly after 1880: they surpassed 1 million tonnes in 1895, reached 1.33 Mt in 1900, and were nearly 2.5 Mt in 1913 (fig. 3.4; appendix G).[18] Although foreign companies initially controlled most of the output (the British ones produced about 60% of all nitrates before the end of the century, and the Chilean stake became dominant only after 1915), a large share of nitrate profits had always remained in the country as taxes accounted for about half of the production cost.[19]

But not all of the imported $NaNO_3$ went to replenish nitrogen in cultivated soils: the compound also became the richest known source of oxygen for use in explosives. Since the invention of gunpowder in the eleventh century, its main ingredient, KNO_3, came either from relatively rare mineral deposits (the easiest ones to harvest

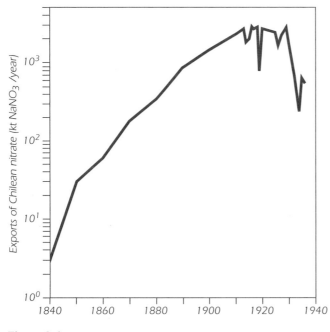

Figure 3.4
Exports of Chilean nitrates, 1840–1940.

were mineral efflorescences on soil surfaces in the Ganges Valley) or from nitrates obtained by bacterial conversion of N-rich organic matter in so-called nitrate plantations.[20]

$NaNO_3$ could be used directly for producing blasting powder (its typical composition being 74% $NaNO_3$, 16% C, and 10% S), or it could be converted to KNO_3 for uses in gunpowder (75% KNO_3, 15% C, 10% S). This conversion relied on potassium carbonate (K_2CO_3) which was initially derived from the aqueous residue of beet sugar molasses (after fermentation and distillation to remove the ethanol); later it was produced by reacting $NaNO_3$ with sylvite, KCl, whose large deposits were discovered in 1857 near Stassfurt in Germany.[21]

Ready availability of rich nitrogen sources was clearly a factor in the invention and commercialization of a variety of modern explosives introduced during the latter half of the nineteenth century.[22] By 1900 the United States consumed nearly half (48%) of its imports of $NaNO_3$ in making explosives. Munroe concluded that

it may, therefore, be safely asserted that but for the discovery and exploitation of the nitrate fields of Chile the explosives industry, as it is known to-day, would have been impossible, and the developments in mining and transportation which have characterized the last half century could not have been made.[23]

As annual $NaNO_3$ shipments topped 1 million tonnes concerns arose about the long-term availability of the resource. But, as is common with ultimately recoverable resource estimates, a widely publicized 1903 forecast—which concluded that at the rate the Chilean nitrates had been mined between 1840 and 1903 their deposits would be exhausted by the year 1938—was too conservative.[24] In 1908 Alejandro Bertrand put the nitrate resources of the Tarapacá and Antofagasta regions at about 220 Mt, and a later detailed appraisal put the country's total probable resources at 920 Mt.[25] If completely exploited, this resource would have been enough to support an annual output of 2.7 Mt, the rate reached just before the beginning of World War I, for 340 years—but those 920 Mt of $NaNO_3$ contain less nitrogen than is now fixed in just two years by the synthesis of ammonia!

Two obvious strategic concerns arose from the world's increasing dependence on Chilean nitrates regardless of the size of the resource: a nation's ability to produce adequate harvests of food and to wage a protracted modern war would be greatly diminished by disruptions of regular nitrate exports or by an outright loss of $NaNO_3$ imports. These fears were particularly acute in Germany. By the beginning of the twentieth century not only was that country the world's largest importer of Chilean nitrates, but it would have also been the easiest target of a maritime blockade, which could deprive it of that crucial import in wartime.[26] Not surprisingly, German efforts to develop new sources of fixed nitrogen were particularly intensive.

By-product Ammonia from Coking

Coal usually contains a small amount of nitrogen (typically between 1 and 1.6%, recorded extremes range from 0.4 to 2.5%) derived from degradation of proteins that were present in the biomass whose eventual transformation by pressure and heat produced the solid fuel. During coal combustion most of this nitrogen is oxidized to NO_x and escapes into the atmosphere. But when coal is heated in the absence of air—during production of coke needed for pig iron smelting, or during generation of coal gas, which was so widely used for lighting in nineteenth-century cities—part of the fuel nitrogen, typically 12-17%, is released as ammonia.[27]

For decades this was an unwanted by-product of coking, as all gases from inefficient beehive ovens, which had been used since the substitution of coke for charcoal began in the United Kingdom after 1750, were simply released into the atmosphere.[28]

Commercial capture of the gas became possible only with the introduction of by-product recovery coking ovens. The first models of by-product recovery coke ovens were introduced in Western Europe during the 1860s and 1870s, and their more efficient versions spread rapidly during the 1880s (fig. 3.5).[29] In contrast, their diffusion in the U.S. iron industry was very slow: in 1905 the Census of Manufacturers showed that less than $1/10$ of all U.S. coal coked was processed in by-product recovery ovens.[30] By far the most common process for recovering by-product ammonia was, and remains, the production of ammonium sulfate, $(NH_4)_2SO_4$ (fig. 3.6).

Raw coke-oven gas is first cooled and washed by countercurrent sprays condensing tar and ammonia liquor. The two substances are separated in a decanter, and the liquor goes into an ammonia still where most of the free ammonia is dissociated by heat, and the fixed ammonia is released by adding calcium hydroxide, $Ca(OH)_2$. Ammonia recovers from the gas in a saturator (originally built of lead-lined cast iron, now of stainless steel) partially filled with an ammonium sulfate solution containing about 5% H_2SO_4. The gas bubbles through the liquid, and the sulfate precipitates in small crystals after the solution becomes saturated.[31] Spray-type absorbers have been also used to recover the gas in diammonium phosphate, $(NH_4)_2HPO_4$.

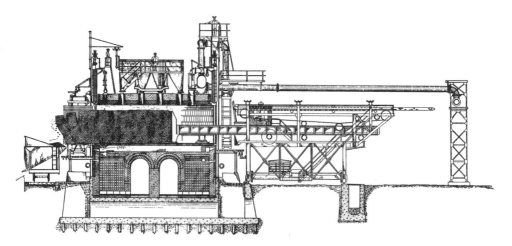

Figure 3.5
Section through a Semet–Solvay by-product recovery coking oven. This was one of the most common types of coking ovens during the late nineteenth and early twentieth centuries. Pusher machine is shown discharging coke into a coke car; gases are recovered at the oven's top. Reprinted, by permission, from H. C. Porter, *Coal Carbonization* (New York: The Chemical Catalog Company, 1924), 174.

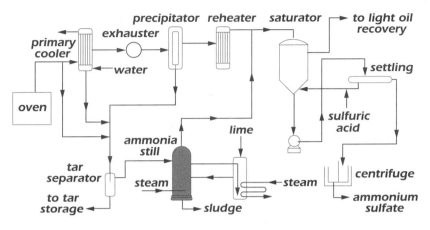

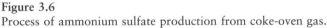

Figure 3.6
Process of ammonium sulfate production from coke-oven gas.

With typical conversions of 15–20% of the original nitrogen into NH_3, ammonia yields in the coke-oven gas are between 2.75 and 3.25 kg per tonne of charged coal.[32] Although a considerable amount of research went into the means to increase the recovery yield, no satisfactory solutions were found. Such a low yield meant that even the potential output of by-product ammonia was relatively limited. Global production of pig iron was almost 40 Mt in 1900, and at that time even the most efficient blast furnaces needed at least 2 kg of coal to smelt 1 kg of the metal.[33]

Consequently, more than 80 Mt of coal a year were used for coking at the beginning of the twentieth century—but even if all of it were processed in by-product recovery ovens the global capture of ammonia would have been equivalent to no more than about 200,000 t N.[34] But given the continuing dominance of beehive ovens in the United States, and the cost of setting up sulfate-making plants at existing coking batteries, the actual output was only a small fraction of the potential rate even in countries that were major producers of metallurgical coke or coal-derived town gas.

In 1905 the U.S. used about 33 Mt of its coal (10% of the total) for coking, a potential source of between 75,000 and 88,000 t of fixed N in recovered NH_3—but the actual output of ammonium sulfate was only about 14,000 t, equivalent of less than 4,000 t N.[35] The United Kingdom was the largest producer of ammonium sulfate until the first decade of the twentieth century: in 1900 it synthesized about 45%

of the global output, with almost half of the total coming from its gas works and only a third from coking ovens.[36]

At the beginning of the twentieth century the global output of nearly 500,000 t of ammonium sulfate contained about 120,000 t N, compared to about 200,000 t N in that year's export of Chilean nitrates. By 1910 German production of ammonium sulfate surpassed the British total, and in 1913 it accounted for nearly 2/5 of the global output, which was equivalent to about 270,000 t N. But the shipments of Chilean nitrates were still larger, providing almost 410,000 t of fixed N (appendix F).[37]

Synthesis of Cyanamide

Synthesis of a cyanide from a carbonate was actually the earliest means of fixing atmospheric N_2 by a chemical reaction in a laboratory: in 1860 F. Margueritte and A. de Sourdeval heated a mixture of barium carbonate, nitrogen, and carbon to 1,200 °C and produced barium cyanide:[38]

$$BaCO_3 + N_2 + 4\,C \rightarrow Ba(CN)_2 + 3\,CO.$$

At lower temperatures and in the presence of water vapor the compound broke down, producing ammonia and barium hydroxide, which could be reused in the synthesis.

In 1895 Adolf Frank (1834–1916) and Nikodemus Caro (1871–1935) substituted barium carbide for the carbonate and reported that the reaction proceeded better with addition of water vapor.[39] In contrast, shortly afterward F. Rothe, employed by Beringer Söhne in Charlottenburg found that the reaction proceeds best with very pure N_2 and in the driest possible conditions. In 1898, after Rothe began working for Frank and Caro in Hamburg, he discovered that cyanamide, rather than cyanide, was the main product. Once it was shown that a much cheaper calcium carbide reacts in much the same way at temperatures between 1,000 and 1,100 °C, it was possible to contemplate a commercial venture based on the reaction

$$CaC_2 + N_2 \rightarrow CaCN_2 + C.$$

An agreement signed in July 1899 in Frankfurt between Frank and Caro and Siemens & Halske established the Cyanid-Gesellschaft. The first field experiments with the product as a nitrogenous fertilizer began in 1901, and in the same year F. E.

Polzenius discovered that addition of calcium chloride ($CaCl_2$) enabled the reaction temperature to be lowered by more than 300 °C. But it produced a highly hygroscopic product that was difficult to store.

The first commercial plant, with an annual capacity of almost 4,000 t N/year, was built in 1905 at Piano d'Orta in the Apennines east of Rome. It was powered by hydroelectricity from a dam on the Pescara River. In 1907 a plant with an annual capacity of 20,000 t a year was completed at Knapsack near Cologne (powered by local lignite), and in 1908 Frank and Caro finished a plant with an annual capacity of 30,000 t of $CaCN_2$ near Trostberg in Bavaria (powered by hydroelectricity generated from the Alz River).

The first step in the commercial process was the synthesis of calcium carbide by fusion of lime and coke, a reaction carried out at high temperature in an electric furnace:

$$CaO + 3\,C \rightarrow CaC_2 + CO.$$

The molten carbide was drawn off, cooled, finely ground, and reacted with nitrogen (prepared by air liquefaction using the recently invented Linde process)[40] in vertical iron retorts heated to about 1,000 °C.

Pure N_2 is necessary to avoid oxidation of the carbide to CaO and CO. The mixture of $CaCN_2$ and carbon was sold as *Kalkstickstoff* (nitrogen chalk): the sintered hard product was milled, and the gray-black powder could be used directly as fertilizer. Pure $CaCN_2$ contains 35% N, but the commercial product was always mixed with 9–13% C and with 20–28% CaO), containing 18–22% N. $CaCN_2$ could be also easily converted to ammonia by reaction with superheated steam:

$$CaCN_2 + 3\,H_2O \rightarrow CaCO_3 + 2\,NH_3.$$

What was generally known as Frank and Caro process thus became the first commercial synthesis of fixed nitrogen, but its high energy requirements—initially more than 200 GJ/t N, and later at least 130 GJ/t N, mostly for the high-quality coke needed to heat the retorts and for electricity required to produce calcium carbide—limited its commercial diffusion. Still, plants were built in a number of European countries, in Japan, and at Niagara Falls in Canada. Total output of cyanamide, mostly in German plants, reached about 100,000 t of $CaCN_2$ in 1913, and in 1918 thirty-five operating cyanamide plants fixed about 325,000 t N.[41]

Electric Arc Process

Transformation of nitrogen compounds by passing electric sparks through air was first noted by Priestley in 1775.[42] A decade later Henry Cavendish (1731–1810) described in detail what happens when the mixture of oxygen and nitrogen is blown through an electric spark: "We may safely conclude that in the present experiments the phlogisticated air [nitrogen] was enabled, by means of the electrical spark, to unite to, or form a chemical combination with, the dephlogisticated air [oxygen], and was thereby reduced to nitrous acid."[43]

But the seemingly simple, and reversible, reaction

$$N_2 + O_2 \leftrightarrow 2\ NO$$

produces very small amounts of the oxide above 2,000 °C, and significant volumes are generated only at very high temperatures above 3,000 °C. Consequently, only a sustained supply of inexpensive electricity could be the basis of a mass-produced commodity. Although there were many additional laboratory experiments with formation of NO during the nineteenth century, particularly during the 1890s, commercial ventures could be contemplated only once relatively cheap electricity became available with construction of larger hydrostations during the last decade of the nineteenth and the first decade of the twentieth centuries.[44]

The first commercial attempt by Charles S. Bradley and D. R. Lovejoy, relying on power from a newly built Niagara Falls hydrostation, was launched in 1902. Their patented method produced more than 400,000 elongated electric arcs per minute inside a furnace made up of a fixed iron shell surrounded by a rapidly revolving cylinder. Nitrogen oxide generated from the passing air enriched with O_2 was converted to nitric acid, but the operation was uneconomical and their Atmospheric Products Company folded in 1904.[45]

High generating heads, obviating construction of large dams, made Norwegian hydroelectricity particularly inexpensive. Kristian Birkeland (1867–1917) and Samuel Eyde (1866–1940), a physicist and an engineer, took advantage of this opportunity by introducing and perfecting a viable commercial generation of nitrogen oxides in intense electric arcs.[46] Their first experimental prototype tested in 1903 at the Christiania (Oslo) University drew just 2.2 kW of electric power, but by May 1905 their Notodden plant had a capacity of 1.875 MW.

An alternating current arc was produced in the Birkeland–Eyde furnaces (their capacity rose eventually to 4 MW/unit) between water-cooled copper electrodes

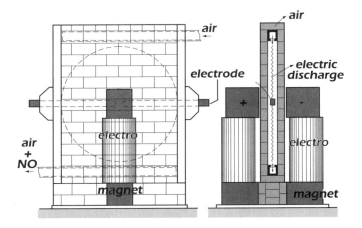

Figure 3.7
Birkeland–Eyde furnace for NO production in the electric arc.

placed between the poles of a strong electromagnet. This arrangement generated a disk-shaped plasma of up to 3 m in diameter and 3,500 °C in which the nitrogen oxides were formed from N_2 and O_2 in the circulating air (fig. 3.7). The enlarged Notodden operation, commissioned in May 1907, used just over 30 MW; the first stage of the Rjukan plant, using electricity generated from a waterfall and completed in November 1911, had the rated capacity of 150 MW, and its second stage, completed four years later, rated almost 250 MW.

In all of these plants highly dilute oxides produced by sparking (the air leaving the chamber had just 1.5–2% NO by volume) had to be cooled rapidly (to 800–1000 °C) in order to prevent the reverse reaction converting the gas back to N_2 and O_2. Then the gases were cooled further and condensed to produce nitric acid, which was reacted with lime to yield calcium nitrate—$Ca(NO_3)_2$, a compound with 17% N—which could be used directly as a fertilizer:

$$2 \, NO + O \rightarrow N_2O_3,$$

$$N_2O_3 + H_2O + 2 \, O_2 \rightarrow 2 \, HNO_3 + O_2,$$

$$2 \, HNO_3 + CaO \rightarrow Ca(NO_3)_2 + H_2O.$$

The process was highly energy-intensive: its initial requirements were in excess of 700 MJ/kg N, while later its most efficient versions still needed between 50 and 75 kWh of electricity per kilogram of nitrogen, or 180–270 GJ/t N.[47] By 1913 the annual output of the three Norwegian plants using the Birkeland–Eyde process reached 70,000 t of calcium nitrate. This was the equivalent of about 12,000 t of fixed nitrogen, a small fraction of either the ammonia recovered from by-product coke ovens (equal to less than 5% of the fixed N derived from that process) or of the Chilean nitrate extraction (equivalent of about 3% of nitrogen in that year's $NaNO_3$ exports; appendix F).[48]

Plant Nutrients and Future Food Supply

The industrializing world was undergoing an unprecedented concatenation of socio-economic changes at the turn of the century, and the rising demand for food and changing diets were among the key ingredients of this great transformation. The steady population increase was obviously the main reason for this change. Between 1850 and 1900 the total population of industrializing countries of Europe and North America grew by two-thirds, from 300 to 500 million.[49] Rapid urbanization had an even greater impact: it was accompanied by rising disposable incomes and by growing female employment in manufactures and services; these changes brought a wide-ranging dietary transition from traditional patterns of eating dominated by grains.[50]

On the most general level, per capita food energy supplies rose during early stages of modernization from preindustrial means of less than 2,500 kcal/day to satiation levels of between 2,800 and 3,000 kcal/day.[51] After an initial period of dietary improvement when the average consumption of staple grains tended to rise slightly, this overall increase of food energy intake included substantial declines in direct consumption of cereals and major gains in the supply of animal foods, and hence also in intakes of animal proteins and lipids. Sugar intakes had also increased appreciably.[52]

Yet this rising demand for food was accompanied by steadily diminishing opportunities for extending the area of land under staple grain cultivation. At the beginning of the twentieth century it was obvious that the next fifty years could not see a replication of that unique feat of the previous fifty years when pioneer farmers converted some 200 Mha of grasslands into grain fields. During that period America's Great Plains and the Canadian Prairies were the sites of the most extensive transformation, followed by the grasslands of Russia, South America, and Australia.[53]

These huge extensions of cultivated land brought much larger grain harvests during the last decades of the nineteenth century when average wheat yields remained fairly low.[54] In 1900 only the best European producers (Denmark, Netherlands, England) harvested more than 1.5 t of wheat per hectare, and they already did so with appreciable help from inorganic fertilizers. Wheat yields in the Austro-Hungarian empire were just around 1 t/ha, in Italy about 800 kg/ha, and rates around 800 kg/ha were also the U.S. mean. With average yields of less than 600 kg/ha, harvests in European Russia were not much higher than was the typical medieval mean in Western Europe.

The global average for the second half of the 1890s was almost certainly below 800 kg/ha, and the single most important step aimed at improving the productivity from a basically fixed amount of land was to increase the supply of the three plant macronutrients. Nitrogen is almost invariably the nutrient required in the largest quantity per unit of a harvested crop. For example, in a good harvest of grain corn there will be nearly four times as much nitrogen as the total mass of potassium and about 4.5 times as much as all forms of phosphorus. Proportions of P and K depend on specific crops and soils, and by the end of the nineteenth century it was clear that a variety of minerals should be able to provide reliable long-term supplies of these two macronutrients. The modern fertilizer industry was launched with the production of phosphatic compounds based on Liebig's idea that bone phosphorus would be more soluble if treated with H_2SO_4.[55]

It was not Liebig, however, but an Irish doctor, James Murray (1788–1871), who became the first commercial vendor of inorganic fertilizers with his 1841 offer of liquid superphosphate made by dissolving bones in H_2SO_4.[56] Two years later Lawes set up a superphosphate factory at Deptford on the Thames. His fertilizer, patented in May 1842 on the same day Murray got his patent, was a virtually identical product, containing "Super-Phosphate of Lime, Phosphate of Ammonia, Silicate of Potass, etc."[57]

The first source of inorganic phosphates were coprolites discovered in Gloucestershire during the 1820s; extraction of much larger deposits of coprolites began a generation later in East Anglia, and these provided Lawes's main supply of raw material until the end of the century. By 1870 ordinary superphosphate—OSP, $Ca(H_2PO_4)_2 \cdot H_2O$—was produced in seventy plants in the United Kingdom.[58] Expansion of the OSP industry stimulated a search for suitable minerals; extraction of high-quality apatite started in 1851 in Norway. The first U.S. rock phosphates came from North Carolina in the late 1860s, but Florida extraction became dominant in 1888.[59]

Depending on the mineral used, OSP contained between 7 and 10% of available phosphorus (8.7% was the standard), an order of magnitude more than most recycled organic wastes.[60] OSP was also a richer source of the nutrient than the basic slag available as a by-product of smelting phosphatic iron ores. This new phosphatic fertilizer, commercially introduced during the 1870s, contained 2–6.5% P. By 1900 the world output of OSP surpassed 4.5 Mt, an equivalent of nearly 400,000 t P, and abundant reserves of phosphate rock obviated any concerns about the adequacy of long-term supply.[61]

Several major crops, including hard spring wheat and rice, actually assimilate more potassium than nitrogen.[62] But because potassium is mostly retained in crop residues rather than in the harvested grain, organic recycling returns a much larger share of the nutrient than in the case of nitrogen or phosphorus.[63] For example, a recent account of total nutrient inputs (organic recycling and inorganic fertilizers) in U.S. agriculture showed that potassium recycled in crop residues and manures contributed almost 55% of all inputs, while the shares for P and N were, respectively, 40 and 30%.[64] Because traditional crops had higher residue/grain shares than modern cultivars, organic recycling could have returned even higher shares of the assimilated nutrient.

Chilean *caliche* contains potassium salts, but a specialized, large-scale potash industry began in 1861 with the exploitation of the Stassfurt deposit discovered in Saxony four year earlier. With the mining infrastructure already in place, only refining and upgrading of the raw ore was needed before the product could be marketed as a commercial fertilizer. German production dominated the emerging world market in potassium fertilizers until the beginning of World War I.[65]

In contrast to abundant supplies of phosphorus and potassium, the availability of inorganic nitrogen remained marginal. In spite of a decades-long search for new sources of fixed nitrogen there were only two options for expanding the nutrient supply by the end of the nineteenth century: imports of Chilean nitrates and the recovery of ammonia from coke ovens. Their combined output in 1900 was equivalent to about 340,000 t of assimilable nitrogen, or on the order of 2% of all nitrogen removed by that year's crops and their residues.[66]

As about 75% of that year's nitrogen inputs, totaling some 30 Mt N, were supplied by recycling of organic wastes and cultivation of legumes, inorganic fertilizers provided no more than about 1.5% of all managed inputs of nitrogen at the beginning of the twentieth century. Another way to illustrate the marginality of this input is to realize that even if the atmospheric deposition at that time would have averaged a mere 2 kg N/ha a year, its worldwide total would have been nearly seven times as large as the supply of nitrogen in inorganic fertilizers.

Moreover, applications of available inorganic nitrogen were highly concentrated in just a few countries. During the last decades of the nineteenth century Europe imported 65–70% of all exports of the Chilean nitrates, with the United States taking at least 20%. In 1900 Germany and the United Kingdom were their largest importers, as well as the leading producers of by-product ammonia from coking.[67] On the other hand, the marginal role of inorganic nitrogen is shown by the American consumption total: in 1900 U.S. imports of Chilean nitrates used in agriculture and recovery of by-product ammonia added up to less than 20,000 t of fertilizer nitrogen, an equivalent of the nutrient incorporated in just 1 million tonnes of spring wheat.[68]

The challenge of finding a reliable, long-term source of fixed nitrogen appeared particularly acute in the United Kingdom and in Germany: besides being, as just noted, the two largest importers of Chilean nitrates as well as the world's largest producers of by-product ammonia, they had already put virtually all of their arable land into relatively intensive production, and in spite of their fairly high wheat yields they were also already large grain importers.[69]

Consequently, it is not surprising that it was an English scientist who made the most eloquent, and a widely publicized, appraisal of long-term food prospects and appealed for a radical shift in securing the world's fertilizer needs. Given the excellence of the country's chemical industry, it is fitting that German chemists found, soon after that call for action, an effective solution of this great challenge.

William Crookes (1832–1919) was a chemist as well as physicist whose work embraced many aspects of both sciences, from identifying a new element (thallium) to pioneering the research on cathode ray tubes (fig. 3.8).[70] Knighted in 1897, he became the president of the British Association, and for his address to the Association's annual meeting in Bristol in September 1898 he chose the topic of the wheat problem. He was aware that some of his statements would sound alarmist ("our wheat-producing soil is totally unequal to the strain put upon it. . . . [A]ll civilised nations stand in deadly peril of not having enough to eat."),[71] but he was actually an optimist who saw his remarks as a warning, not a prophecy, and who pointed out a way out of this existential dilemma. The presentation caused a furor among economists and politicians but elicited much agreement as well as vigorous criticism among agricultural experts. It became one of the best known speeches at the turn of the century.

Figure 3.8
William Crookes (1832–1919). Courtesy of the E. F. Smith Collection, Rare Book and Manuscript Library, University of Pennsylvania.

Crookes's reasoning ("founded on stubborn facts") was straightforward. Larger populations in the twentieth century would need much more wheat; only a limited amount of soils suitable for wheat farming remained uncultivated; in the future the land then under wheat would be better farmed, and hence would produce larger harvests from the identical amount of land. But such improvements could be pushed only so far, and relying on a steady flow of much higher exports by such producers as Russia or Australia might be unwise.[72] Crookes concluded that were the low wheat yields, prevalent during the late 1890s, to continue, the rising demand would result in global wheat deficiency as early as 1930.

Higher yields would inevitably require higher nutrient inputs, and the need for nitrogen would be particularly high: "It is now recognised that all crops require what is called a 'dominant' manure. Some need nitrogen, some potash, others phosphates. Wheat pre-eminently demands nitrogen, fixed in the form of ammonia or nitric acid."[73] But the quantities of the nutrient recovered in ammonia from coking ovens were comparatively small, green manuring has a limited potential in intensive farming, and when applying Chilean nitrates "we are drawing on the earth's capital, and our drafts will not perpetually be honoured."[74] In contrast, the amount of nitrogen in the atmosphere is practically unlimited, and hence the key to solving the world's food challenge was to find the way in which this enormous resource could be tapped to produce fixed nitrogen assimilable by plants.

Near the end of his lecture Crookes offered an eloquent summation why the fixation of atmospheric nitrogen was one of the great discoveries awaiting the ingenuity of chemists:

This unfulfilled problem, which so far has eluded the strenuous attempts of those who have tried to wrest the secret from nature, differs materially from other chemical discoveries which are in the air, so to speak, but are not yet matured. The fixation of nitrogen is vital to the progress of civilised humanity. Other discoveries minister to our increased intellectual comfort, luxury, or convenience; they serve to make life easier, to hasten the acquisition of wealth, or to save time, health, or worry. The fixation of nitrogen is a question of the not far-distant future.[75]

The conclusion was obvious: "It is the chemist who must come to rescue. . . . It is through the laboratory that starvation may ultimately be turned into plenty."[76] But I suspect that even Crookes, who so clearly envisaged this development, was surprised by the speed with which his urgings were turned into a commercial reality.

4

A Brilliant Discovery
Fritz Haber's Synthesis of Ammonia

Major discoveries in science rarely arise de novo, and the admired breakthroughs that may be later rewarded by prizes commonly amount to just a last few steps on a long intellectual journey that could not have been pursued without many earlier, and all too quickly largely forgotten, pilgrims charting the course. This was very much the case with the synthesis of ammonia from its elements: although none of the experiments preceding Haber's efforts resulted in commercial applications, many of them had either contained procedures, or resulted in suggestions, that proved not merely helpful but turned out to be essential for the eventual solution of the challenge.

Haber's success, which came after he abandoned his early experiments for a few years, owed a great deal to pioneering research on gas equilibria that was carried out by Walther Nernst and his pupils, as well as to experiments done in Haber's Karlsruhe laboratory by his colleagues Gabriel van Oordt and Robert Le Rossignol. Haber was well aware that it was his privilege to take those last few steps on a long path to a major discovery. In closing his Nobel lecture he noted that "the synthesis of ammonia from the elements is a result which physical chemistry was bound to reach. The notion of the reversibility of the breakdown of ammonia was already held by Deville, Ramsay and Young, and by 1901 Le Châtelier had already given thought to the effects of temperature and pressure."[1] The value of Haber's creative contributions is not thereby diminished; we merely see once again that modern science is a cooperative enterprise par excellence.

Haber's Predecessors

In 1784, three years before Claude-Louis Berthollet joined Lavoisier, de Morveau, and de Fourcroy in naming the noxious part of the air *azote*, Berthollet became aware that the element joins with hydrogen to make ammonia, and he was even able to establish the approximate ratio of the two constituents.[2] The first recorded failed attempt to combine nitrogen and hydrogen—a simple trial done by Georg Friedrich Hildebrand (1764–1816) that involved mixing the two elements in sealed flasks under atmospheric pressure—came just eleven years later.[3] Nor did the pressure of 3.1 MPa—achieved by submerging a bottled mixture of N_2-H_2 into the ocean in an 1811 experiment—help.

The first attempt at synthesis using high temperature dates to 1788, and more than a dozen chemists published results of similar experiments during the nineteenth century. These included widely cited tests by Otto Erdmann (1804–1869) and Richard Marchand (1813–1850), who passed N_2 and H_2 (or water vapor) over red-hot iron or glowing coals and reported formation of trace amounts of NH_3. The first use of catalysts in ammonia synthesis goes back to 1823, when Johann Wolfgang Döbereiner (1780–1849) experimented with platinum, and by the 1870s many different elements (pure metals) and compounds (mixtures of metals with coal, or asbestos) were tested and even patented. Work published in 1865 by Henri Saint-Claire Deville (1818–1881), who was seemingly successful in fixing NH_3, received particular attention.

A new, and ultimately successful, research direction was launched in 1884 when William Ramsay (1852–1916) and Sydney Young (1857–1937) came up with a consistent finding during their study of the decomposition of ammonia in temperatures around 800 °C: the decomposition was never complete, because a trace of undecomposed gas had always remained.[4] Yet they could not produce such a trace from the two elements brought together at that temperature in the presence of iron. More fruitless attempts at ammonia synthesis from its elements followed during the 1880s and 1890s, and then the new century brought a surprising claim.

As Anglo-German antagonism increased because of the Boer War, Wilhelm Ostwald (1853–1932), one of Germany's most famous scientists, realized that a military conflict with the superior naval power would cut off nitrate imports from Chile and deprive Germany of nitrogen needed for its fertilizer and explosives. Ostwald thought he had found an excellent solution to circumvent this potential threat. On March 12, 1900, he notified his friend Heinrich von Brunck (1847–1911), General Director of the BASF in Ludwigshafen, that he had succeeded in synthesizing NH_3 from its elements in the presence of an iron catalyst at elevated temperature and atmospheric pressure. (He also notified Farbwerke Hoechst, another large German chemical company.) Ostwald believed that the low costs of the process, both in material and energy terms, would revolutionize the supply of fixed nitrogen.[5]

Ostwald was one of the founders of physical chemistry (together with Svanté Arrhenius and Jacobus van't Hoff) and an expert on catalysis, and so his announcement had to be taken very seriously (fig. 4.1).[6] The BASF, mindful of the enormous commercial potential of such a discovery, rapidly concluded an agreement to investigate Ostwald's claim, although it could not foresee where it would conceivably

Figure 4.1
Wilhelm Ostwald (1853–1932). Courtesy of the E. F. Smith Collection, Rare Book and Manuscript Library, University of Pennsylvania.

obtain large volumes of pure H_2 needed for any eventual commercialization of the synthesis.[7]

The researcher entrusted by von Brunck with checking Ostwald's claim was Carl Bosch, a young Leipzig-trained chemist who joined the company in 1899. His report, dated April 7, 1900, concluded that the small amount of ammonia obtained by Ostwald was due not to the fixation of atmospheric N_2, but to the hydrogenolysis of iron nitride (Fe_3N) formed on the iron catalyst during its treatment before the experiment.[8] Ostwald was unaware that such trace quantities of nitrogen were present in all commercial iron catalysts at that time. Bosch was unable to detect any catalysis even when he tested the reaction under a pressure of 0.5 MPa.

Ostwald's first, and in view of Bosch's subsequent brilliant accomplishments an irresistibly quotable, reaction was to complain to the BASF directorate that "when you entrust the task to a newly hired, inexperienced chemist who knows nothing, then naturally nothing will come out of it." But by April 7, 1900, he had acknowledged his mistake and withdrawn his patent application.[9] Still, Ostwald's patent contained all of the basic ingredients of what eventually became the successful Haber–Bosch method: high temperature and elevated pressure, a metal catalyst, and gas circulation.[10] Not surprisingly, Ostwald later called himself the intellectual father of the ammonia industry. In his autobiography he wrote that an expert can readily recognize his March 1900 patent submission as "the unequivocally and clearly expressed basic thinking" behind the ammonia synthesis.[11]

Yet another failure came during the next year when Henry Louis Le Châtelier (1850–1936) attempted ammonia synthesis at high pressure. He calculated the required temperature, pressure, and amount of iron catalyst in 1900, but in 1901, when testing his theoretical results, his small high-pressure apparatus blew up, and he abandoned his research.[12] Haber found out about his work only a long time after he had completed his successful experiments. This then was the state of the field when Fritz Haber (1868–1934) began his work on ammonia synthesis (fig. 4.2). Some two decades later, during his lecture celebrating the centenary of the Franklin Institute, Haber summarized the state of affairs thus:

At that time, the possibility of producing chemically even traces of ammonia from its elements was disputed. Only the method of electrical production by silent discharge was proven. . . . Of course, the stability of the compound at ordinary temperatures seemed probable, in view of the amount of heat necessary for its formation. But after the numerous fruitless experiments of the previous generation, it seemed beyond doubt that nitrogen and hydrogen would not unite spontaneously at this temperature. Experiments at high temperatures, on the other hand, had up to that time proved nothing but its rate of decomposition, and

Figure 4.2
Fritz Haber (1868–1934) lecturing at the Technische Hochschule in Karlsruhe in 1909, the year he successfully demonstrated the synthesis of ammonia from its elements. Courtesy of Bibliothek und Archiv zur Geschichte der Max Planck Gesselschaft, Berlin, Germany.

alleged synthetic results turned out to be experimental errors, as the action of water upon nitrogen compounds or similar unintended by-reactions were answerable for the occurrence of ammonia.[13]

Fritz Haber

Haber was neither a child prodigy, nor was his career single-mindedly devoted to the pursuit of one all-consuming goal.[14] His mother died three weeks after he was born on December 9, 1868, in Breslau in Prussia (present-day Wrocław in Poland). His father, Siegfried, who married a cousin, Paula Haber, in 1867, remarried in 1876, but Fritz remained the only son in the family, which by 1881 included three girls. The Habers were not among the richest merchant families in Breslau, but Siegfried's wholesale business with natural dyes, paints, lacquers, and chemicals secured a comfortable living.

Haber's birth coincided with the creation of a new, Prussian-dominated Germany, a united Reich proclaimed in Versailles after the defeat of France in the war of 1870–1871.[15] As a child and as a young man he lived through the period of an unprecedented economic expansion that made Germany the most powerful state in Europe and that transformed its social structure.[16] These changes were driven, for the first time in history, by ceaseless technical and scientific advances. Germany was a latecomer as a modern state, but it had a long-standing tradition of scientific excellence.[17] German contributions to the creation of modern chemistry—inorganic, organic, or physical chemistry, as well as electrochemistry and biochemistry—were particularly outstanding.[18]

The family's affluence, as well as the advancing integration of well-educated Jews into privileged German society, made it possible for Haber to take over the family business or to strike out on his own in industry or in academia. Eventually he dabbled in the first two endeavors before choosing the third one, but even than he kept many close ties with industry, and during the last two decades of his life he became one of the most influential managers of scientific research in the country, a career shift from research to administration that was followed by many outstanding twentieth-century scientists.[19]

Haber's academic *Lehrjahre* actually resembled the *Wanderjahre* of an apprenticing craftsman: doubts about his career choices led him to three different universities in four years. After graduating from a classical gymnasium he left Breslau at the age of eighteen to study chemistry. His studies began at the University of Berlin under August Wilhelm von Hoffmann (1818–1892), one of the century's leading organic chemists, and continued at Heidelberg under Robert Wilhelm Bunsen (1811–1899).[20] In October 1889 he returned to Berlin and registered at the Technische Hochschule (Technical University) in Charlottenburg where he studied mainly under Carl Liebermann (1842–1914), an organic chemist best known for the synthesis of a red dye, alizarin.[21]

Because Haber decided that he wanted a doctorate and not just a diploma granted by the Hochschule, he had to submit his thesis at the Friedrich-Wilhelms-Universität. This was possible because science students could move freely between the Hochschule and the university, as long as they had fulfilled the requirement of six completed semesters at a university before they registered at the Hochschule. Consequently, Haber graduated, cum laude, from the university on May 29, 1891. His doctoral thesis, assigned by Liebermann, dealt with derivatives of piperonal.[22]

After graduation Haber began wandering once more: after a year of compulsory military service (in the artillery in Breslau) he went through a rapid succession of jobs and locales. He worked in his father's business and held three brief industrial practicums, in a distillery in Budapest, in a soda factory near Kraków, and in a cellulose plant in Feldmühle in the foothills of the Riesengebirge, before spending a term at the Institute of Technology in Zurich. Finally deciding on research as his career, he secured an assistantship at the University of Jena.[23]

Shortly after his arrival in Jena he converted to Protestantism and was baptized in the Michaeliskirche in November 1892.[24] Yet another move followed in 1894 when Hans Bunte (1848–1925) offered him an assistantship at the Technische Hochschule in Karlsruhe. Haber stayed there for seventeen years, becoming a *Privatdozent* in 1896 and a professor of physical chemistry and electrochemistry in 1906.[25] This recital of Haber's academic career does not sound unusual for an ambitious scientist in Wilhelmine Germany. What set Haber apart was his range of experience and the depth of expertise. He started out as an organic chemist: his first published paper, coauthored in 1890 with Liebermann, was on methylene indigo; in Jena he and Ludwig Knorr (1859–1921) published a paper on ethyl diacetosuccinate.[26]

In Karlsruhe his superiors turned his research in different directions. Bunte's main interest was in combustion chemistry, and Carl Engler (1842–1925) introduced him to the study of the decomposition and combustion of hydrocarbons (the topic of Haber's *Privatdozent* habilitation).[27] Shortly after his arrival in Karlsruhe Haber also began working on electrochemistry, and by 1898 he had published a textbook on the subject featuring his preferred combination of theory and application.[28]

The first five years of the new century were Haber's most productive period, both in terms of the total number of publications (almost fifty) or the variety of topics he researched. He pursued different kinds of electrochemical studies (ranging from electrolysis of solid salts to the invention of the glass electrode for determining the acidity of liquids), researched the loss of energy by various prime movers (steam engines, turbines, internal combustion engines), and probed the luminous inner core of the Bunsen flame. In 1905 he published a book on the thermodynamics of technical gas reactions, which was soon translated into English.[29]

In 1902 Haber was chosen by the German Electrochemical Society, with van't Hoff's recommendation and financial support, to make a sixteen-week fact-finding and study tour of the United States. He left on August 18 and traveled extensively around the country to be able to report on the country's chemical education and electrochemical industry.[30] After returning home in the third week of December

1902, he lectured around Germany about his impressions of American chemistry and engineering, and several months later he was involved in preparing the chemical part of the German exhibition at the St. Louis Exposition of 1904.

On August 3, 1901, Haber married Clara Immerwahr (1870–1915), whom he had known since his student years in Breslau.[31] Her career was much more unusual, and much more straightforward, than Haber's. Born into an affluent Breslau family, she enrolled at the University of Breslau and, like her father and several other members of her well-educated family, she chose chemistry. While at the university she too converted to Christianity.

Eventually she became a protégé of Richard Abegg (1869–1910), Haber's former classmate and a newly appointed professor of chemistry in Breslau, and she earned her magna cum laude doctorate as the first woman at the University of Breslau in December 1900.[32] Marriage followed just half a year later. Their only child, Hermann, was born in 1902, and she was transformed from an eager young scientist to a dissatisfied housewife who now shared little of her husband's preoccupation with his research.[33]

Haber's First Experiments with Ammonia

Haber's work on ammonia synthesis began during the summer of 1904 not because of any planned, determined effort to solve an elusive task, but because of an unexpected request from the Österreichische Chemische Werke in Vienna, a company set up by the brothers O. and R. Margulies.[34] Haber did not immediately take up the offer: he advised the brothers to take what was at that time the easiest route and recover by-product ammonia from coking. As he knew something about Ostwald's earlier work on the problem (but not about his unsuccessful deal with BASF and about the withdrawal of Ostwald's 1900 patent application), he wrote to him hoping that he would cooperate with the Margulies brothers.

Eventually Haber decided to proceed with fundamental experiments, and he received generous financial support from the Viennese company for his work. As he summed it up in his Nobel lecture, the challenge was both simple and exceedingly difficult;

We are concerned with a chemical phenomenon of the simplest possible kind. Gaseous nitrogen combines with gaseous hydrogen in simple quantitative proportions to produce ammonia. The three substances involved have been well known to the chemist for over a hundred years. During the second half of the last century each of them has been studied hundreds of times in its behaviour under various conditions during a period in which a flood of new chemical knowledge became available.[35]

Indeed, few reactions appear simpler than

$$N_2 + 3H_2 \leftrightarrow 2NH_3.$$

But because none of the many attempts to achieve a spontaneous union of the two gases to form ammonia had ever succeeded with absolute certainty, a widely held belief arose that such a synthesis might not be possible—and such feelings deterred further attempts at the necessary theoretical research.

But when Haber decided to use Ramsay and Young's investigations of ammonia decomposition in the vicinity of 800 °C as his point of departure, he did not feel that he was facing an impossible task. Rather than seeing those results as a cul-de-sac, Haber saw them as an intriguing opportunity for further investigations: "there was a point of uncertainty here, and if this could be cleared up it would indicate the possibility of a direct synthesis of ammonia from the elements."[36]

His first experiments, which he conducted with his assistant Gabriel van Oordt, were to determine the approximate position of NH_3 equilibria in the vicinity of 1,000 °C.[37] They slowly passed pure ammonia over heated iron, removed any ammonia that did not decompose, and introduced the remaining mixture of N_2 and H_2 into a second reactor also filled with finely divided iron. Almost immediately they found that the amount of ammonia formed in the second reactor was about as much as the volume of the undecomposed gas leaving the first reactor.

The equilibrium was thus confirmed by approaches from both sides of the equation, but the ammonia content was very low—between 0.005 and 0.0125%.[38] Haber believed, incorrectly as it turned out, that the higher value would eventually prove to be the correct rate. Haber obtained the same results with nickel as the catalyst, and he found that calcium and manganese allowed the two gases to combine at even lower temperatures.

Haber's broad research experience and eagerness to apply his findings was of great help in facing the challenge of ammonia synthesis: as he was not narrowly specialized, he was not easily deterred by a discouraging result. As Appl noted, his great initial accomplishment was to overcome his colleagues' excessive preoccupation with unfavorable equilibrium concentrations.[39] Although the reaction yield was far too small to be of any economic interest, Haber's circulation of gases—which brought the hot gas in contact with the catalyst and then washed out the ammonia at normal temperature—made it possible to gradually convert a relatively large volume of the gas.

Theoretical considerations allowed Haber to predict increased yields with decreasing temperature and the need for a stoichiometric gas ratio to get the highest NH_3 content. Haber realized that once the temperature surpasses 600 °C (the dark red heat stage) no catalyst will produce anything but a trace of NH_3 from the optimum gas mixture at normal pressure. Practical results under normal pressure would thus require catalysts working in temperatures no higher than 300 °C. This, rather than the very low reaction yield, spelled the temporary end of Haber's effort; he considered the discovery of any effective low-temperature catalysts unlikely. The Margulies brothers generously paid him for the negative results, and Haber considered the matter closed.

Although he knew that a reaction under high pressure was an obvious choice, and although he had the technical means to proceed with such experiments, he "did not think it worth the trouble; at that time I supported the widely-held opinion that a technical realization of a gas reaction at the beginning of red heat under high pressure was impossible. Here the matter rested for the next three years."[40]

Nernst and Haber

In yet another series of chance events, Haber's return to work on ammonia synthesis was prompted by a peculiar case of scientific rivalry, by one man's publicly tactless behavior injuring other man's pride. Hermann Walther Nernst (1864–1941) was only four years older than Haber, but by the time of their clash in 1907 he was certainly better established and better known (fig. 4.3). One of Ostwald's pupils from Leipzig, he became the first professor of physical chemistry in Göttingen in 1894, and his pioneering contributions ranged from technical inventions (Nernst's lamp and electrical piano, neither of which was a commercial success) to the theory of heterogeneous reactions.[41]

In 1906 he published his heat theorem, which soon became widely known as the third law of thermodynamics and which earned him a Nobel Prize for physics in 1920.[42] A discrepancy arising from this thermodynamic research led indirectly to Haber's eventual success. In his review of existing equilibria measurements Nernst discovered that the values for ammonia published in 1905 by Haber and van Oordt were the only ones seriously at variance with his heat theorem. He entrusted his assistant Fritz Jost to replicate the experiment, using a small iridium apparatus heated in an iridium furnace to more than 2,000 °C and an ingenious microbalance of his own design.[43] In order to increase the equilibrium ammonia concentration he

Figure 4.3
Walter Nernst (1864–1941). Courtesy of the E. F. Smith Collection, Rare Book and Manuscript Library, University of Pennsylvania.

also decided to conduct these experiments under pressure and built an apparatus withstanding up to 7.5 MPa (75 atm).

When recalculated for a stoichiometric mixture at 1,000 °C and 0.1 MPa (Haber's experimental temperature and pressure), Nernst's results (obtained at 5 MPa and 685 °C) showed 0.0032% ammonia yield, only about a quarter of Haber's high value (0.0125%) but close to Nernst's theoretical prediction (0.0045%). Nernst wrote to Haber, who, with his new assistant, Robert Le Rossignol, an Englishman trained in Ramsay's laboratory, almost immediately repeated the 1904 experiment (again at 1,000 °C and at atmospheric pressure) and ended up with 0.0048%, a good confirmation of Nernst's theory though still a bit above Nernst's results.

Haber and Le Rossignol published these new results one week before the annual meeting of the German Bunsen Society, held in 1907 in Hamburg between May 9 and 12.[44] During the discussion on May 12 Nernst maintained that Haber's new figures were, once again, incorrect and that they should be withdrawn like the previously published results because the measurements were made at normal pressure and hence at low ammonia concentrations. This public criticism by one of the world's most respected physical chemists was a serious setback for a freshly appointed professor.

Nernst concluded his critique with a judgment that spurred Haber to prove that the older man was wrong not only about the particulars of ammonium equilibria but also as far as the very possibility of ammonia synthesis is concerned:

It is very unfortunate that the equilibrium is more displaced towards the side of very low ammonia formation than the strongly inaccurate figures of Haber had formerly led us to assume, since one had inferred from them that it might be possible to synthesize ammonia from nitrogen and hydrogen. Now, however, the conditions are much less favorable, the yields being about a third what was thought previously.[45]

This was a puzzling statement in several ways: Haber publicly corrected his old figures just before the meeting at which Nernst attacked him so bluntly (calling his results not just unlikely or somewhat incorrect, but strongly inaccurate: *stark unrichtigen Zahlen*). Haber's new figures were accurate, agreeing better with Nernst's theorem than Nernst's own experimental values did; based on his first, incorrect, determinations, Haber did *not* conclude that ammonia synthesis is feasible, just the opposite.

Although Svante Arrhenius endorsed Haber's findings, Nernst's disparaging public remarks clearly damaged Haber's reputation, and his reaction was anything but a shrug-off: Clara Haber's letter to Abegg written in July mentioned Haber's being upset and suffering from digestion and skin problems.[46] At the same time Haber, who abandoned his earlier work because of what he perceived to be yields too low to justify further experiments, was now determined to vindicate himself, and he and Le Rossignol prepared to conduct a new set of more careful experiments, this time under a pressure of 3 MPa. Soon their results confirmed the accuracy of the revised data and showed the expected advantage of high pressures: at 500 °C the ammonia yield at 3 MPa was twenty-eight times as large as at 101.3 kPa (fig. 4.4).[47]

But conversion of these findings into an economically viable process had to reckon with the contradictory requirements of catalytic synthesis and operational profitability. The lowest possible pressure is desirable in order to minimize the energy cost

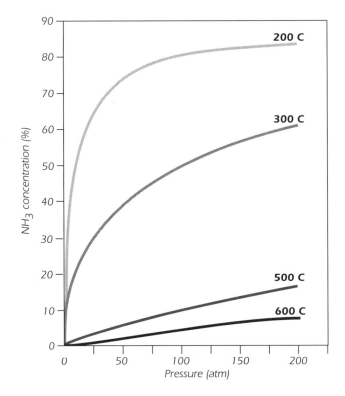

Figure 4.4
Dependence of ammonia yield on pressure and temperature as determined by Haber and Le Rossignol in 1907.

of the synthesis, but this shift in equilibrium conditions must be compensated for by lower reaction temperatures. This combination will result in lower reaction rates—and, unless a more active catalyst is deployed, in longer reaction times, and hence in larger catalyst volumes. Lower pressure also results inevitably in reactors with larger volumes.

Calculations based on Haber's and Le Rossignol's new measurements showed that under a pressure of 20 Mpa, at that time the limit of the high-pressure compressor in their laboratory, and at 600 °C it would be possible to have an appreciable NH_3 yield (about 8%)—providing that a new catalyst effective at such a relatively low temperature could be found (fig. 4.4). Manganese and nickel, the best catalysts employed at that time, were effective only at temperatures above 700 °C, and at that level their action was slow and the ammonia yield was low. Not surprisingly, Haber was eager to do further experiments with combinations of higher pressures and high temperatures and to find a better catalyst.

Also, immediately prior to this reexamination, Haber familiarized himself with the newly invented process of liquefaction of air and had, as he put it, "simultaneously caught a glimpse of the formate industry," which reacted flowing CO with alkali under heat and increased pressure. Alert as he was to a system approach in solving difficult problems he "no longer considered it impossible to produce ammonia on a technical scale under high pressure and at high temperature. But the unfavourable opinion of colleagues taught me that an impressive advance would be needed to arouse technical interest in the subject."[48] In 1908 he and Le Rossignol set out to make that very advance. Their task was made easier by a valuable partnership that Haber formed with the country's, and indeed the world's, leading chemical company.

BASF and Haber

Determining ammonium equilibria was not the only challenge related to nitrogen fixation that preoccupied Haber at that time. The Birkeland–Eyde process was emerging as a technically promising means of large-scale fixation of nitrogen, but as its energy costs were so extraordinarily high, its economic viability remained questionable. Any practical possibility of its refinement or of a modification leading eventually to higher conversion efficiencies had an obvious attraction for any electrochemist.[49]

In 1907 Haber and his pupil A. König published their first paper on the topic of NO formation in a high-voltage electric arc.[50] They argued, contrary to the prevailing opinion (which they, too, shared before they began their experiments), that it was desirable to proceed with the lowest possible temperature in order to minimize the decomposition of the formed NO and hence to increase the overall efficiency of the conversion. Haber and his assistants conducted the experiments under different pressures, cooled the electric arc from the wall as well as from the anode, and determined the relationship between energy consumption and frequency up to about 50 kHz. Their best results yielded concentrations of about 10% NO in air at decreased pressure, and conversion efficiencies 10–15% higher than the standard procedure.

Not surprisingly, their work attracted the attention of BASF in Ludwigshafen. BASF was at that time the world's largest chemical enterprise with a rich history of pioneering accomplishments and a keen interest in virtually any kind of chemical innovation.[51] The company was set up in 1865 by Friedrich Engelhorn, who had been producing aniline and fuchsin from coal tar, a by-product from coking bituminous coal in his Mannheim gas plant, but who wanted to expand the business to include a full range of tar-based products. A decade later the company offered a complete range of synthetic dyes for the expanding textile industry.

The first great diversification came in 1888 when Rudolf Knietsch (1854–1906) introduced the process for catalytic production of sulfuric acid; his discovery made BASF the world's largest H_2SO_4 producer at that time. The same year Knietsch also discovered the liquefaction of chlorine.[52] But dyes remained at the core of the company's business (accounting for more than 90% of the company's earnings), and in 1897 its researchers succeeded—after an unprecedented effort lasting seventeen years—in synthesizing indigo; just four years later René Bohn (1862–1922) discovered a better blue dye, indanthren.[53]

These advances assured the continuing success of BASF's dye business, but they kept the company less diversified than Hoechst, Bayer, or Agfa. But BASF also had a considerable experience with by-product ammonia from its coal gasification plant and with synthesizing HNO_3 by using imported Chilean nitrate, and it was trying to develop both of the two newly introduced ways of nitrogen fixing. Since 1903 Carl Bosch had been investigating the possibility of barium cyanide synthesis from barytes, coke, and nitrogen. Experimental production of $Ba(CN)_2$ began in August 1907, but it was closed a year later; in 1899 BASF had already set up a research program into nitrogen fixation in electric arc.

The BASF researchers Otto Schönherr (1861–1926) and Johannes Hessberger (1871–1934) eventually experimented with devices using 500–700 kW of electricity. In Schönherr's furnace (patented in 1905) a long (5–7 m) electric arc was confined within the calm core of a vortex created in the grounded tube by tangentially blown air; NO-laden gas was cooled by water as well as in the counterstream of the incoming gas, which was preheated to 500 °C (fig. 4.5).[54] As the commercialization of this process required the vicinity of cheap sources of energy, preparations began for construction near the Alz hydrostation in Bavaria.

As BASF also decided that it should not miss the opportunity to participate in the commercial development of the electric arc process getting under way in Norway, it concluded a cooperative agreement with Norsk Hydro, proprietors of the Birkeland–Eyde design, in December 1906. In the fall of 1907 the company set up an experimental plant at Kristianssand in southern Norway where it tested electric arc furnaces with power requirements of up to 450 kW.[55] Any patentable increase

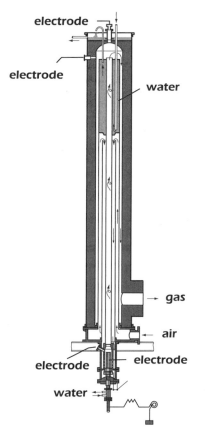

electrode

electrode

water

gas

air

electrode

electrode

water

Figure 4.5
Cross-section of Schönherr's electric arc furnace.

in the efficiency of nitrogen fixation in the electric arc could be obviously of a great benefit for the company.

At the very beginning of 1908 Carl Engler, Haber's senior colleague at Karlsruhe and a member of BASF's board of directors, asked Haber if he would like to enter into a binding contractual research agreement with BASF. A month later BASF director August Bernthsen (1855–1931) visited Haber in Karlsruhe, and Engler wrote to the company's directors recommending Haber as

a very energetic man, from whose talent and enterprising attitude I expect important results and whose solutions of disputed matters in the field of electrochemistry show that he is as much a fundamentally sound expert as he is a sharp and clever dialectician. As he is not unaware of his worth, and, like Ostwald and his pupils, he would like to earn something, he does not come naturally quite cheap.[56]

Two agreements were concluded on March 6, 1908. The first one between Haber and König and BASF concerned the synthesis of nitrogenous gases from nitrogen and oxygen, and it would have earned them 10,000 marks after approval of the patent. The second agreement was between Haber and BASF, and it guaranteed him 6,000 marks a year until February 1911 (later extended to October 1914) to use as he saw fit in support of his nitrogen fixation research. All patentable discoveries would become BASF's property, and Haber was to receive 10% of the earnings during any patent's duration.[57]

In retrospect, it is remarkable how eager BASF was to support electric arc research and how reluctant its leadership was to pursue ammonia synthesis. Twenty years later Le Rossignol remembered how frustrating the beginnings of that cooperation were,[58] and Haber himself noted in his Nobel lecture that the company

thought so highly of my efforts to obtain improved efficiency from electrical energy in the combining of nitrogen and oxygen, as to get in touch with me in 1908 and—by providing their resources—to facilitate my work on the subject; whereas they agreed with every caution to the proposal to back me in the high-pressure synthesis of ammonia as well, approving it only with hesitation.[59]

BASF, at that time committed to the improvement of the Birkeland-Eyde process, looked for advances in nitrogen oxidation, and Haber's contracted work in that field continued until 1910, when he and three of his assistants published a series of papers summarizing several years of their experiments.[60] But there is no doubt that by 1908 he was already certain that oxidation in the electric arc was not the way to large-scale nitrogen fixation. He felt that any improvements he was able to achieve "were not conclusive, being moreover achieved by methods which were hardly suited to adaptation to mass-production."[61] In contrast, Haber was encouraged by his new equilibria measurements obtained after his disagreement with Nernst. He believed that demonstrating a commercially appealing synthesis of ammonia from its elements was a definite possibility.

High-Pressure Catalytic Synthesis

The quest proceeded along two main lines: building a high-pressure apparatus for bench-testing the process, and coming up with a better catalyst whose deployment in the newly built experimental setup would give results convincing enough to attract commercial interest. Decades later Paul Krassa, Haber's student between 1906 and 1909 and Le Rossignol's friend, recalled the extraordinary difficulties at that time facing any researchers wishing to work with high pressures and high temperatures. Le Rossignol's exceptional skills were required to manufacture special valves for controlling the flow of gases; he made them during summer vacations in his own workshop.[62]

The reaction proceeded in an iron tube under a pressure of 20.3 MPa; inside a nickel heating coil was used to raise the temperature of gases to a desired level, and there was a thick-walled tube of transparent quartz—which was drawn on one side into a fine capillary—that continued to a point outside of the furnace, as the converter was called at that time (fig. 4.6). The catalyst was placed just in front of the tube's narrowed outlet so the gases, once they passed over the catalyst, would leave the oven rapidly. Their immediate cooling allowed the maintenance of the equilibrium according to the catalyst's temperature, which was measured with a thermoelement.

A new apparatus also needed an improved circulation system so that the synthesized ammonia could be separated from the flowing gas at a constant high pressure,

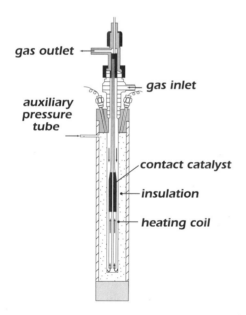

Figure 4.6
Haber's experimental converter for ammonia synthesis (1909).

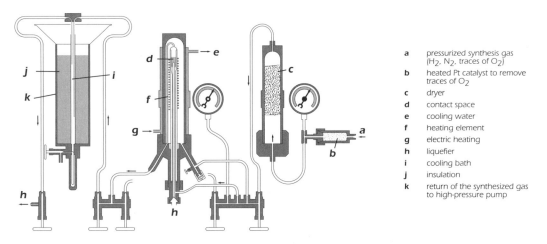

a	pressurized synthesis gas (H_2, N_2, traces of O_2)
b	heated Pt catalyst to remove traces of O_2
c	dryer
d	contact space
e	cooling water
f	heating element
g	electric heating
h	liquefier
i	cooling bath
j	insulation
k	return of the synthesized gas to high-pressure pump

Figure 4.7
Haber's experimental laboratory apparatus for ammonia synthesis (1909).

and the heat from the exothermic reaction could be removed from exhaust gases and used to preheat the freshly charged gas replacing the synthesis gas removed by the conversion (fig. 4.7). A small double-acting steel pump, also of Le Rossignol's design, circulated the gases; its efficiency was limited because of leather packing on its piston, an indication of the simple means with which the breakthrough apparatus was made to work. Recycling of the gas over the catalyst made it possible to sustain relatively high production rates even though the thermodynamic equilibrium was not particularly favorable.[63]

This setup was outlined in Haber's first patent concerning ammonia synthesis—No. 235421, valid since October 13, 1908, and issued on June 8, 1911, later widely described as the circulation patent—and its principle is still used in every ammonia plant today.[64] A twelve-line patent claim summed up the invention:

Process for synthetic production of ammonia from its elements, whereby an appropriate mixture of nitrogen and hydrogen is continuously subjected to both the production of ammonia under the influence of heated catalysts, and continuous removal of the resulting ammonia, characterized by constant pressure and the transfer of process heat from the ammonia-containing reaction gases to the incoming ammonia-free gas mixture.[65]

In order to achieve yields that would attract commercial attention, it was imperative to discover catalysts inducing rapid conversion at temperatures between 500 and 600 °C. Elements of the sixth, seventh, and eighth group of the periodic system—containing the well-known catalytic metals chromium, manganese, iron, and nickel—were the most promising candidates. Then Haber brought a small amount of what turned out to be his best catalyst, osmium, from one of his Berlin visits.[66]

During the months of research on ammonia synthesis Haber was also acting as a consultant to the Deutschen Gasglühlicht-Aktiengesellschaft, established in 1892 by the inventor of the gas mantle light, Auer von Welsbach (1858–1929).[67] The company used osmium for the filaments of their first metal-filament lamps (later it was replaced with much more readily available tungsten), and it had accumulated most of the world's total supply (amounting to some 100 kg) of the rare element, whose stock it maintained by buying back the burnt-out lamps and recycling the metal.

Finely divided osmium remained active in temperatures below 600 °C. When the metal was used as a catalyst at 17.7–20.2 MPa and 550–600 °C, ammonia yields of about 6% and higher were clearly of economic interest.[68] Moreover, the rate of ammonia production per unit volume of contact space—several grams per cm^3 of heated high-pressure converter—made Haber confident that the dimensions of the converter for eventual commercial reaction could be made small enough to allay any engineering objections. When he reported these findings in a letter to BASF on March 23, 1909, he also advised the company to buy up all the osmium in the possession of the Deutschen Gasglülicht-Aktiengesellschaft to prevent any future competitive bidding.[69]

But BASF leadership still had to be convinced that Haber's laboratory success could eventually be translated to a commercial process. During a meeting on March 26, 1909, at which Brunck, Bernthsen, and Bosch represented the company, serious doubts were expressed about running syntheses at pressures above 10 MPa (100 atm). When Haber answered Bernthsen's question about the required pressure by saying, "Well, at least 100 atmospheres," Bernthsen, head of BASF's laboratories, was horrified: "A hundred atmospheres! Just yesterday we had an autoclave under a mere seven atmospheres flying into the air!" Brunck then asked the opinion of the man who was responsible for the company's nitrogen fixation research, and Bosch replied without hesitation: "I believe it can go. I know exactly what the steel industry can do. It should be risked."[70] That settled the matter.

The osmium catalyst patent was filed on March 31, 1909, and Haber proceeded to perfect his bench-top apparatus in order to demonstrate convincingly the potential of high-pressure synthesis with gas recirculation. His search for suitable catalysts also continued, and soon he filed another patent application for using uranium in several forms: as a pure or alloyed metal, as a nitride (UN), or in a mixture with other compounds. Experiments with a uranium catalyst indicated ammonia yields on the order of 10% at 10 MPa and 500 °C, clearly a commercially appealing rate.

Synthesis of ammonia belongs to that special group of discoveries—including Edison's lightbulb or the Wright brothers' flight—for which we can pinpoint the date of the decisive breakthrough.[71] The BASF archives contain a letter sent by Haber to the company's directors on July 3, 1909, describing the events of the previous day when Alwin Mittasch and Julius Kranz witnessed the demonstration of Haber's synthesis in his Karlsruhe laboratory:

Yesterday we began operating the large ammonia apparatus with gas circulation in the presence of Dr. Mittasch and were able to keep its uninterrupted production for about five hours. During this whole time it had functioned correctly and it produced continuously liquid ammonia. Because of the lateness of the hour, and as we all were tired, we had stopped the production because nothing new could be learned from continuing the experiment. All parts of the apparatus were tight and functioned well, so it was easy to conclude that the experiment could be repeated. . . . The steady yield was 2 cm^3/minute and it was possible to raise it to 2.5 cm^3. This yield remains considerably below the capacity for which the apparatus has been constructed because we have used the catalyst space very insufficiently.[72]

Twenty years after the event, Le Rossignol, misdating the experiment to June 1908, remembered that both Bosch and Mittasch came to witness the morning experiment, and Krassa recalled the tension that tends to accompany such momentous events, and the fear of *Tücke des Objekts* (the spite of things). Indeed, one of the bolts of the high-pressure apparatus sprang during the tightening, and the demonstration had to be delayed for several hours. Bosch could not stay any longer and returned to Ludwigshafen.

As Mittasch, later in the afternoon of July 2, witnessed the first synthetic ammonia rising in the water gauge, he excitedly pressed Haber's hand.[73] The yield was on the order of 80 g NH_3/hour. There is yet another, slightly different, account by Julius Kranz, a technician who accompanied Mittasch and who was seventy-three years old when the BASF information office published his memoir on the occasion of the fiftieth anniversary of ammonia synthesis.[74]

Haber's third key patent for ammonia synthesis—No. 238450 filed on September 14, 1909, and issued on September 28, 1911, in Haber's, not BASF's, name, and generally known as the high-pressure patent—stressed the catalysis under pressure and high temperature. Its claim read: "Process for production of ammonia from its elements using catalysis under pressure and elevated temperature, whose characteristic is that the synthesis takes place under very high pressure of about 100 atm, but to be useful at 150–250 atm or higher."[75]

Now there could be no doubt about the fundamental feasibility of synthesizing ammonia from its elements: Haber's research was conclusive, and von Brunck was finally convinced that BASF should commit its full resources to the commercialization of the enterprise. The challenge was to scale up Haber's experimental laboratory convertor—a pressurized tube a mere 75 cm tall and 13 cm in diameter—into a pilot plant device that would withstand the required high pressure and high temperature, fill it with a catalyst as effective as Os or U but considerably cheaper and easily available, and ensure economical supplies of the two reactant gases. Carl Bosch was put in charge of this challenge.

5

Creating an Industry
Carl Bosch and BASF

Haber's laboratory version of ammonia synthesis was indisputably effective, but even after Brunck made the decision that BASF should proceed with the commercialization of the discovery doubts persisted inside the company about the success of such a costly, and seemingly open-ended, exercise. Given the absence of any precedent for running continuous catalytic synthetic reactions at such high pressures and temperatures, it was only reasonable to expect that even with intensive effort it would take a very long time to transform Haber's bench model to a smoothly operating commercial plant.[1]

As with any pioneering endeavor—and particularly with such a complex effort in which both scientific and engineering problems had to be overcome before it was conceivable to go ahead with a full-scale project—there were unforeseeable complications that only inventive and dedicated teamwork could resolve. The contributions of Alwin Mittasch and his colleagues, who elevated the field of catalytic studies to a new level of accomplishment with their prolonged, extensive, and ingenious search for inexpensive but highly effective catalysts, were especially critical. But, as in the case of Fritz Haber and his collaborators, these do not diminish Bosch's extraordinary contributions (fig. 5.1).[2]

As already noted, Bosch's conviction was decisive in persuading the company's leadership to proceed with the commercialization of the process. He was the only expert who knew the capabilities of the steel industry, and he believed that a sizable high-pressure converter could be built; he also believed that cheaper catalysts could be found, and that raw materials could be supplied in requisite amounts and purities. Bosch's inventiveness was decisive in removing a critical obstacle to further progress when the walls of experimental steel converters began failing after only short periods of operation.

Figure 5.1
Carl Bosch (1874–1940) in a 1914 photograph. Courtesy of BASF Unternehmensarchiv, Ludwigshafen, Germany.

In the end, not only was Bosch able to pioneer what has become the world's most essential chemical synthesis, but he was able to turn Haber's and Le Rossignol's experimental design into a commercial success in an extraordinarily short time. In just four years the enterprise went from a 75 cm tall converter sitting on Haber's laboratory bench in Karlsruhe's Technische Hochschule and producing about 100 g NH_3/hour to an 8 m tall converter installed at the first ammonia plant in Oppau and turning out up to 200 kg NH_3/hour.[3]

Construction of the world's first large-scale ammonia synthesis plant at Oppau near Ludwigshafen, and soon afterwards of a much larger Leuna plant near Halle, were not just pioneering steps for ammonia production. They presented manifold engineering challenges—most notably the necessity of handling large volumes of corrosive gases under both high temperature and high pressure—whose solutions had much wider technical implications: they opened up possibilities for a new field of large-scale high-pressure synthesis.[4]

Commercialization of ammonia synthesis also required a wide variety of associated high-pressure components (gaskets, joints, fittings), as well as of quick-acting magnetic shutoff valves and continuously operating instruments measuring and recording flows, densities, and compositions of gases and liquids, components later used in other continuous syntheses. Successful operation of the first two ammonia plants was also a turning point on the way from traditional, small-scale batch processing to modern continuous catalytic synthesis operating on a massive scale. And industrial catalysis was transformed by the introduction of painstakingly optimized mixed materials.[5]

Carl Bosch

During the first thirty years of its existence the Nobel Prize in chemistry was never given to work aimed directly at the achievement of technical improvements and practical progress. As Professor Palmaer of the Royal Swedish Academy of Sciences noted in his Nobel Prize presentation speech in 1932, a contributing reason for the absence of such awards was the fact that technical progress tends to be the outcome of cooperation among several individuals, and hence it is usually difficult to decide who is the most deserving person. But 1931 was a different year: "This year, however, the Academy of Sciences believes it has discovered a technical advance of extraordinary importance and in respect of which it is also quite clear to which persons the principal merit is to ascribed."[6] Carl Bosch shared the prize with Friedrich Bergius, the inventor of coal hydrogenation, for "their services in originating and developing chemical high-pressure methods."[7]

Bosch had what in retrospect seems to be nearly perfect preparation for achieving such a success in applied science and chemical engineering. His father was a mechanic and a businessman in Cologne, a founder and partner in what eventually became a large wholesale company for plumbing and gas installation supplies whose sales and attached workshops employed seventy-five people. Carl Bosch, born on August 27, 1874, was the oldest of six children.

He grew up playing with all kinds of tools—one day his mother found her sewing machine disassembled, and another time, to her consternation, her son planed the family's mahogany furniture with a tool he had received as a gift—and experimenting with chemicals.[8] He did not enter the university right after finishing the *Oberrealschule,* but spent a year in Kotzenau, a small town north of Liegnitz in Silesia, where his father's business associate owned a smelting company. There Bosch apprenticed in a number of trades, as a molder, mechanic, carpenter, and locksmith.

Bosch began his university studies at the Technische Hochschule in Charlottenburg in 1894, just over three years after Haber had left the school, not as a chemist but as a metallurgist and mechanical engineer. During these studies he received a great deal of further practical experience in German iron and steel mills. After two years he changed his subject and school: he moved to the University of Leipzig, where he studied organic chemistry under Johannes Wislicenus (1835–1902).[9] After passing his oral doctoral examinations in May 1898, he stayed for nearly a year in his professor's laboratory as an analytical assistant. In January 1899 he wrote an application letter to BASF in which he listed not only his academic qualifications but also all the crafts he had learned in Kotzenau. He was promptly hired and joined the company in April 1899.

He was first assigned to work under Rudolf Knietsch on synthetic indigo, and then on an expansion of the phthalic acid plant. His first independent task was the already described replication of Ostwald's March 1900 proposal for catalytic synthesis of ammonia.[10] From that period comes an anecdote of his future wife Else Schilbach (they were married in May 1902) of how, as they were walking one day though the old part of Cologne, Bosch stopped suddenly and said, "I will solve the nitrogen problem!"[11] As it turned out, a long period of detours and cul-de-sacs lay on the road toward fulfilling that ambitious goal.

Bosch began working on the nitrogen problem as a result of BASF's interest in fixation using electric arc. His task was to research the ways to convert the generated nitrogen oxides into nitrates, but he soon became convinced that the very high energy requirements of the arc process would not make it a major source of fixed nitrogen. Starting in 1903 he turned instead to research on barium cyanide, and Alwin Mittasch joined him in that effort in 1904. After years of extensive experiments a pilot plant began operating in August 1907, and its daily output of about 5 t of $Ba(CN)_2$ was converted to 300–350 kg of ammonia, rates 30–40% lower than expected on the basis of laboratory work. Low yield and high cost brought the abandonment of the project just ten months later, after it had produced only some 90 t of NH_3.[12]

Bosch's concurrent, and subsequent, experiments with titanium nitride (TiN, produced at 1500 °C by reacting TiO_2 and N_2) and silicon nitride (Si_3N_4) were no more successful. After seven years of work on an array of nitrogen fixation–related problems, Bosch's research had yielded no commercial promise. Then came Haber's unexpected breakthrough, his letter of March 23, 1909, informing BASF that his combination of high pressure and osmium catalyst produces appreciable yields of liquid ammonia. Bosch the experimental chemist lost the race to invent a commercially promising nitrogen fixation method.

But when Brunck asked for Bosch's opinion, it was Bosch the metallurgist who decided the immediate fate of Haber's invention by saying that the company should go ahead with the commercial development of Haber's laboratory design. His negative judgment would not have been questioned by the company's apprehensive and far from enthusiastic leadership, and because of the outbreak of World War I the eventual development of the process could have been delayed by many years.

But Bosch's recommendation to take the risk would have been just an empty gesture if he had not proceeded to do a masterful job of managing the complex process of developing the synthesis and personally resolving a key technical obstacle that was preventing large-scale synthesis of ammonia. His organizational skills, his perseverance as an experimenter, and his ingenuity were critical for the success of that unprecedented enterprise.[13] When in 1914 Carl Duisberg (1861–1935), speaking to a Reichstag commission, used for the first time the term *Haber–Bosch Verfahrern*— Haber–Bosch process—he was merely putting the correct label on the century's most far-reaching invention.[14]

In 1909, when Bosch was given extraordinary powers by Brunck to go ahead with the task of commercializing Haber's ammonia synthesis process, he set up his own technical workshop headed by Franz Lappe (1878–1950), a young engineer he knew from the phthalic acid plant. Bosch realized that he had to settle three key problems before any full-size plant could be built: the supply of the two feedstock gases, nitrogen and hydrogen, at a lower price than had been possible so far; the identification and production of effective and stable catalysts; and the construction of durable high-pressure converters.[15] These challenges had to be dealt with simultaneously, but they were not equally difficult: an unexpected event in the early stage of experiments showed that building a high-pressure converter would be the greatest challenge.

Designing High-Pressure Converters

One of the first steps taken in Ludwigshafen in 1910 was to build two experimental converters appreciably larger than Haber's laboratory model (fig. 5.2). Then Mannesmann produced the contact tubes 2.5 m long with a 15 cm inside diameter and wall thickness about 30 mm for the first technical trials (fig. 5.3). They were heated electrically from the outside, filled with the catalyst, and housed in reinforced concrete chambers. Both tubes burst after about eighty hours of operation under high pressure.[16]

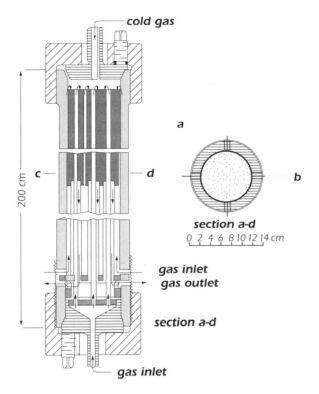

cold gas

a

c — d

b

200 cm

section a-d

0 2 4 6 8 10 12 14 cm

gas inlet
gas outlet

section a-d

gas inlet

Figure 5.2
BASF experimental converter (1910).

The converter's precautionary placement in a concrete container prevented any wider damage, but that was a small consolation. Examination of the burst tubes revealed that their structure had totally changed: their inner part had totally lost its elasticity, it contained countless line fractures, and it had become hard and brittle, very much like cast iron. Once the undamaged perimeter metal became too thin, the tubes gave in to the internal pressure.

Haber's experiments had not provided even an inkling of such a severe problem, and its cause was not obvious: hydrogen should have been harmless to the metal, and tests showed no buildup of embrittling iron nitride. There were no precedents to consult; at that time the only technique involving high pressures was Linde's liquefaction of air, which was done in an apparatus made of soldered copper, a metal unsuitable for use in temperatures of up to 600 °C.[17] Owing to its mechanical strength, carbon steel was the only practical choice to be used for the ammonia converter.[18]

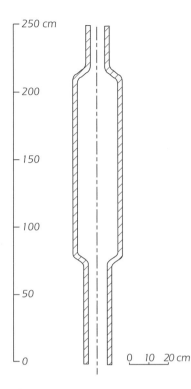

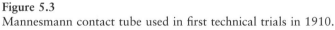

Figure 5.3
Mannesmann contact tube used in first technical trials in 1910.

Bosch discovered the cause of the failure by performing a metallographic analysis of a thin section of the burst converter wall, a procedure he was familiar with from his metallurgical studies. Although steel decarburization took place, the loss of carbon did not produce, as might be expected, soft iron, but a hard and embrittled material prone to cracking.[19] Eventually it became clear that hydrogen diffusing into the steel wall causes the decarburization through the formation of methane; then the combination of the mechanical stress and of the gas trapped under high pressure within the metal loosens the structure and eventually results in a complete wall failure. External heating of these experimental converters put further stress on their walls.

Immediate laboratory investigations showed that at high pressure and high temperature all carbon steels will be invariably weakened by hydrogen and will fail in just a matter of hours or days.[20] Obvious solutions did not work: all the suitable

metals and materials that could possibly protect the steel either were also destroyed
or allowed easy transdiffusion of hydrogen into the structural metal. Bosch solved
the problem by accepting that hydrogen diffusion into the iron, decarburization of
perlite, and embrittlement of the metal are unavoidable, and by devising a solution
that rendered them harmless.

His fundamental insight was to separate the two functions played by the converter
wall—providing a gas-tight seal, and withstanding the pressure of highly compressed
gases—by designing a converter made up of two tubes (fig. 5.4):[21]

The long-sought solution consisted in fitting a pressure-bearing steel jacket internally with a
quite thin lining of soft steel in such a way that as the hydrogen, which of course only diffuses,
passes through the thin lining it is able to escape without building up pressure before it can

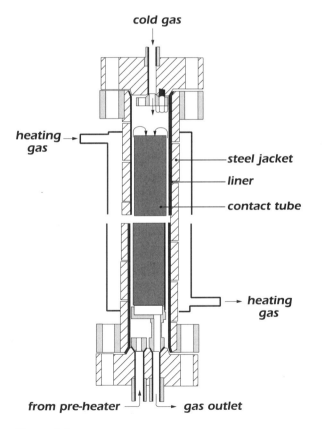

Figure 5.4
Bosch's first converter with a liner and external heating (1911).

attack the outer steel jacket at the high temperature. This is readily achieved by the grooves produced on the outside when the tube is being turned, and by a large number of holes drilled in the steel jacket through which the hydrogen is free to emerge. Right at the outset the thin lining tube fits tightly against the jacket under high pressure and later, when it has become brittle, there is no way of expanding it again so that no cracks occur. Diffusion losses are minimal.[22]

As Bosch concluded, this "solution appeared simple and was in fact so, yet the entire development of the process depended on it to a greater or lesser extent."[23] But even when not attacked by hydrogen, steel jackets of the converters, exposed to high-pressure and high-temperature gradients from the external converter heating, tended to buckle and crack, causing more explosions and requiring costly repairs.[24] Bosch came up with another ingenious solution: interior heating done by igniting air forced into the hydrogen-rich mixture inside the converter and producing the so-called reverse flame.[25] Even then, additional heating had to be applied to the converter by forced air.

Once the internal heating was introduced, it was possible to further reduce the attack of hydrogen on the converter's iron walls by flushing the space between the lining tube and the steel jacket continuously with nitrogen, which is harmless to the metal (fig. 5.5).[26] The nitrogen flow also helped to reduce the pressure differences between the inner and the outer side of the lining. The next critical step after resolving the problem of converter wall failure was to come up with the most energy-efficient design of the conversion process.

This required an efficient and even removal of heat from the exothermic synthesis, which was done by heating the incoming synthesis gas mixture by countercurrent exchange from the exhaust gases.[27] Knietsch's design for the H_2SO_4 contact process served as the obvious model. As the converters grew larger it was possible—once their inside temperature reached the initial reaction level, and with good heat exchange—to offset heat losses completely by the heat of the highly exothermic reaction.[28]

Pioneering high-pressure design was also needed for gas compressors and circulation pumps. Leaks could be tolerated in modestly sized compressors such as those used at that time in mine locomotives and air liquefaction, but not in machines feeding the synthesis gas to ammonia converters: the risk of explosion and financial losses from possible damage and lost production required reliable units that could be operated for months before requiring scheduled maintenance. After years of trials Bosch's group introduced 3,000 hp (2.2 MW) compressors operating for six months without failure.[29]

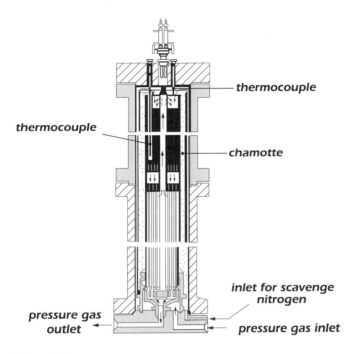

Figure 5.5
Converter with nitrogen flushing (1916).

It was also necessary to ensure the tightness of extensive networks of steel tubing and of numerous gaskets for high-pressure flange connections and valves. Safety considerations resulting from high pressure, high-temperature flows led to the design of magnetically operated quick-acting valves, self-closing valves with a spring release, and slide valves for rapid closing in the event of pipe breakage (fig. 5.6).[30]

Bosch also believed that anything that could be measured should be measured, and he championed the introduction of extensive process controls in the chemical industry. Continuous operation of ammonia synthesis required constant monitoring of temperatures, pressures, and gas and liquid flow rates, densities, and compositions. No such commercial monitoring instruments were available at that time, and Bosch's team had to devise, build, and test them.[31] Particular attention had to be paid to the monitoring of O_2 and CO in the reaction gases: to O_2 because of the explosion risk and because of the reduced yield due to the steam generated in the

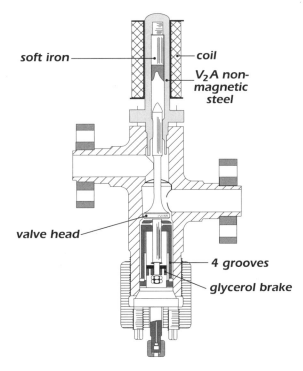

soft iron

coil

V_2A non-magnetic steel

valve head

4 grooves

glycerol brake

Figure 5.6
Magnetically operated quick-acting safety valve (1915).

hot catalyst, to CO because the gas is, next to sulfur, the greatest poisoner of iron catalysts.

Finding New Catalysts

Haber achieved excellent results with osmium, but the metal could not be used in large-scale synthesis for two reasons: in contact with air it was readily converted to volatile osmium tetroxide (OsO_4); more importantly, as already noted, its worldwide supply in 1909 (most of it, as Haber advised, was actually acquired by BASF for 400,000 marks) was limited to some 100 kg, and in October 11, 1910, BASF estimated that this would suffice to produce no more than about 750 t NH_3 a year.[32] The other effective catalyst, uranium, was also expensive, but it was, even at that time, obtainable in larger quantities. Tests showed, however, that it too was not a viable technical choice: it was extremely sensitive to oxygen and water, and the necessity to use it in a very fine powder prevented its application in a mass-production process.

Testing of catalysts by Alwin Mittasch (1869–1953; fig. 5.7) and his colleagues Hans Wolf (1881–1937) and Georg Stern (1883–1959) began soon after they replicated Haber's Karlsruhe experiments in their Ludwigshafen laboratory. In order to go through a massive winnowing process in the shortest possible time, the team resorted to a systematic approach suited to a highly repetitive task.[33] Stern was responsible for designing a small-scale, high-pressure apparatus in which catalysts could be tested continuously by easily inserting and removing cartridges containing 2 g of catalyst samples. The apparatus consisted of a vertical high-pressure tube with internal electric heating and external air cooling (fig. 5.8).[34] Eventually as many as thirty of these minireactors were operating simultaneously, and in just a few months this strategy led to the choice of an inexpensive yet highly effective catalyst.

Figure 5.7
Alwin Mittasch (1869–1953) in his BASF office in Oppau in 1929. Courtesy of BASF Unternehmensarchiv, Ludwigshafen, Germany.

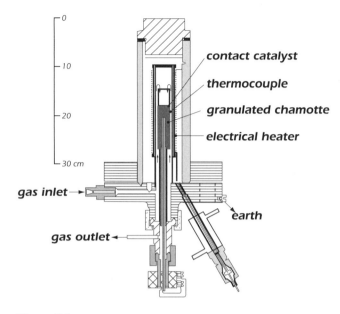

Figure 5.8
Laboratory apparatus designed by Georg Stern and used by Alwin Mittasch for testing catalysts (1910).

Because of the low cost and ready availability of the metal, the team paid special attention to iron in all of its many forms.[35] Although not altogether discouraging, yields with iron catalysts lagged behind those achieved with osmium and uranium. Then on November 6, 1909, one of Wolf's samples, magnetite (Fe_3O_4) from the Gällivare mines in northern Sweden, showed an unusually high yield. The combination of the material's high density and porous structure appeared to be the most obvious explanation, but tests with other magnetites could not replicate the yield obtained with the Gällivare mineral.

Mittasch eventually concluded that the presence of additional elements, acting as promoters of the reaction, must determine the effectiveness of a metallic catalyst, and he launched a systematic testing of a large number of promoters, both individually as well as in combinations with differing amounts of various ingredients, added to a pure iron catalyst. The first experiments with mixtures were done just before the end of 1909. NaOH and KOH were the first tested additions, and the first patent concerning the mixed catalysts (DRP 249 447) was filed by Bosch, Mittasch, Stern, and Wolf on January 9, 1910.[36]

Nothing was overlooked in the search for the best possible catalyst: all known metals with catalytical properties—Co, Fe, Mn, Mo, Ni, Os, Pd, Pt, U, W—were tested singly in pure form as well as binaries (such as Al-Mg, Ba-Cr, Ca-Ni, Fe-Li, Na-Os, or W-Zr) or ternary elemental catalysts (ranging from Al-Cr-Na to Mn-Si-Na), as well as even more complex mixtures.[37] Soon the testing identified a number of substances that act, even at very low concentrations, as catalytic poisons and whose presence must be strictly avoided, as well as a number of compounds that act as strong promoters of catalysis. Sulfur, phosphorus, and chlorine are the three most common elements in the first category.[38]

In contrast, numerous metal oxides—above all alumina (Al_2O_3) and magnesium oxide (MgO)—were identified as catalytic promoters. Moreover, combinations of several compounds may transform materials that by themselves have a neutral or even negative effect into promoters; perhaps most notably, when combined with alumina, oxides of alkaline metals act as promoters.[39] Appl marvels at how well Mittasch identified the optimal catalyst for ammonia synthesis for temperatures up to 530 °C and pressures up to 35 MPa.[40] Most commercial catalysts on the market today are just slight variations on Mittasch's basic theme: they use magnetite with 2.5–4% Al_2O_3, 0.5–1.2% K_2O, 2.0–3.5% CaO, and 0–1.0% MgO (as well as 0.2–0.5% Si present as a natural impurity in the metal).[41]

Nobody was as surprised as Haber when he was informed about the discovery of mixed iron-based catalysts: "And so it is iron, with which Ostwald worked at first, and which was then tested hundreds of times in pure form, which now acts only in impure forms. I see again how one should follow every trace until the end."[42] A most welcome property of these catalysts was their extraordinary stability. As long as no catalyst poison is present in the synthesis gas, no deactivation of catalysts has been observed in the lower layers of modern large-volume converters even after more than a decade of operation, and with extremely pure synthesis gas catalysts can even serve up to twenty years.[43]

By January 1910 the search for an inexpensive yet effective catalyst was thus basically over, but further testing and refining continued to make sure that the optimal catalytic material had been found.[44] By the beginning of 1912 Mittasch's laboratory performed 6,500 tests with 2,500 different substances, and the effort did not cease with the selection of a mixed catalyst for the first commercial plant. By the time the testing ended in 1922, the tally was up to 20,000 runs of more than 4,000 different substances, and the team selected other new effective catalysts for the oxidation of ammonia.

Producing the Feedstocks and Oxidizing Ammonia

In order to prevent catalyst poisoning ammonia synthesis requires hydrogen and nitrogen feedstocks of high purity. By 1909 nitrogen could be obtained relatively cheaply by Linde's fractional distillation of air, but the production of hydrogen was the single most expensive item in the total cost of ammonia synthesis.[45] Initial experiments and the first pilot plant in Ludwigshafen used hydrogen from chlorine-alkali electrolysis and obtained the stoichiometric volume of nitrogen by the partial combustion of the generated hydrogen with air.[46] But neither this method of hydrogen generation, nor any other commercial process available at that time, was suitable for large-scale operation: they were either too expensive, or they yielded excessively contaminated gases.

Given BASF's dependence on coal, Bosch decided that the sole eligible source of hydrogen was water gas produced by the reaction of glowing coke with water vapor:

$$C + H_2O \leftrightarrow H_2 + CO + 118.7 \text{ kJ/mol.}$$

The gas contains about 50% H, 40% CO, and 5% each of N and CO_2, and the cryogenic process was used initially to condense CO out of the gas at -200 °C and 2.5 MPa. As the volume of ammonia synthesis increased, Bosch sought a better solution. In the summer of 1911 Wilhelm Wild (1872–1951), one of Nernst's pupils working at BASF, proposed a cheaper method of hydrogen production from water gas: a catalytic process during which the reaction of CO with water shifts the gas to CO_2 and produces additional H_2 for ammonia synthesis.

In the ammonia plant this was combined with concurrent generation and processing of the feedstock nitrogen. In 1912 Wild and Bosch patented this catalytic process for mass production of purified hydrogen.[47] BASF's *Wasserstoffkontaktverfahren* began with production of water gas and coke-oven gas, the latter mixture being the product of the reaction of hot (in excess of 1,000 °C) coke with air and containing 62% N, 32% CO, 4% H_2, and 2% CO_2.[48] As is the case with water gas, the presence of sulfur in H_2S (derived originally from the low-sulfur coal used) is the most objectionable trace impurity in the coke-oven gas.

The next step was to wash both gases with water in towers and mix them before circulating them over activated charcoal to remove H_2S. Running the gas production processes together with the appropriate air:steam ratio will produce synthesis gas with the correct stoichiometric ratio. The key segment of the process came next: a catalytic conversion, a shift reaction transforming CO and steam into CO_2 in the presence of iron oxide and a chromium oxide catalyst at 250–450 °C, producing more hydrogen:

$$CO + H_2O \leftrightarrow CO_2 + H_2 - 41.2–91 \text{ kJ/mol.}$$

CO_2 was then removed in a water scrubber at 2.5 MPa. Removing the residual amount of CO (2–3%) turned out to be an unexpected challenge. Bosch chose scrubbing under a pressure of at least 20 MPa in a solution of cuprous oxide (Cu_2O), but it was corroding any iron surface with which it came in contact. Replacing numerous iron parts with more expensive metal was unthinkable.

Bosch, for once dejected and under strong pressure by the company's directors, turned to Carl Krauch (1887–1968), a young chemist on his team. Krauch solved this last complication on the road to commercial ammonia synthesis by adding ammonia to the solution. Cuprous ammonium formate proved to be an excellent absorber of CO, and Krauch's solution remained in use for decades.[49] The remaining moisture was not removed from the mixture used in the first two commercial plants, but it is taken out in modern processes.[50] If the purified mixture contained excess H, its content was adjusted with nitrogen from liquefied air to the exact stoichiometric ratio of 1:3.

BASF planned to convert most of the synthesized ammonia into ammonium sulfate to be used directly as crop fertilizer. This was to be done in a straightforward manner by reacting ammonia with sulfuric acid:[51]

$$2NH_3 + H_2SO_4 \rightarrow (NH_4)_2SO_4 - 283.5 \text{ kJ/mol.}$$

The second most important route to a final product was to oxidize ammonia into nitric acid, which, in turn, can be used to produce nitrates, be they for fertilizers or for explosives.

This conversion also did not need any fundamentally new research by Bosch's team, as the requisite catalytic process, which begins with converting NH_3 to NO in the presence of hot platinum, had been known for a long time. The reaction sequence is

$$4NH_3 + 5O_2 \rightarrow 4NO + 6H_2O,$$

$$2NO + O_2 \rightarrow 2NO_2,$$

$$2NO_2 + 1/2H_2 \rightarrow HNO_3 + NO.$$

Ostwald and Eberhard Brauer investigated the oxidation of ammonia shortly after Ostwald's failed attempt to synthesize the compound, and on November 18, 1901,

Ostwald offered the process to the Farbwerke Hoechst, noting that commercialization of the process would remove the threat of an English naval blockade denying Germany access to Chilean nitrates in the case of a war.[52] As soon as Ostwald's attempt to patent the process became known in 1903, BASF disputed the claim. The company pointed out that the reaction was first described by Charles Frédéric Kuhlmann (1803–1881) in 1838, and then in many of his subsequent papers, and it also noted its own research in related areas of high-temperature catalysis. In April 1907 Ostwald was forced to withdraw his claim.[53] By the time BASF decided to go ahead with the oxidation there were no legal problems, but it still had to find a cheaper catalyst.

The first ammonia oxidation plant, using by-product ammonia from coking in Gerthe in the Ruhr region since 1906, was equipped with a platinum catalyst that was too expensive for BASF's intended large-scale production. Once again, an intensive search rapidly came up with an effective solution: after just a few months of work, in February 1914, Christoph Beck (1887–1960) found that a mixed catalyst, which had previously been found ineffective for ammonia synthesis, gave impressively good results. The combination of iron, bismuth, and manganese oxides thus superseded the expensive platinum.[54] Within a year, with Germany's increasing need for explosives, this process was incorporated into the world's first ammonia synthesis plant.

The First Ammonia Plant at Oppau

The Bosch-directed, indeed Bosch-driven, process of converting a bench-top process into a large-scale commercial reality proceeded at an impressive pace.[55] The first ammonia produced at the experimental site (Lu 398) in the company's main Ludwigshafen compound began to flow on May 18 and 19, 1910, just ten months after Haber's Karlsruhe demonstration. Two months later, on July 19, there was enough ammonia to fill a 5 kg container.[56]

Soon afterward the work shifted to a new site, Lu 35. All the machinery for gas preparation, high-pressure synthesis, and gas cleaning was tested and improved in this prototype plant, which began operating on August 10, 1910. During the first three months of its operation the plant, using single-wall converters, produced only about 100 kg NH_3, but during December 1910 it managed an average daily rate of almost 10 kg NH_3, and about 18 kg NH_3/day during the first two weeks of 1911.[57]

After the first converter with Bosch's soft iron insert was put into experimental operation in March 1911 its ammonia output kept rising steadily, and the daily rate of 100 kg NH_3 was surpassed in July 1911. Total NH_3 output during 1911 reached 11 t (averaging 30 kg/day); by the end of the year internally heated ammonia converters had a length of 4 m and a diameter of 15 cm. In February 1912 a new 8 m tall converter was introduced, and two months later its daily output topped 1,000 kg NH_3. Another production milestone came in July 23, 1913, when the aggregate synthesis at Lu 35 reached 1 million kg of NH_3.[58]

Planning for a plant with a daily capacity of 10 t NH_3/day began in April 1911 when the pilot plant was not even producing 0.1 t NH_3/day. By the time the decision to build the first commercial plant was made in November 1911, its capacity had increased to 30 t NH_3/day; about 70% of this output was to be converted to ammonium sulfate to be used directly as fertilizer.[59] As there was insufficient space next to the main Ludwigshafen plant, a new site was chosen between the small village of Oppau and the Rhine River, 3 km north of BASF's main plant. But while the preparations for the first plant's construction were proceeding as fast as possible, BASF was fighting in courts.

Several companies (Kunheim, Griesheim-Elektron, Hoechst), as well as Fritz Jost, Nernst's collaborator, contested Haber's basic patents. BASF's adroit handling of its principal opponent was key to resolving the problem: Walther Nernst, allied with Griesheim, was offered an annual "honorarium" of 10,000 marks for five years, and he promptly pledged "as a national duty" to defend Haber's patent. Hoechst recruited Ostwald to its cause, but what contemporary observers saw as an astonishing union of former rivals—Haber and Nernst marching side by side into the courtroom—carried the day: all counterclaims were dismissed on March 4, 1912.[60]

Construction at Oppau began on May 7, 1912, and Johannes Fahrenhorst headed the building project. The plant's 500,000 m^2 site was to contain a completely integrated operation beginning with deliveries of coal and coke and the production of water gas and coke-oven gas, proceeding to purification and compression of gases and the ammonia synthesis proper, and ending with the subsequent conversion of ammonia to ammonium sulfate and storage of the sulfate in silos and shipping facilities.

Because of the unprecedented volumes of pressurized gases, which required the construction of lengthy pipelines carried on elevated structures, the plant acquired a look unlike that of any other previous chemical enterprise (fig. 5.9). Bosch was

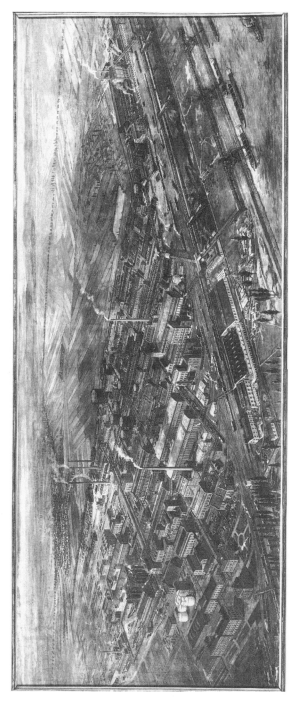

Figure 5.9
Ammonia plant in Oppau in a painting by Otto Bollhagen (1920). Courtesy of BASF Unternehmensarchiv, Ludwigshafen, Germany.

confident of the project's success: on October 15, 1912, he assured the company's directors that "our ammonia process is a good, winning project full of future promise."[61]

Work on the site proceeded without major problems, and the plant produced its first ammonia on September 9, 1913, after less than fifteen months of construction.[62] By October 24 the plant was synthesizing more than 10 t of NH_3 a day. Fahrenhorst became Oppau's first director, Philipp Borchardt was responsible for Linde liquefaction of air producing nitrogen, Wilhelm Wild ran the purification of the feedstock gas, and Hans Keller supervised the coal gasification and production of ammonium sulfate.

Quite surprisingly, the synthesis of ammonium sulfate, a straightforward process of neutralizing the sulfuric acid with ammonia that was being carried out without any complications in many plants using by-product ammonia from coke ovens, was the source of repeated and costly leaks of ammonia, corrosion of metal surfaces, and acid spills during the initial stage of the plant's operation in early 1914. Their cause was eventually traced to the high purity of synthetic ammonia.[63]

Scaling-up of the high-pressure converters continued both during the plant's construction and after its completion. Size went up from the 1912 pilot plant unit—4 m long, 23 cm in diameter, and weighing 1 t—to 8 m long units with inner diameters of 28.5 cm and weighing 3.5 t, which were installed at Oppau and whose daily production capacity was 3–5 t NH_3 (appendix H). Soon they were widened to 45 cm (weighing 8.5 t), and in 1914 a 67.5 cm converter weighed 25 t. In 1915 two 6 m units were joined to form a 12 m long converter with 1 m diameter weighing 75 t. This unit had to be forged from a 100 t block, the maximum ingot mass that could be handled by German forges at that time.[64]

During the first full year of Oppau's operation in 1914 the plant fixed about 20 t of nitrogen daily, and its production of ammonium sulfate reached 26,280 t, the equivalent of 5,500 t N. In the summer of 1914 plans were made to expand its capacity to 40 t N/day (150,000 t of ammonia sulfate a year). In 1914 Bosch also realized his plan to establish an agricultural research station, at Limburgerhof near Ludwigshafen, where the company began an expanding program of studies on fertilizers and plant physiology. Then the war broke out, and the role of ammonia synthesis changed in just a matter of months from producing fertilizers to sustaining Germany's munitions industry.

Ammonia Synthesis for War

On August 4, 1914, all Reichstag parties voted for war credits.[65] In October Ludwig Fulda's manifesto *To the Civilized World*—which made an astonishing claim that "[w]ere it not for German militarism, German civilization would long since have been extirpated. . . . The German army and the German people are one"—was signed by 93 scientists.[66] Among the prominent names were three past (Paul Ehrlich, Emil Fischer, and Wilhelm Ostwald) and three future (Richard Willstätter, Fritz Haber, and Walther Nerst) Nobel Prize winners, who were convinced that Germany bore no responsibility for the war and that it simply had to defend itself.[67] Albert Einstein's pacifist proclamation attracted four signatures, including his own!

But the blind enthusiasm for war was shortlived as it became obvious that Germany would not achieve a lightning victory on the Western front. After the horrendous loss of life brought by an enormous expenditure of munitions during the first Battle of the Marne (September 6–9, 1914) it was realized that trench warfare would require more explosives than foreseen before the war. In order to be able to wage a prolonged position war, the country had to restructure much of its resource base—and the Haber–Bosch process made a critical contribution to the success of this task, which so greatly increased the war suffering.

An allied naval blockade deprived BASF of its worldwide export market for textile dyes as effectively as it cut off Germany from its imports of Chilean nitrates needed for producing increasing quantities of munitions. German producers of sulfuric acid had also lost access to imports of foreign pyrites that they used as feedstock for their synthesis, a loss that affected the availability of the acid used by BASF to produce ammonium sulfate. By the end of 1914 it looked as if most of BASF's large Ludwigshafen plant would be out of production, and, as Emil Fischer and Walther Rathenau had pointed out, unless some extraordinary steps were taken, the country would largely run out of munitions by the spring of 1915.[68]

So, on the one hand, BASF faced seriously declining revenues due to the loss of its traditional markets, but, on the other hand, it had the capacity to convert its ammonia output into nitrates, the commodity in critical demand by the country's military. Here is clearly one of the origins of a phenomenon that marked so much of Germany's, and the world's, subsequent history: the rise of a military-industrial complex.[69] For BASF and the Second Reich this was a natural partnership: the country's leading company looking for new markets, a state at war desperately needing the company's newly developed product.[70] Although their interests coincided, Bosch found an astonishing degree of ignorance about the chemical prerequisites of modern war during his first meeting in the War Ministry in September 1914.[71]

The War Ministry was eager to expand significantly the output of nitric acid based on the two established processes (by-product ammonia and cyanamide, whose output was also to rise sharply) by adding BASF's ammonia oxidation. But as of September 1914 the company had done nothing but small laboratory experiments on converting ammonia to HNO_3. Consequently, switching Oppau's production from ammonium sulfate to nitric acid presented another large engineering challenge. As Alwin Mittasch recalled, when Bosch returned from Berlin, "I was suddenly asked if I thought it would be possible, on the basis of our laboratory experiments, and without using platinum . . . to begin the production of nitric acid as early as possible. My answer could be, after a short evaluation, positive."[72]

The commitment to commercial operation without the benefit of any existing pilot plant process was yet another calculated gamble on Bosch's part. Although the company already had a suitable catalyst to replace scarce and expensive platinum, new contact chambers of unprecedented size had to be built, and all the valves, pumps, coolers, and pipes had to be fashioned for the first time from a newly developed chromium-nickel stainless steel, which at that time could not be produced in large pieces. But by using soda from the Solvay process, Oppau could produce $NaNO_3$ and eliminate the country's dependence on Chilean nitrates.[73]

Once again, Bosch's gamble paid off. The plant, with a daily capacity of 150 t HNO_3 was finished on time, and the first deliveries for military use began in May 1915.[74] It became obvious very quickly that BASF's nitrate synthesis would have to be greatly expanded: output of by-product ammonia from coking ovens had a limited scope for expansion, and, in spite of the cyanamide industry's aspirations to become the premier supplier of fixed nitrogen (and to secure monopoly prices), it was clear that even its expanded synthesis would still leave the country short of the fixed nitrogen needed for the war, which was showing no signs of ending soon.

The German High Command favored the expansion of the Oppau plant as the fastest way to boost nitrate supplies, but the first French air raid on Ludwigshafen on May 27, 1915, led Bosch to suggest that the new plant should be located deep inside Germany.[75] In a meeting on February 4, 1916, Lieutenant Hermann Schmitz from the *Kriegsrohstoffabteilung* (the War Ministry's Department for Raw Materials) in Berlin asked Bosch if BASF was ready to build a large ammonia plant in central Germany and how long it would take. Bosch made yet another unprecedented commitment: he promised to build a plant with an annual capacity of 38,500 t NH$_3$ (31,700 t N) in just one year.[76]

Bosch's choice for the plant's location was in the heart of Germany's large brown-coal (lignite) mining region in Saxony, near the small village of Leuna on the southern outskirts of Merseburg on the Saale River about 20 km south of Halle.[77] The Saale was to supply the needed water, and the gypsum for the production of ammonium sulfate (to which the plant would largely convert after the war) was to come from a nearby mine in the Harz region.[78] Final agreement between BASF and the state treasury was signed on April 10, 1916, with the government contributing 12 million marks and BASF pledging to maintain a fixed price for its product. Construction at Leuna began on May 19, 1916. Initially it progressed well, but then it hit snags during the harsh winter of 1916/1917. In March 1917 Bosch sent Krauch to resolve the difficulties.[79]

As with Oppau, there was a big difference between the initially planned capacity (31,700 t N/year) and the eventual size of the plant. In December 1916 the decision was made to more than quadruple the annual production to 130,000 t N/year. In spite of mounting wartime difficulties and material shortages, the construction proceeded rapidly, and the first converter was turned on on April 27, 1917, just eleven months after construction began. The first ammonia was produced two days later.[80]

Closely patterned after the Oppau plant, Leuna's most distinctive feature was the use of steam engines, instead of gas engines, to run the synthesis gas compressors. This gave the plant its readily recognizable silhouette of thirteen large boiler buildings with adjoining tall stacks (fig. 5.10). In July 1918 plans were made for another expansion, to 200,000 t N/year. The plant reached that capacity only in 1923, becoming by far the world's largest ammonia synthesis enterprise, extending 2 km along the Halle-Erfurt rail line. A decade later, at that time a part of the I. G. Farben, Leuna occupied an area of 3.5 × 1.3 km, and it was the largest chemical plant in the world.[81]

Oppau and Leuna made the decisive difference to Germany's capacity to produce munitions: their success had undoubtedly delayed the Second Reich's collapse. In spite of the vigorous lobbying by the cyanamide industry to secure the largest share of the fixed nitrogen business during the war, the Haber–Bosch process emerged rapidly as the preferred method of nitrogen fixation. By 1918 ammonia synthesis produced 60% more fixed nitrogen than the cyanamide process, and it also surpassed the combined production of ammonia in Germany's numerous coking plants using by-product recovery.[82]

Figure 5.10
Ammonia plant in Leuna in a painting by Otto Bollhagen (1920). Courtesy of BASF Unternehmensarchiv, Ludwigshafen, Germany.

Oppau and Leuna also provided the country's chemical industry with postwar prospects to replace the dye markets, which Germany lost to foreign competitors during the years of the Allied blockade, with a profitable commodity: by the mid-1920s nitrogen fixation accounted for 65% of the I. G. Farben's huge profits.[83] But the cost of developing the new industry was enormous. Beike estimated that the total cost of BASF's development of ammonia synthesis was on the order of $100 million (using the January 1919 exchange rate), with half of the sum borrowed from the government.[84]

But the admirably rapid German commercialization of Haber–Bosch synthesis was not followed by a similarly rapid conquest of the world fertilizer market. Synthesis of ammonia from its elements did not become the dominant means of providing fixed nitrogen for agriculture until almost two decades after Oppau's startup, nor did the commercial availability of the process lead to a rapid increase in average applications of nitrogen fertilizers during the first half of the twentieth century.

6

Evolution of Ammonia Synthesis
Diffusion and Innovation

Nearly a century after its invention the Haber–Bosch process of ammonia synthesis from its elements remains fundamentally unchanged, yet at the same time radically transformed (fig. 6.1). This is only a seemingly contradictory statement: while Haber and Bosch would have the satisfaction of recognizing the essentials of their creation in today's modern ammonia plants, they would be amazed at the size of individual operations and at their efficiency, and they would certainly admire many innovations, large and small, that have made ammonia synthesis into one of the most remarkable industrial undertakings of modern civilization.[1]

These advances have not been a matter of steady, incremental progress. The innovation process in ammonia synthesis has been uneven, with periods of consolidation and stagnation clearly distinct from spells of fundamental and rapid change. The technique's development and diversification proceeded rapidly during the 1920s, but average plant outputs, engineering improvements, and energy efficiencies stagnated during the next two decades.[2] Global output of ammonia began rising rapidly after 1950—first thanks to the rapid growth of U.S. production, later because of a broader-based expansion—but by far the most far-reaching technical transformations were introduced only during the 1960s (fig. 6.2). Higher energy prices during the 1970s and the early 1980s aided the continuing quest for higher efficiencies of ammonia synthesis.

A steady increase of the worldwide output of fixed nitrogen had accompanied these innovations until the end of the 1980s. Average fertilizer applications already began leveling off in a number of affluent countries during the early 1980s, and the beginning of the 1990s saw the first decline in the worldwide synthesis of ammonia since the end of World War II. Its most important cause was the demise of the Soviet Empire and the great political and socioeconomic transformations in its wake rather than any worldwide weakening of the inherent long-term need.[3]

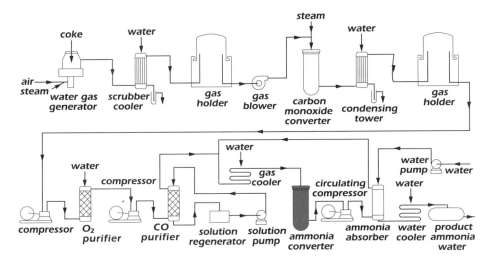

Figure 6.1
Classical Haber–Bosch process of ammonia synthesis.

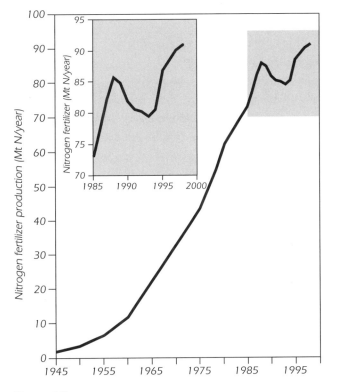

Figure 6.2
Global production of ammonia-based nitrogen fertilizers, 1945–1998.

By the end of the 1990s continuing strong growth of Asian production restored the global ammonia synthesis to the record levels of the late 1980s, but pinpointing the longer-term future demand remains uncertain (fig. 6.2). As I will explain in some detail in chapters 8 and 10, future demand is not just a simple function of continuing population growth: it will also depend on the composition of prevailing diets and on our success in making the use of fertilizers more efficient.

Slow Diffusion of Ammonia Synthesis: 1918–1950

Oppau did not produce any nitrogen fertilizer during the last year of World War I because all of its ammonia was used to synthesize HNO_3 for Germany's munitions industry. Ammonium sulfate production at the plant resumed only in June 1919, and output reached almost 13,000 t during the remainder of the year, increasing to more than 92,000 t in 1920. On September 21, 1921, a major explosion of ammonium nitrate and ammonium sulfate stored in a silo destroyed a large part of the plant and killed 561 people; in spite of this setback, that year's output of ammonium sulfate surpassed 190,000 t and in 1922 rose to 303,000 t.[4]

In 1922 BASF also introduced the first large-scale commercial process to synthesize urea from NH_3 and CO_2.[5] In 1926, a year after BASF joined the five largest German chemical companies in forming I. G. Farben, it patented *Nitrophoska*, the first mixed fertilizer containing balanced amounts of all three macronutrients.[6] Most of Leuna's capacity was also converted shortly after the war from producing nitrates to the synthesis of ammonium sulfate. After 1926 Leuna—which was, like Oppau, originally equipped with discontinuous production of water gas from coke—acquired a continuous supply of synthesis gas from the gasification of lignite, and Germany's output of synthetic ammonia, destined mostly for fertilizers, remained above 400,000 t N/year.[7]

The first transfer of the Haber–Bosch process abroad was a result of the Versailles Treaty, which the defeated Germany had to sign in 1919. By its terms, BASF was obliged to license construction of an ammonia plant with an annual capacity of 100,000 t in France. A newly set-up Office National Industriel de l'Azote located the plant in Toulouse and built it partially with French equipment; BASF supplied the water gas production process and high-pressure converters.[8] The British Ministry of Munitions planned to build Britain's first ammonia synthesis plant, and in 1918 it acquired a site at Billingham in North Yorkshire, which was bought in 1919 by Brunner, Mond & Company, who eventually developed the project.[9]

After three years of delays the U.S. government finally contracted with the General Chemical Company to build the first modified Haber–Bosch plant: U.S. Nitrate Plant No. 1 at Sheffield, Alabama. Only a single small unit was finished before the war ended, and many technical problems, above all rapid poisoning of a poor catalyst, prevented its successful operation. The first commercial synthesis of ammonia in the United States began in 1921 in a small plant (just 5,000 t N/year capacity) built by Atmospheric Nitrogen Corporation at Syracuse, New York; the second plant, built by Mathieson Alkali Works, was completed at Niagara Falls a year later.[10]

A steady increase of ammonia synthesis capacity brought the worldwide output to 930,000 t N in 1929. With that year's $CaCN_2$ synthesis total at 255,000 t N and fixation via electric arc at about 20,000 t N, ammonia synthesis accounted for roughly ¾ of all industrially fixed nitrogen, but still for less than half (43%) of all nitrogen in inorganic fertilizers (appendixes F and L; fig. 6.3).[11] Not surprisingly,

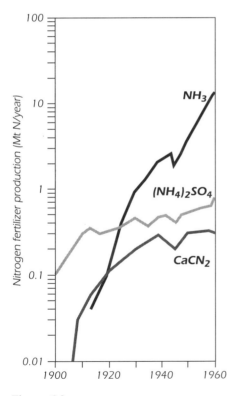

Figure 6.3
Ammonia synthesis became the dominant means of nitrogen fixation by the late 1920.

German output of Haber–Bosch ammonia was by far the largest. Continuing research done at Ludwigshafen also brought what in retrospect is clearly the most important innovation of the process introduced during the 1920s, the steam reforming of light hydrocarbons.

Pure methane (CH_4) is the most economical feedstock for reforming, because it has the highest H:C ratio of all hydrocarbons. Most of the natural gases are mixtures dominated by methane but also containing a few other gaseous and light liquid homologues of the alkane series; they, too, are excellent candidates for reforming.[12] It is also possible to use naphtha, a mixture of relatively light liquid hydrocarbons separated by fractional distillation of crude oil at temperatures between 30 and 260 °C. The general reforming reaction for alkanes is

$$C_nH_{(2n+2)} + nH_2O \rightarrow nCO + (2n + 1)H_2.$$

During the early decades of the twentieth century, Germany had no access to natural gas, but Bosch believed, even before Oppau went into production, that methane should be investigated as the source of hydrogen, and Mittasch began preliminary experiments in 1912. The gas for these experiments was catalytically synthesized in Ludwigshafen from CO and H_2.[13] During the 1920s Georg Schiller was eventually able to reform CH_4 in an externally heated tube oven in the presence of a nickel catalyst.

A license agreement transferred this technique to Standard Oil of New Jersey, which began using it for production of hydrogen in its Baton Rouge, Louisiana, refinery in 1931.[14] Two years earlier, the Shell Chemical Company's new plant at Pittsburgh, California, was the first ammonia plant in the world to use methane as a feedstock, but it relied on thermal cracking of methane, rather than on its steam reforming, to produce hydrogen and solid carbon.[15] The first ammonia plant based on steam reforming of methane was built by the Hercules Powder Company in California in 1939.

The 1920s also brought the introduction of several rival processes of ammonia synthesis. Two designs attempted to increase the reaction yield by raising the synthesis pressure far above the Haber–Bosch standard of 20 MPa at 500 °C. Georges Claude (1870–1938) demonstrated his process in 1917, and the first plant was built in Montereau, about 100 km south of Paris, in 1921. Instead of a single large converter it had several arrayed reactors through which the synthesis gas passed sequentially without recirculation under the pressure of 90–101 Mpa, the highest level ever employed in ammonia synthesis (fig. 6.4).[16] Naturally, more compact converters could be used with higher pressures, and expensive heat exchangers, needed for operations at lower pressure, could be eliminated.

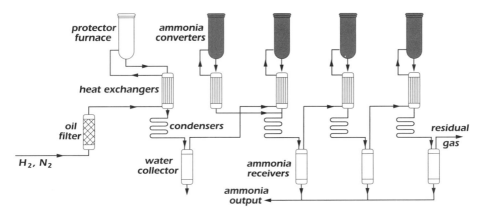

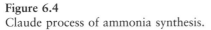

Figure 6.4
Claude process of ammonia synthesis.

Luigi Casale (1882–1927), whose experiments began in 1916, chose pressures around 20 Mpa and temperatures between 500 and 550 °C. The first commercial operation started in Terni in southern Umbria in 1921, and plants using the process were eventually built not only in Italy but also in France, Belgium, and Japan.[17] The other Italian variation—the process designed by Giacomo Fauser and adopted by the Montecatini Company—worked at 20–30 MPa and used H_2 from hydroelectricity-powered electrolysis; it became the country's most important means of nitrogen fixation before World War II (fig. 6.5). The process introduced by the U.S. Nitrogen Engineering Company was very similar to the Haber–Bosch design.[18]

In contrast, the Mont Cenis–Uhde process—developed by Friedrich Uhde to use by-product gas from the coke ovens of the Mont Cenis coal mine at Herne in the Ruhr area, and put into full operation in 1928—circumvented all high-pressure patents by using highly purified gas and new catalysts under only 9.1 MPa and 350–430 °C (fig. 6.6).[19] These rival designs achieved considerable commercial success before World War II: by 1939 about 90% of the world's 110 operating synthetic ammonia plants used the Casale, Fauser, Nitrogen Engineering, Mont Cenis, or Claude process (in that order of frequency), but considerably larger Haber–Bosch plants accounted for nearly half of all installed production capacity. After World War II Haber–Bosch plants, or their various modifications, became dominant.[20]

As with so many technical developments, the global financial collapse at the end of the 1920s and early 1930s and the subsequent economic depression brought a long period of stagnation and slow growth in the fertilizer industry. Oppau's

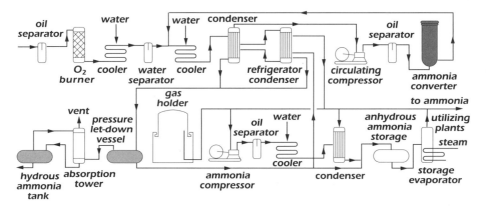

Figure 6.5
Fauser process of ammonia synthesis.

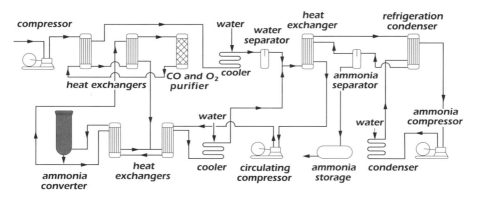

Figure 6.6
Mont Cenis–Uhde process of ammonia synthesis.

ammonia sulfate plant had to close so suddenly in December 1929 that it could not even use up all the gypsum stored in its silos.[21] Germany's pre–World War II output of ammonia peaked in 1928 at almost 850,000 t N; it was cut nearly in half by 1931, and it surpassed the 1928 peak only in 1939 (appendix I).[22] Pre–World War II U.S. ammonia output was only a fraction of German production (appendix J).[23]

The anticipated wartime demand for explosives led the U.S. government to embark on a large-scale construction of ammonia plants. The first one was completed in

1941; by 1945 the annual capacity of ten new plants added to about 880,000 t NH_3.[24] But this production gain could not make up for the wartime destruction of European and Japanese plants, which pushed the global ammonia output below the level of the late 1920s during the immediate post–World War II period. Worldwide production of fixed nitrogen began to rise sharply by the beginning of the 1950s: the end of food rationing in Europe, increasing disposable incomes and growing consumption of animal foods throughout the Western world, and the diffusion of hybrid corn in the United States (the first kind of high-yielding grain cultivar dependent on higher fertilizer applications) were the major reasons for the exponential growth of ammonia synthesis.[25]

Expansion and Changes since 1950

The United States emerged from World War II as by far the world's strongest economy, and its expansion of ammonia synthesis accounted for nearly ⅓ of the tripling of the global output during the 1950s (appendix J).[26] Construction of new ammonia plants was one of the early examples of a truly global business endeavor: a relatively small number of American, European, and Japanese engineering firms that held patents and licenses on various processes needed for a modern ammonia synthesis plant were eager to design new plants, oversee their construction, and help in launching their operation anywhere in the world.

Several trends characteristic of American ammonia synthesis in the 1950s have since become nearly universal features of the nitrogen-fixing industry: dominance of natural gas as the source of hydrogen, spatial concentration of new large plants near gas wells (or along natural gas pipelines) and near water transportation (Texas, Louisiana, and Oklahoma emerged as the preferred U.S. locations), and integration of ammonia synthesis with production of various solid and liquid fertilizers, including not only urea and urea-ammonia solutions but also complex compounds containing phosphorus.[27]

During the 1960s the worldwide capacity of ammonia plants grew about 2.3 times to 42 Mt N/year, and this rate of growth was almost matched during the 1970s: the rising demand for nitrogen fertilizer in Asian and Latin American agricultures adopting new high-yielding cultivars of rice, wheat, and corn was the single largest reason for that expansion. Expansion of Soviet synthesis, based on abundant natural gas resources discovered in Central Asia and Siberia, was a major contributing factor. Considerably slower growth prevailed during the 1980s and continued during the 1990s. The global capacity of ammonia plants surpassed 120 Mt NH_3 in 1990, and by 1999 it topped 160 Mt NH_3.[28]

The four principal steps of the Haber–Bosch process have remained constant throughout this period as the global ammonia synthesis capacity expanded more than twentyfold: production of feedstock gases, shift reaction, which nearly eliminates CO and yields more H_2 and CO_2, removal of CO_2 and CO, and the high-pressure catalytic synthesis itself. A diminishing number of ammonia plants, accounting for about one-tenth of the global production capacity, still use either bituminous coal or lignite as the feedstock. During the 1970s and the early 1980s, when OPEC's actions brought a sharp rise in world oil prices, there was a renewed interest in coal-based ammonia synthesis in North America and Europe, but the return to low-cost oil after 1985 made it a short-lived phenomenon.[29]

Relatively large modern coal-based syntheses now operate only in several Asian countries (mostly in India and Pakistan) and South Africa, but most of the world's coal-based ammonia is synthesized in hundreds of small plants in China. Most of these coal-based ammonia plants were built between 1963 and 1975; their typical annual capacities were just between 3,000 and 5,000 t NH_3/plant, and they were highly inefficient, consuming in excess of 120 GJ/t NH_3.[30] Their numbers peaked in 1979 (with 1,539 units), and the subsequent upgrading reduced the total to 750 units, with an average annual capacity of 25,000 t NH_3, in 1997.

Their contribution to China's overall ammonia synthesis peaked in 1990 at almost 58% and then fell to 49% by 1997. Many of these plants now produce urea, aqua ammonia, ammonium chloride (NH_4Cl), or ammonium sulfate, but most of their output is still converted to ammonium bicarbonate (NH_4HCO_3), a highly volatile compound that should be vacuum-packed for distribution and incorporated deep into soil to eliminate large storage and postapplication losses (even when properly packed it will lose 10–20% of its weight in a year).

A new process of partial oxidation, first developed in Ludwigshafen between 1936 and 1939, made it possible to use almost any hydrocarbons regardless of their molecular weight; because of their low cost, heavy fuel oils have been the most common choice.[31] After preheating they are mixed with high-pressure steam and oxygen, and the partial combustion proceeds at 1,200–1,500 °C without any catalysts. Shell and Texaco developed the two widely used partial oxidation techniques of heavy hydrocarbons. Dozens of plants, including some large operations in India and China, are based on partial oxidation of hydrocarbons. But because the heavier hydrocarbons subjected to partial oxidation have a lower H:C ratio than does natural gas, the process uses more feedstock and requires larger facilities to handle larger volumes of CO that has to be converted to CO_2 and then removed.[32]

The two leading coal gasification processes, Lurgi and Koppers-Totzek, are also partial oxidations done under pressure and high temperature (the first process at 560–620 °C, the other at more than 1,400 °C). Since the mid-1970s ammonia synthesis has also been using by-product hydrogen or waste gases rich in hydrogen, including streams from the production of caustic soda (99.86% H_2), CO (97% H_2), ethylene (84.4% H_2), and methanol (80% H_2).[33] But most of the hydrogen must come from steam reforming of natural gas.

Both the U.S. and Russian expansions of ammonia capacities since the 1950s have been based overwhelmingly on natural gas. European ammonia syntheses remained coal-based until the 1950s, but then the rising crude oil imports led to new plants based on heavier hydrocarbons. Natural gas–based expansion was made possible first by the development of the giant Groningen field in the northern Netherlands, then by natural gas deliveries from rich North Sea fields, and finally by massive exports of Russian gas, which began during the mid-1980s.[34]

New Asian, Middle Eastern, and Latin American ammonia plants built since the 1960s have been based overwhelmingly on natural gas. Consequently, reforming of natural gas has come to dominate the global synthesis of ammonia: gas-based plants, which also use the methane as the principal source of their process energy, now account for about 80% of the world's ammonia production capacity.

Natural Gas–Based Ammonia Synthesis

Operation of natural gas–based plants follows a basic sequence whose details differ among various proprietary processes (fig. 6.7).[35] Natural gas must first be purified to remove H_2S and any dust or liquid particles. Initial filtering is followed by compression to reformer pressure and preheating, and H_2S sulfur is removed by absorption by zinc oxide (ZnO).[36] The gas is then mixed with superheated steam and reformed in two stages. Primary reforming, producing a mixture of H_2 and CO, is done inside alloyed stainless steel tubes packed with a nickel catalyst; in most ammonia plants these tubes are placed in a furnace heated by gas (or liquid fuel) to 750–850 °C.

The first thermal reforming units operated at an only slightly elevated pressure. Subsequently increases of pressure, heat fluxes, and temperatures, as well as better

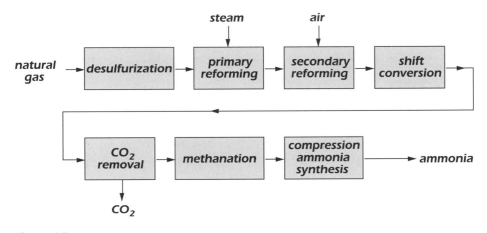

Figure 6.7
Principal stages of natural gas–based ammonia synthesis.

heat recovery, raised the efficiency of reforming. The process remains the single largest consumer of energy in ammonia synthesis. Traditional primary reformer designs differ most obviously in the arrangement of tubes and burners, in the gas distribution, and in the choice of tube alloys.

Tubes are made either of chromium-nickel or chromium-niobium alloys, and they are expected to last at least for 100,000 hours (more than eleven years). The latest primary reformer designs have dispensed with the furnace altogether, and the process is heated by gases from the secondary reformer. The reforming reaction is always the same

$CH_4 + H_2O \leftrightarrow CO + 3H_2 + 206$ kJ/mol.

The exit gas is oxidized in the secondary reformer, a cylindrical vessel lined with a refractory material whose lower part is filled with the catalyst; the exothermic reaction results in temperatures on the order of 1,000 °C. Oxidation is necessary in order to reduce the amount of unconverted methane (anywhere between 5–15% of the gas is still unreacted at that point) and to isolate the appropriate amount of N_2 for the synthesis gas containing the stoichiometric $H_2:N_2$ ratio of 3:1:

$CH_4 + 2O_2 \leftrightarrow CO_2 + 2H_2O - 35.6$ kJ/mol.

After leaving the secondary reformer the gas has 56% H_2, 23% N_2, 12% CO, 8% CO_2, and less than 0.5% CH_4, as well as excess steam. After passing through a heat exchanger the cooled mixture is subjected to a two-step shift reaction in the presence of catalysts. Fe_3O_4 and CrO are used to catalyze the first, high-temperature, stage, and CuO, ZnO, and Al_2O_3 are used for the low-temperature segment:

$$CO + H_2O \leftrightarrow CO_2 + H_2 - 41.2 \text{ kJ/mol}.$$

After the shift conversion the synthesis gas must be stripped of its residual amounts of CO (about 0.3%) and CO_2 (about 0.1%). Starting in the early 1960s methanation—the reverse of the reforming—replaced absorption by cuprous ammonium solutions as the means of final CO removal; it also removes the remaining traces of CO_2:

$$CO + 3H_2 \rightarrow CH_4 + H_2O - 206.1 \text{ kJ/mol},$$

$$CO_2 + 4H_2 \rightarrow CH_4 + 2H_2O - 164.9 \text{ kJ/mol}.$$

Methanator temperatures are between 300 and 400 °C, and the strongly exothermic reactions go almost to completion, leaving behind just a few ppm of oxygen-containing compounds that could act as catalyst poisons.

Synthesis gas, containing 74% H_2, 24% N_2, 0.8% CH_4, and 0.3% Ar, is then compressed to the design pressure, stripped of any moisture in molecular sieves (in order to prevent degradation of the catalyst) and of any excess nitrogen, and fed into the converter to undergo a strongly exothermic ammonia synthesis:

$$N_2 + 3H_2 \leftrightarrow 2NH_3 - 91.8 \text{ kJ/mol}.$$

Metallic iron catalysts, based on magnetite (Fe_3O_4) and promoted with Al_2O_3, KCl, and Ca, speed up the reaction at temperatures mostly between 400 and 450 °C in converters with different gas-flow arrangements (axial, radial, crossflow).

Modern large-capacity ammonia plants invariably use multiple bed converters. In order to keep the converter wall cool and to prevent its embrittlement by hydrogen, the incoming synthesis gas usually flows through an annular space between the converter shell and the catalyst cartridge. Only then does the gas enter an internal heat exchanger, where it is preheated to synthesis temperature before it flows through two or three vertical or horizontal catalyst beds (fig. 6.8). Large cross-sectional areas ensure a very low pressure drop even when very small catalyst particles are used.

The exit gas, containing 12–18% NH_3, is first cooled by heat exchange with the incoming synthesis gas, then by air or water, and finally by refrigeration, whose temperature depends on the mode of storage. If the gas is to be stored at atmospheric pressure, it must be cooled to −33 °C, ammonia's boiling point. The unreacted synthesis gas is recompressed and recycled into the converter. A large number of innova-

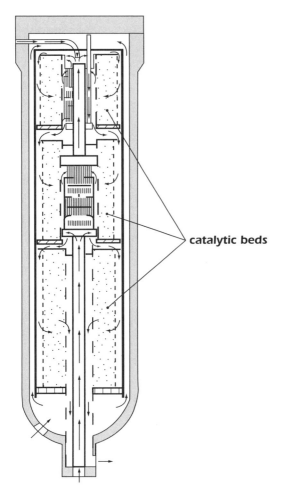

catalytic beds

Figure 6.8
Haldor Topsøe's three-bed radial flow converter. Arrows indicate the flow of the synthesis gas.

tions in every stage of gas preparation and ammonia synthesis has added up to impressive gains in unit capacities and overall process efficiencies.[37]

Until the early 1950s natural gas reformers worked only under low pressure (less than 1 MPa); successively higher reforming pressures (first over 2, and then even up to 4 MPa), made possible by better alloys and better fabrication techniques, reduced the size of equipment and the volume of catalysts, lowered the downstream energy requirements for gas compression, eliminated the need to compress the process gas for the purification stage, and recovered more waste heat.[38]

Removal of CO_2 has moved through three different stages. Absorption in water, the original method of the CO_2 removal, was abandoned during the 1940s because of its high energy intensity and the considerable solubility of H_2 and N_2 in water under high pressure. Until 1960 a 15–20% solution of monoethanolamine was used; other reaction systems use triethanolamine and methyl diethanolamine. But since then the dominant means of CO_2 removal, available in more than a half dozen proprietary versions, has been the use of hot potassium carbonate (K_2CO_3) solution containing additives promoting the process and inhibiting corrosion.[39] The compound absorbs CO_2, and when the resulting potassium bicarbonate ($KHCO_3$) is heated it releases the gas and regenerates K_2CO_3.

Although modern iron-based catalysts are clearly similar to mixtures selected by Mittasch, there have been many improvements in their preparation and efficacy. The smaller particle size of catalysts—1.5–3 mm used in radial or horizontal flow converters instead of 6–10 mm used in older axial converters—has increased their catalytic surface area per unit volume and hence improved the overall catalyst efficiency.[40] But there is no doubt that the technical innovation with the greatest impact on the modern ammonia industry has been the introduction of large single-train plants using centrifugal compressors, a development that brought not only much larger plant average capacities but also much improved energy efficiencies.

Single-Train Plants with Centrifugal Compressors

During the first fifty years of their existence ammonia plants were set up as parallel synthesis loops, called trains, with the gas compression requirements of each unit supplied separately by a reciprocating compressor (during the last year of its operation in 1990 Leuna had eighteen synthesis loops). These compressors were powered initially either by coke-oven gas (as in Oppau) or by steam (as in Leuna), but electric motors soon became the prime movers of choice. Reciprocating compres-

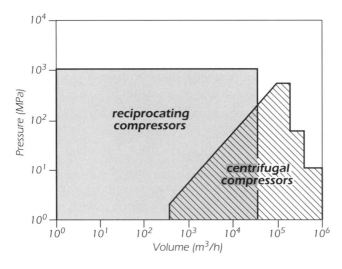

Figure 6.9
Comparison of reciprocating and centrifugal compressor performances.

sors are the only machines that can produce extremely high pressures (up to 100 MPa) through several stages of gas compression (fig. 6.9). They are also excellent energy convertors (with efficiencies up to 87%), but they are expensive both to install and to maintain, and they are not suitable for working with very large volumes of air.[41]

By the beginning of the 1960s the largest design capacity of an ammonia synthesis train was just over 300 t NH_3/day. With insufficient storage capacity needed to hold the output during the periods of low demand, these plants operated year-round by simply shutting down one or more trains.[42] As the development of refrigerated units enabled the storing of ammonia under atmospheric pressure, and as the long-distance transport by pipelines and vessels became possible, year-round synthesis at high capacity could be realized.

The M. W. Kellogg Company of Houston, an engineering firm active in oil and petrochemical industries and engaged in ammonia synthesis since 1943, pioneered a radically new design that optimized the entire production process rather than merely improved some of its key segments and that could produce ammonia in much larger quantities at a much lower cost.[43] All the compression needs in a plant were provided by a single centrifugal machine.[44] Because these compressors could be powered by steam turbines, it has been possible to integrate the plant's energy use and maximize the energy conversion efficiency of the whole energy cycle of ammonia synthesis by using steam as the working medium.[45]

A great deal of attention has been thus paid to minimizing the major causes of energy losses throughout the process, above all from surface condensers associated with steam turbines, from CO_2 removal (due to the regeneration of absorbing solutions), and from furnace stack gas. As a result, waste heat recovered throughout the plant is sufficient not only to drive the compressor but also to energize endothermic reactions, pumps, and reheating needs. Kellogg designers have also made changes in the reforming and converting of gases.[46]

Depending on the process design, plants with reciprocating compressors used pressures up to 80 MPa, with typical pre-1963 compressors operating with synthetic loop pressures of 30–35 MPa. In contrast, most modern large plants use pressures just around 15 MPa, and the latest ones operate at below 10 MPa.[47] Operations at higher pressure—with more favorable position of the equilibrium, higher reaction rates, and lower refrigeration costs during ammonia recovery—require the use of high-activity catalysts, while lower pressures make a large economic difference due to lower energy requirements for gas compression. Much lower compressions reduce both plant investment and operating costs per tonne of ammonia—but because centrifugal compressors can be throttled back to no less than about 70% of capacity, they could be installed only in plants that maintain high-volume production throughout the year.[48]

In December 1963 M. W. Kellogg completed the world's first single-train ammonia plant for the American Oil Company at Texas City. This plant was designed to produce 600 t NH_3/day. Shortly afterwards the company built another plant with identical capacity for the Monsanto Company at Luling, Louisiana, which used only centrifugal compressors, as well as the world's first ammonia plant rated at 1,000 t NH_3/day for the Coastal Chemical Company at Yazoo City, Mississippi. Four large single-train ammonia plants were operating in the United States by 1966, ten by 1967, and by the end of the decade twenty-eight such plants accounted for half of total U.S. ammonia synthesis capacity.[49] Kellogg's innovation ushered in the greatest capacity buildup in the history of the ammonia industry, the process led by large orders placed by China and Indonesia as they began their quest for self-sufficiency in nitrogen fertilizer production.[50]

Synthesis of ammonia has been a highly competitive business since the 1920s, and so it was inevitable that other engineering firms soon followed Kellogg's lead in designing, building, and perfecting large ammonia plants. Every leading designer and builder of these plants—British ICI, Danish Haldor Topsøe, German Krupp Uhde, Japanese Chiyoda, Kinetics Technology India—has introduced a number of proprietary improvements, ranging from new catalysts to modification of reforming and shift processes and better converters.[51] Still, Kellogg Brown & Root remains the most successful provider of advanced, large-scale, low-energy ammonia synthesis plants. By 1999 more than 150 Kellogg-designed plants were operating in some thirty countries, accounting for more than half of the world's NH_3 capacity added since the mid-1960s.[52]

The most obvious consequence of widespread adoption of the single-train design was the rapidly rising average capacity of ammonia plants.[53] In 1962 the largest U.S. plant produced about 800 t NH_3/day, but the average installed capacity of sixty U.S. ammonia plants was just short of 200 t NH_3/day. The typical size of a new plant surpassed 300 t NH_3 day by 1965, and 500 t by the early 1970s. The modal size of the world's new ammonia plants rose from 200–400 t/day in the early 1960s to 800–1,000 t/day a decade later, and to 1,200–1,400 t/day during the 1980s, when the largest plants surpassed 1,600 t/day.

During the late 1970s there seemed to be no technical obstacles to constructing a single-train plant producing 2,700 t NH_3/day—but no such plant has been actually built. Most of the new large ammonia plants completed or commissioned during the 1990s have capacities between 1,000 and 1,500 t NH_3/day, and the largest ones—Topsøe designs for P. T. Kaltim Pasifik Amoniak of Indonesia and Profertil S.A. of Argentina—rate, respectively, 2,000 and 2,050 t NH_3/day.[54]

Several ammonia complexes containing three to seven large single-train plants have annual capacities in excess of 1 Mt: Togliattiazot in Togliatti, Russia, is the world's largest (2.46 Mt N), followed by complexes in Veracruz, Mexico (Cosoleacaque, 1.7 Mt N), Donaldsonville in Louisina (1.55 Mt N), and Point Lisas in Trinidad (1.4 Mt N).[55] Perhaps the most obvious outward demonstration of this growth has been the increasing size of ammonia converters, illustrated by comparing Haber's first experimental unit (fig 4.6) and BASF's first designs (fig. 5.3) with such modern assemblies as Haldor Topsøe's 23 m tall unit (fig. 6.10).

Ammonia production is now one of the world's two largest chemical syntheses. Worldwide capacity of ammonia plants rose from Oppau's 9,000 t N in 1913 to about 2.5 Mt on the eve of World War II; it surpassed 20 Mt in 1958 and 100 Mt N in 1979, and in 1998 it stood at 156 Mt NH_3, or 128 Mt N.[56] Modern ammonia

Figure 6.10
Haldor Topsøe's ammonia converter installed at the Aonla plant in India (height 23 m, diameter 3 m, annual capacity 500,000 t NH$_3$). Courtesy of Haldor Topsøe, Lyngby, Denmark.

plants operate typically at 83 (80–85)% of their installed capacity, which means that during the past half century global ammonia output has risen from less than 4 Mt during the late 1940s to 129 Mt (106 Mt N) by 1998. Ammonia and sulfuric acid are thus the world's two most important synthetic chemicals in terms of annually produced mass: during the late 1990s their mass output was virtually identical, but ammonia (molecular weight 17 vs. 98 for H_2SO_4) was number one in terms of synthesized moles.[57]

Construction of much larger, and spatially more concentrated, ammonia plants necessitated requisite advances in storage and transportation. Traditional high- and medium-pressure storage tanks were replaced by fully refrigerated vessels able to hold liquid ammonia at $-33\ °C$ at atmospheric pressure; these tanks are now also used on barges and ships. Pressurized tanks (at 2.75 MPa) are still used on jumbo railroad cars carrying between 63 and 72 t NH_3. In the United States the major agricultural regions with the highest demand for nitrogen fertilizer are served by ammonia pipelines (the first one, from northern Texas to northern Iowa, was built during the late 1960s).[58]

Continuing Innovation

Performances of single-train plants with centrifugal compressors have continued to improve since the early 1960s: a combination of incremental improvements and radical redesigns has been behind these gains. Several fundamental design changes came during the 1980s. In 1984 Chiyoda proposed eliminating the primary reformer furnace and energizing the first, endothermic reforming step by the heat from the second, highly exothermic, reaction. This idea was commercialized for the first time by the Imperial Chemical Industries as a part of the company's Leading Concept Ammonia (LCA) Process. The process was designed to replace the company's two oldest, and least efficient, ammonia plants at Severnside near Bristol with two relatively small units whose performance would match efficiencies previously achievable only in large plants.[59]

The LCA Process diverged from the traditional design of ammonia plants in a radical way, as it replaced the integrated process with a core unit containing all the key gas preparation and synthesis processes (gas desulfurization, reforming, shift conversion, methanation, and synthesis) and a separate utility area. LCA hydrodesulfurization of natural gas is done at a low temperature, and the primary reformer furnace is replaced by a tubular, compact, pressurized gas-heated reformer (GHR), which operates at 3.7 MPa and uses hot secondary reformer outlet as its heat source. Ammonia is synthesized in the converter under a pressure of 8 MPa.

Improved alloy selection has reduced the frequency of tube failures exposed to high temperatures and pressures inside primary reformers.[60] KBR's Advanced Ammonia Process (KAAP) is the first ammonia synthesis design that is not based on the classical magnetite catalyst but on a precious metal, ruthenium, with co-promoters.[61] This catalyst, supported on a graphite structure and developed jointly with British Petroleum, in some applications has up to twenty times the activity of magnetite-based materials, and it makes it possible to achieve high ammonia yields (18–22%) under milder operating conditions: converter pressures are just 7–9 MPa, and exit temperatures are 350–420 °C. However, the first bed of the four-bed KAAP reactor uses the traditional magnetite catalyst in order to benefit from high reaction rates achievable with this material at low NH_3 concentrations.

Another key feature of this new process is the patented KBR Reforming Exchanger System (KRES), which eliminates the need for the directly fired primary reformer furnace (fig. 6.11). This reduces capital costs by 5–8%, and emissions of NO_x and CO_2 by 70–75%. The first KAAP process was installed for the Ocelot Ammonia Company's (now Pacific Ammonia Inc.) retrofit of its plant in Kitimat, British Columbia, in 1992; and when the first two new KAAP plants (for Farmland Misschem and PCS Nitrogen) were completed at Point Lisas, Trinidad, in July 1998, they became the world's largest ammonia facilities (each at 1,850 t/day) to date.[62]

All advanced plant designs deliver high operational reliability: design capacities are commonly available 98–100% of the time. But by far the most impressive result of these continuing improvements of natural gas–based ammonia synthesis has been a steady decline in energy intensity, that is, in the total amount of fuels and electricity needed to produce a tonne of the compound (appendix K). The energy requirements of chemical syntheses are made up of the heat equivalents of fuels and electricity used in the synthetic process, and of the energy embodied in the feedstocks. Ammonia's lower heating value is 18.6 GJ/t, and the stoichiometric energy requirement for ammonia synthesis is 20.9 GJ/t (fig. 6.12).

After accounting for the purge gas, net feedstock inputs of modern ammonia plants are very close to this value. A purge stream must be drawn in order to control the concentration of inert gases in the recycled flow; recovery of H_2 from the purge gas (cryogenically, by membrane separators, or by pressure swing absorption)

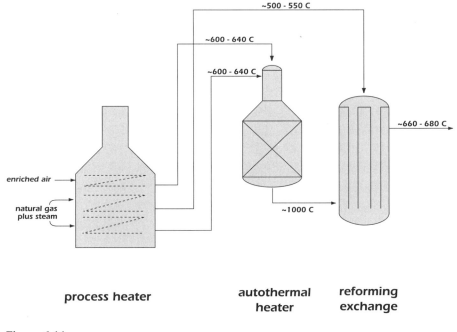

~500 - 550 C

~600 - 640 C

~600 - 640 C

~660 - 680 C

enriched air

natural gas
plus steam

~1000 C

process heater

**autothermal
heater**

**reforming
exchange**

Figure 6.11
KBR's reforming exchanger system.

and its recycling reduces the amount of the gas produced by steam reforming. But while the exothermic ammonia synthesis should supply heat (about 580 MJ/t of NH_3), the industrial process actually requires a substantial input of energy. Reduction of this energy demand during the eight decades of NH_3 synthesis has been impressive.[63]

Coke-based Haber–Bosch synthesis in Oppau initially needed more than 100 GJ/t NH_3, and typical pre–World War II plants required around 85 GJ/t NH_3 (fig. 6.12). During the 1950s natural gas–based synthesis with low-pressure reforming (0.5–1 MPa) and reciprocating compressors needed between 50–55 GJ/t NH_3. In the early 1960s total energy consumption of a typical plant working with reformer pressure of 1 MPa and synthesis loop pressure of 35 MPa would have added up to around 45 GJ/t NH_3.

A decade later a large single-train plant—with high-pressure reforming (above 3 MPa) and centrifugal compressors producing converter pressure 15 MPa—reduced the energy need to 35 GJ/t NH_3.[64] Because their compressors are powered by

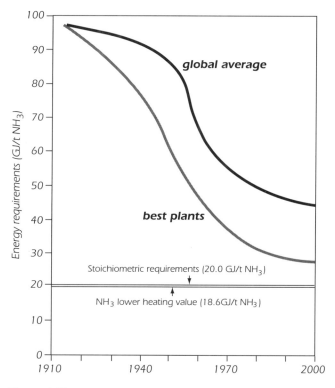

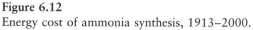

Figure 6.12
Energy cost of ammonia synthesis, 1913–2000.

steam, the electricity demand in these plants is a small fraction of the need in the pre-1963 units. Reciprocating compressors powered by electric motors needed 520–700 kWh/t NH_3 (the typical U.S. rate was about 630 kWh/t), while with steam turbine drives in much larger post-1963 plants, the overall electricity consumption fell to as little as 20–35 kWh/t NH_3.[65]

By the late 1970s further gradual efficiency improvements brought the average energy needs of ammonia plants down to just around 30 GJ/t NH_3 (fig. 6.12). Then the early 1980s brought another substantial efficiency gain: M. W. Kellogg and Krupp Uhde began offering new designs requiring less than 28.9 GJ/t NH_3.[66] Kellogg's latest KRES design saves another 1 GJ/t NH_3, and Topsøe's HER (Heat Exchange Reformer) cuts the need for reformer fuel by 70% and the total natural gas consumption by 15%: the company's best designs now need about 30% less energy than in the early 1970s.[67] The most efficiently operating plants in the 1990s needed only about 27 GJ/t NH_3, and rates of 25–26 GJ/t appear to be technically feasible. The lower rate would be less than 20% above the stoichiometric natural gas requirements.

Naturally, average worldwide energy efficiency of ammonia synthesis—which is affected by more energy-intensive reforming of heavier hydrocarbons and coal—has improved as well: before 1955 the global mean of energy consumption in all ammonia plants (natural gas, oil, and coal-based) was at least 80 GJ/t NH_3, by 1965 it declined to just over 60 GJ/t, and by 1980 it was a bit more than 50 GJ/t.[68] The best estimate for the late 1990s would be between 40–45 GJ/t.[69] Even with these huge improvements the synthesis of ammonia remains an energy-intensive process, particularly when compared to the production of other inorganic fertilizers.

Energy requirements for producing phosphorous fertilizers add up to as little as 15 GJ/t P for phosphate rock and as much as nearly 30 GJ/t for complex P-containing compounds.[70] Mining, crushing, and beneficiation of potash (KCl) usually does not cost more than 10 GJ/t of the nutrient.[71] As we will see in the next chapter, only a small part of synthesized ammonia is used directly as a fertilizer: conversion of NH_3 to liquid and solid nitrogen fertilizers that dominate the field applications requires a considerable amount of additional energy.[72]

Using rather generous means of 50 GJ/t N, 20 GJ/t P, and 8 GJ/t K indicates that during the mid-1900s the global production of these macronutrients would have claimed, respectively, almost 100 Mt, just over 7 Mt, and no more than 4 Mt of oil equivalents.[73] Global production of nitrogen fertilizers thus requires about nine times as much energy as the combined output of P and K compounds. Even if all of the energy needed to fix the fertilizer nitrogen would come from natural gas, it would be just 5% of the recent annual global consumption of the fuel, and about 1.3% of all energy derived from fossil fuels.[74]

Clearly, there is little reason to be anxious either about the current needs or the future supplies of energy for nitrogen fixation. We could release energy equivalent to today's global fixation demand merely by burning natural gas in the world's industrial boilers and household furnaces with 7% higher efficiency, or by improving the current average gasoline consumption of private cars by about 12%.[75] And although the world has abundant natural gas resources, nitrogen fixation could proceed easily (albeit more costly) even in a world devoid of any gas by tapping the earth's enormous coal deposits.[76]

Looking ahead, there is no doubt that higher absolute energy needs brought about by rising demand for nitrogen fertilizers (see chapter 8) will be partially offset by lower requirements for nitrogen fixation and by higher efficiencies of fertilizer use (see chapter 10). Moreover, today's low-income countries will see much faster growth of energy needs in sectors other than the fertilizer industry.[77] As a result, the share of global fossil fuel consumption claimed by nitrogen fixation by the middle of the twenty-first century may be only marginally higher than it is today.[78] Clearly, the future nitrogen fixation needed to produce enough food for the additional 2–4 billion people the planet will house by the year 2050 will not be limited by energy and feedstock supplies.

Synthetic Fertilizers
Varieties and Applications

Anhydrous ammonia can, and in the United States and Canada actually is, applied directly to fields—but most of the ammonia destined to become crop fertilizer is used indirectly, as the key feedstock in a complex industry producing a large variety of solid and liquid compounds. Although their diverse properties make individual nitrogen fertilizers, singly or in combination with other essential nutrients, more or less suitable for particular field applications, appropriate agronomic practices, rather than specific compounds, generally have a much greater effect on the efficiency with which the fixed nitrogen is taken up by crops. Urea became the world's most important fertilizer because of its combination of high nutrient content and the ease of applications.

Ever since the end of World War II the synthesis of ammonia has been driven predominantly by the rising demand for nitrogen fertilizers. Between 1950 and the early 1980s their average applications kept increasing even in small Western European countries that already had high field cultivation intensities, but beginning in the 1960s much higher gains were made in many populous, low-income countries, particularly in Asia. A major shift has been evident since the late 1980s: average applications have kept rising, albeit less rapidly, in nearly all populous low-income countries, but they have either stagnated or have significantly declined in most of the industrialized nations.

A comprehensive account of nitrogen inputs reaching the world's farmlands shows that during the 1990s synthetic fertilizers provided about half of the total nutrient available to annual and permanent crops. Expectedly, there are large regional departures from this mean: some export crops aside, nitrogen applications throughout sub-Saharan Africa are either absent or remain extremely low, while the world's highest rates prevail in the countries of northwestern Europe and East Asia. Synthetic nitrogen fertilizers are now the dominant sources of the nutrient in all of the world's most intensively cultivated agroecosystems, be they the cornfields of the U.S. Midwest, wheat lands of Atlantic Europe, or paddies of South China. Managed nitrogen inputs commonly add up to 90–95% of the nutrient reaching intensively farmed fields.

Nitrogen Fertilizers

About 7% of synthesized ammonia is lost in production, transport, and storage, and nearly 10% of ammonia's global output recently has been going into products other than fertilizers. The long-established use of the compound as the key feedstock for the production of explosives has been joined by applications in organic syntheses now producing such common materials as nylon, perlon, plexiglass, polyurethane, urea formaldehyde, amines, and amides. Other relatively large users of ammonia include oil refining, metallurgical, rubber, and pulp and paper industries, and the compound remains an important industrial refrigerant.

Only a negligible fraction of today's synthetic nitrogen fertilizers does not start with ammonia (appendix L).[1] And while during the first forty years after the commercialization of ammonia synthesis the compound served only as the feedstock for making more complex fertilizer compounds, since the 1950s it has also been applied directly to fields, although this practise has been restricted to North America. Another usage largely confined to North America has been the spreading use of nitrogen in fertilizer solutions. Certainly the most important global trend has been the increasing conversion of ammonia to urea. As a result of these changes, applications of traditional synthetic nitrogen fertilizers have been declining.

Ammonium sulfate—the first nitrogenous fertilizer made from ammonia by BASF and the leading form in which fixed nitrogen was supplied before World War II—now accounts for a small and falling share of the world fertilizer market (about 4% in 1985, 3% in 1997).[2] The compound is chemically stable, it is not highly hygroscopic, and it is, of course, particularly suitable to providing nitrogen in sulfur-deficient soils (fig. 7.1).[3] But its low nitrogen content (21%) increases its packaging, storage, and distribution costs, and its decomposition in the soil is strongly acid-forming, a disadvantage in most nonalkaline soils; consequently, the sulfate's use has been in prolonged decline. However, increasing sulfur deficiencies in fields have brought some renewed interest in the compound (it contains 24% S) during the 1990s.[4]

Compound	Formula	Nitrogen content (%)		Solubility at 20 C (g/100g H_2O)	Density (g/cm^3)
		Pure	As marketed		
Ammonium nitrate	NH_4NO_3	35.0	30.0 - 34.5	187	1.725
Ammonium sulfate	$(NH_4)_2SO_4$	21.2	21.0	75	1.769
Calcium nitrate	$Ca(NO_3)_2$	17.0	15.5	172	2.230
Urea	$CO(NH_2)_2$	46.6	45.0	107	1.335

Figure 7.1
Properties of major solid nitrogen fertilizer compounds.

The opposite is true about calcium nitrate, which is produced mostly by neutralization of HNO_3 by ground limestone. The compound's decomposition helps to improve acidified soils, its calcium displaces the sodium that is absorbed by soil clays, and it is also perhaps the best compound to use on saline soils (fig. 7.1).[5] Only a few European countries (the Netherlands, Norway, Portugal, and the Ukraine) and Egypt produce appreciable amounts of calcium nitrate. Calcium ammonium nitrate (CAN), produced by the rapid mixing of a concentrated ammonium nitrate solution with ground or precipitated $CaCO_3$ and containing between 21 and 27.5% N, is a more popular fertilizer than calcium nitrate (it had about 4% of the world nitrogen market in 1997), but its production has been stagnating.[6]

In contrast, ammonium nitrate (35% N) remains a leading fertilizer. It had about 10% of the world nitrogen market in 1997, largely because of its widespread use in Europe (fig. 7.1).[7] Its advantages come from the combination of two forms of nitrogen, which provide extended and immediate nutrient supply. The ammoniacal nitrogen must be first nitrified to become readily available to plants, while the highly soluble nitrate nitrogen can be assimilated almost immediately after the fertilizer application. Nitrate nitrogen is thus useful in regions with a short vegetation period, while in cooler and wetter climates ammoniacal nitrogen is not nitrified rapidly, and a large part of it is not then lost to leaching. And the application of ammonium nitrate is not as acid-forming as that of ammonium sulfate, nor does it release volatilized ammonia, which may damage seedlings, as does urea.

But ammonium nitrate is a far from ideal fertilizer. Unless the soluble nitrate is promptly assimilated, the leaching losses of nitrogen are higher than with ammoniacal fertilizers; it is not the best choice in rice fields; it is very hygroscopic; and although it does not burn itself and is difficult to detonate, it is a powerful oxidizing agent, and heating it in confined space can result in highly destructive explosions.[8] The last concern goes beyond safe storage and handling of the compound before its application to fields: ammonium nitrate is also a main ingredient of the ammonium nitrate fuel oil (ANFO) mixture that has been widely used not only as an industrial explosive but also in powerful car and truck bombs deployed by terrorists around the world.[9]

Urea, a highly water-soluble white crystalline compound, has by far the highest nitrogen content among solid fertilizer compounds (46.6%); as noted, its commercial synthesis was introduced by BASF in 1922 (fig. 7.1). Its good physical properties and lack of fire and explosion hazards make it the most commonly used solid nitrogen fertilizer, while liquid urea is commonly used in nitrogen solutions, and the prilled compound is a common part of mixed fertilizers.

All of today's industrial processes for urea synthesis still rely on the same basic reaction combining liquid ammonia with CO_2 to synthesize ammonium carbamate, which is then dehydrated to produce solid urea:

$$2NH_3 + CO_2 \rightarrow NH_2CO_2NH_4,$$

$$NH_2CO_2NH_4 \rightarrow CO(NH_2)_2 + H_2O.$$

The first reaction is exothermic and goes almost to completion, while the second, endothermic, reaction converts less than 70% of feedstocks to urea. Consequently, the once-through process of urea synthesis had to be combined with production of other nitrogen compounds. Modern syntheses use a more expensive total recycle process in which all the unreacted CO_2 and NH_3 is separated from the effluent solution and returned to the reactor.[10] Conversion rates range between 65 and 67% for each pass, and the overall reaction yield is on the order of 99%. Three commercial processes—Dutch Stamicarbon, Italian Snamprogetti, and Japanese Toyo—dominate today's world market for large (up to 2,000 t/day) urea plants using the total recycle process.[11]

Expensive synthesis—it takes at least 35% more energy to supply fixed nitrogen as urea than as NH_3, and both ammonium sulfate and ammonium nitrate also require less heat and electricity (about 25% and 15% less) to produce—was only one reason for urea's slow diffusion as field fertilizer.[12] Some earlier experiments indicated that urea may be less effective than other sources of nitrogen, largely because under some conditions the compound releases its nitrogen much more slowly than do nitrates.

The hydrolysis of urea releases ammonia, which is then converted by bacterial nitrification to nitrite and finally to highly soluble nitrate ions:[13]

$$CO(NH_2)_2 + H_2O \rightarrow 2NH_3 + CO_2,$$

$$2NH_4^+ + 4O_2 \rightarrow 2NO_3^- + 2H_2O + 4H^+.$$

Nitrification rates decline significantly in lower temperatures, reducing the availability of nitrogen in cooler climates.

Two other agronomic considerations that prevented urea's wide adoption in cooler regions of Europe were its possible phytotoxicity when placed close to seeds, and volatilization losses of ammonia produced by hydrolysis. But today there is no doubt that a properly used urea is as effective for cereal fertilization in cool climates as ammonium nitrate, even when it is used in alkaline soils.[14] Although urea has found a somewhat greater acceptance in North American farming, its importance in Asia's rice fields has made it the world's leading solid nitrogen fertilizer.

Results based on nine to eleven years of studies showed that the nitrogen uptake, its agronomic efficiency and grain yield in tropical rice, were highest with applications of relatively expensive urea super granules, and only slightly lower when using prilled urea.[15] Unlike other staple grains, rice can efficiently assimilate ammonium nitrogen, which becomes available without long delays in warm tropical and subtropical climates where urea's hydrolysis (and ammonia nitrification) proceeds rapidly. Volatilization losses from properly applied urea in rice fields are also lower than denitrification losses from nitrates in anaerobic flooded paddy soils.[16]

Urea's expanding production has thus been closely tied with increasing synthesis and application of nitrogen fertilizers in the rice-growing countries of Asia. In 1960 less than 10% of the world's nitrogen fixed by the Haber–Bosch process was converted to urea; the share reached 15% in 1968, doubled to 30% by 1973, and in 1997 stood above 46%. China, India, Indonesia, and Pakistan are the world's largest urea producers, and most of the new ammonia capacity is directly tied to urea synthesis in integrated ammonia-urea plants.[17]

In contrast to Asia's dependence on urea, North America's pattern of nitrogen fertilizer use has moved from solids to fluids, a diverse category including anhydrous ammonia, aqua ammonia, and unpressurized nitrogen solutions. The availability of pressurized containers for distribution (by railcars, barges, and trucks), storage and application of ammonia, and the construction of ammonia pipelines carrying the liquid from Texas and Louisiana to the Corn Belt, made anhydrous ammonia the leading nitrogen compound for direct agricultural use during the 1960s and 1970s.[18]

Because of its volatility, the liquid has to be injected 15–30 cm into the soil using special application knives.[19]

During the 1980s the use of liquid ammonia in the United States stabilized, and during the 1990s it declined, but direct applications of liquid ammonia still accounted for almost ⅓ of the U.S. nitrogen fertilizer market (and for almost 6% of worldwide use in 1997).[20] Declining use of liquid ammonia has been accompanied by the rising popularity of various nitrogen solutions. These solutions may use just a single compound (aqua ammonia), but most commonly they are mixtures of two or more nitrogen fertilizers. Aqua ammonia, which is safer to handle than the anhydride, usually contains 20% N.[21]

The most popular nonpressurized nitrogen solutions are prepared from urea and ammonium nitrate and contain between 28 and 32% of nitrogen. No special equipment is needed for their storage and application; they are easier to mix and to handle than solids, have low distribution losses, can be accurately placed, are suitable for foliar and aircraft application, and are excellent carriers of micronutrients as well as of pesticides and herbicides. As a result, about 60% of U.S. fertilizer nitrogen is now applied in fluids, and the use of solutions is becoming also more widespread in Europe.[22]

One other notable worldwide trend has been the increasing share of nitrogen applied in mixed fertilizers. These materials range from such traditional choices combining two macronutrients as potassium nitrate (with 13.8% N) and various ammonium phosphates (containing anywhere between 10 and 21% N) to granulated or bulk-blended materials combining all three macronutrients, often with the addition of selected micronutrients.[23]

Fertilizer Applications: Global Views

No inorganic nitrogen was applied to any crops before Chilean nitrates reached the European and American markets in the 1840s. By 1900 nitrogen exports and by-product ammonia recovered from coking ovens added up to less than 350,000 t N (appendix F).[24] In 1913 Oppau's ammonia synthesis was a marginal addition to nearly 800,000 t of fixed nitrogen in Chilean nitrates, by-product ammonia sulfate, cyanamide, and compounds produced from nitrogen oxides generated in electric arcs (appendix F).[25]

Synthetic ammonia provided half of the world's inorganic nitrogen only in 1931; by 1950 that share was close to 80%, and by 1962 it surpassed 90% (appendix

L).[26] During the late 1990s Haber–Bosch synthesis supplied more than 99% of fixed inorganic nitrogen; the tiny remainder (about 0.7%) comes mostly from Chilean nitrates and by-product ammonia from coke ovens (appendix L).[27]

Pre–World War II nitrogen fertilizer applications, which rose to 3 Mt N by 1939, barely registered on the global scale of nitrogen inputs to farm soils: at best they amounted to less than 1/4 of all nitrogen received by the world's farmland every year from the atmospheric deposition of ammonia and nitrates. Only in a few European countries, Japan, and Egypt did inorganic nitrogen make a significant difference. Dutch agriculture was by far the most intensive user of nitrogen fertilizers before World War II, with applications averaging 50–60 kg N/ha, compared to 20–25 kg N/ha in Germany and less than 3 kg N/ha in the United States.[28]

Global consumption of nitrogen fertilizers rose to 3.63 Mt N during the crop year 1949/1950, it stood at 9.23 Mt N a decade later, and it had more than tripled to 31.7 Mt N by 1970.[29] In spite of sharply higher world energy prices, consumption doubled during the 1970s to 60.7 Mt by 1980, and to about 80 Mt N eight years later (appendix M; fig. 7.2). Most of this spectacular increase was due to the rapid diffusion of new, shortstalked, high-yielding varieties of wheat and rice, an agronomic shift that became commonly known by the evocative but inaccurate name of the Green Revolution.[30]

Norman Borlaug, one of the leaders in developing and diffusing these new cultivars, summed up the importance of fertilizer nitrogen in his speech accepting the Nobel Peace Prize in 1970 by using a memorable kinetic analogy:

If the high-yielding dwarf wheat and rice varieties are the catalysts that have ignited the Green Revolution, then chemical fertilizer is the fuel that has powered its forward thrust. . . . The new varieties not only respond to much heavier dosages of fertilizer than the old ones but are also much more efficient in its use. The old tall-strawed varieties would produce only ten kilos of additional grain for each kilogram of nitrogen applied, while the new varieties can produce 20 to 25 kilograms or more of additional grain per kilogram of nitrogen applied.[31]

Stagnation in the global use of nitrogen fertilizers after 1988 was due to a drastic decline of demand in the former Soviet Union and in other post-Communist European economies and to a more measured reduction of fertilizer use in most countries of the European Union, whose excessive agricultural subsidies led to unsustainably high levels of fertilizer applications. Global consumption of nitrogen fertilizers fell below 73 Mt N in 1993 and 1994, but during the following year it was up, and in 1996 it surpassed 80 Mt N (appendix M; fig. 7.2).[32] Further steady increases should bring the annual consumption to over 85 Mt N by the first year of the new century.[33]

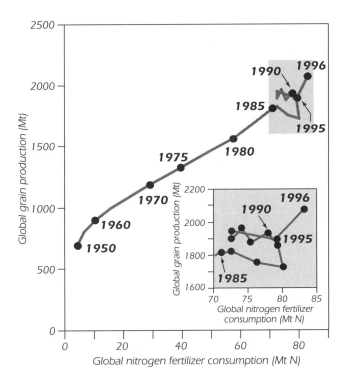

Figure 7.2
After decades of steady growth global consumption of nitrogen fertilizers—as well as the global grain harvest—declined and stagnated during the early 1990s before resuming their growth.

Commonly used national averages of nitrogen applications per hectare of cultivated land are misleading for several reasons. Obviously, different crops do not benefit equally from these applications, and the means are particularly meaningless for countries where a few principal crops (often permanent plantation species such as oil palm and fruit trees) receive the bulk of the nutrient. Some countries use a large share of the nutrient on permanent pastures, and dividing total fertilizer use by cropped area gives exaggerated results. In the Dutch case it more than doubles the amount of nitrogen applied to field crops! Finally, application rates should be adjusted for various degrees of multicropping: high fertilization rates in Asia's monsoonal agricultures, where double- and even triple-cropping of rice is common, are then cut by ½ or ⅔ in order to make proper comparisons with regions producing just a single crop per year.

Consequently, the most revealing way to compare the intensity of fertilization is to use averages for individual crops, but few countries regularly publish such data. The best available global breakdown of nitrogen fertilizer use by individual crops comes from worldwide surveys organized by the IFA, FAO, and IFDC. Fertilizer application rates based on informed estimates, rather than on actual field surveys, are now available for more than ninety countries.[34]

During the mid-1990s the combined total of all fertilizers used for food and feed cereals was 55% (wheat received 20%, corn 14%, and rice 13% of the total); 12% of all fertilizer went to oil crops, 6% to tubers, 5% to fruits and vegetables, and 4% to sugar crops. Pastures and fodder and silage crops received about 11%, and fiber crops (mainly cotton) no more than 4% of all fertilizer. The remaining 3% included the fertilizing of legumes, cocoa, coffee, tea, and tobacco. Consequently, about 95% of all fertilizer is used to produce food and feed crops. The highest relative fertilizer applications go to bananas, sugar beets, citrus fruit, vegetables, and potatoes. The highest national averages are between 120 and 180 kg N/ha for rice and wheat, 120 and 150 kg N/ha for corn, and 30 and 60 kg N/ha for soybeans (appendix N). Surveys also show very low nitrogen applications in many low-income countries, particularly for corn.

When the application rates per hectare of arable land are used just as indicators of global trends, they prorate to negligible values of less than 1 kg N/ha until the early 1920s and by 1950 reach about 4 kg N/ha, ten times the rate in 1900. Subsequent rapid increases brought the means to about 11 kg N/ha in 1960, double that amount in 1970, and to 40 kg N/ha by 1980. During the 1990s the annual means fluctuated around 50 kg N/ha (fig. 7.3).[35] Per capita rates of nitrogen fertilizer consumption rose from 0.2 kg in 1900 to 2 kg in 1950 and 15 kg in 1990 before falling a bit: the rate in 2000 was almost 14 kg N/capita (fig. 7.3).[36]

Consequently, the twentieth century had seen a roughly 125-fold increase in the average global rate of inorganic nitrogen applications per hectare of cropland, and less than a seventy-fold increase in average per capita use. The rapid recent increases of nitrogen applications are perhaps best illustrated by the fact that half of all nitrogen fixed by the Haber–Bosch process between 1913 and 2000 was consumed only during the last two decades of the twentieth century.[37]

Naturally, the often cited global consumption rates hide an enormous amount of spatial variation. By far the most obvious gap used to be an enormous disparity between average applications in affluent and low-income countries.[38] Both the production and consumption of synthetic nitrogen fertilizers used to be concentrated

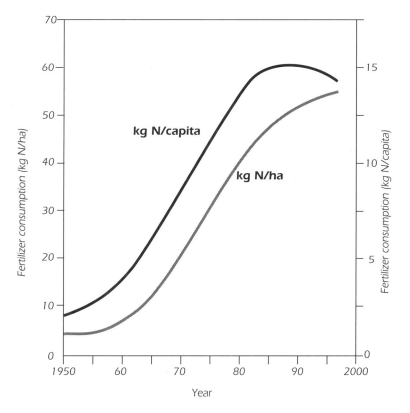

Figure 7.3
Global means of nitrogen applications per hectare of arable land and per capita, 1950–1996.

overwhelmingly in high-income, industrialized countries. Europe, North America, and Japan consumed 95% of all nitrogen fertilizer in 1950, about 82% in 1960, and still almost 75% in 1970. Only in 1989 did agricultures of those two very unequal segments of humanity—at that time roughly 1.2 billion people in high-income nations and 4.1 billion inhabitants of low-income countries—consume an equal mass of nitrogen fertilizers, about 40 Mt N/year each (fig. 7.4; appendix M).[39]

By the late 1990s, with low-income countries applying almost ⅔ of the world's nitrogen fertilizer, the disparity was finally erased at least in terms of fertilizer use per hectare of cropland, but, of course, it remains large in per capita terms.[40] After the recent decline in average applications of fertilizers in North America, Japan, and

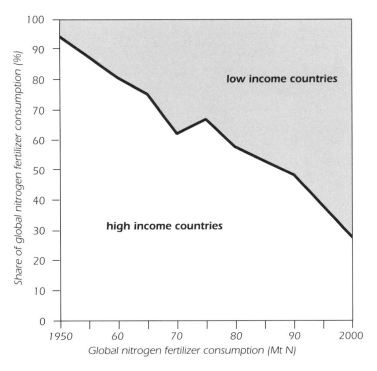

Figure 7.4
Shares of global nitrogen fertilizer consumption, 1950–2000.

Europe, the high-income countries are using annually almost 45 kg N/ha of their cropland, while the mean for low-income countries has surpassed 55 kg N/ha. But the respective per capita means, at 25 kg and 10 kg N, are still far apart—and in spite of the anticipated growth of total applications in low-income countries and stagnation, or even further slight declines, in rich nations, this gap will not close rapidly.[41]

Fertilizer Nitrogen in Global Crop Production

The most accurate way to calculate the nitrogen taken up by the global crop harvest is to multiply the output of every major crop by the nitrogen contents of harvested parts (grains, stalks, leaves, roots, fruits).[42] Global crop harvests of the mid-1990s amounted to about 2.65 Gt of dry matter per year, and they contained 50 Mt N (appendixes O and P). Cereals accounted for 60% of total dry crop phytomass, and they also contained 60% of all removed nitrogen. Sugar crops ranked second in dry mass (16%), but legume crops were second in assimilated nitrogen (20%).

Quantification of nitrogen incorporated by crop residues is not as straightforward, as no country keeps statistics on their production. My calculations resulted in the most likely total of 3.75 Gt of dry residual phytomass.[43] Cereal straws, stalks, and leaves accounted for $^2/_3$ of all residual phytomass, and sugarcane tops and leaves were the second largest contributor. As already explained in chapter 2, substantial inter- and intraspecific variations in nitrogen content result in a rather broad range of the nutrient incorporated annually in crop residues.[44]

Nitrogen taken up by the world's crop residues amounted to about 25 Mt/year during the mid-1990s, or $^1/_3$ of the total taken up by crop biomass (appendix P). Quantifying annual harvests of forages grown on arable land (alfalfa, clovers, vetches, and various legume-grass mixtures) is even more uncertain; the best estimate of the total yield of these forages is about 500 Mt of dry matter containing at least 10 Mt N.[45]

Consequently, the global annual crop harvest of the mid-1990s was about 7 Gt of dry matter, and about 85 Mt N were incorporated in this phytomass (appendixes O and P). This total is very close to the annual mean of almost 80 Mt of nitrogen applied recently in synthetic fertilizers, but the approximate equivalence of the two flows does not, of course, mean that Haber–Bosch fixation now supplies all of the nutrient needed in the world's agriculture. In order to find its share of total supply it is necessary to calculate, and to add up, all the other nitrogen-bearing inputs reaching the world's crops. To avoid an appearance of unwarranted precision, these fluxes should be presented as ranges rather than as single value estimates.

Planted phytomass (seeds and roots) and irrigation water are two minor sources of nitrogen, the first one returning annually only about 2 Mt N, the second one contributing about twice as much.[46] Nitrates (mainly from oxidation of NO_x released by combustion of fossil fuels) and volatilization of NH_3 (from animal wastes, soils, and plant tops) are the two largest sources of fixed atmospheric nitrogen, whose total annual deposition on the world's agricultural land is about 20 (18–22) Mt N.[47]

As no country systematically traces the fate of its crop residues, it is necessary to account for many competing uses of this phytomass—mainly as household fuel and building material, bedding and feed for animals, substrate for cultivation of mushrooms, and feedstock for making paper and extracting organic compounds. In addition, a significant share of straws and stalks is still burned after the harvest in the fields.[48] The most plausible range of crop residue recycling would return 12–16 Mt N annually to the world's croplands.

As detailed in chapter 2, wide variations of manure output and its nitrogen content make calculations of the nutrient recycling in animal wastes highly prone to error. My conservative assumptions result in the global voiding of about 75 Mt N during the mid-1990s, less than several other recently published estimates.[49] About 40% of this total, or some 30 Mt N, is either collected for recycling or is directly returned by grazing animals to harvested fields and cropland pastures. After preapplication losses, close to 20 Mt N would eventually be applied in recycled animal manures.[50]

In order to make the best possible estimate of symbiotic biofixation I have used specific ranges for eight kinds of seed legumes (beans, broad beans, chickpeas, lentils, peas, peanuts, soybeans, and other pulses). This resulted in the most likely global total of 10 Mt N fixed by *Rhizobium* symbiosis with leguminous grains, and a fairly conservative total of N fixation in 100–120 Mha of cropland forages and green manures (averaging 100 kg N/ha) is about 12 Mt N/year.

Biofixation by non-*Rhizobium* diazotrophs is of lesser importance, although cyanobacteria (mainly *Anabaena* and *Nostoc*) in rice fields can fix 20–30 kg N/ha during the growing season. Nitrogen fixation by *Anabaena azollae* may have contributed between 4 and 6 Mt N/year in 150 Mha of the world's paddies. Recently discovered endophytic diazotrophs (*Acetobacter*, *Herbaspirillum*) living inside sugarcane roots, stems and leaves add another 1–3 Mt N for the world's 18 Mha of sugarcane.[51] The total biofixation in agricultural soils would then have been 33 (25–41) Mt N/year during the mid-1990s.

The grand total of all nitrogen inputs reaching the world's croplands during the mid-1990s would then add up to 151–186 Mt N/year, with about 170 Mt N/year being the most likely figure (appendix Q). Haber–Bosch synthesis of ammonia is thus the source of approximately half (the most likely range being 44–51%) of all fixed nitrogen reaching the world's croplands. As seeds, organic recycling, and irrigation water supplied about 33 (29–37) Mt N a year, human management was responsible for 80–85% of all nitrogen received by the world's agricultural land (appendix Q).

Regional and National Perspectives

As must be expected, differences in nitrogen applications go beyond the split between rich and poor countries: regional and national application averages show significant departures from both global and continental means. The key worldwide determinants of high rates of nitrogen applications are high population densities, scarcity of farmland, and a high share of irrigated cropland. Combination of these variables explains Egypt's high use of nitrogen, while the rest of the African continent consumes less than 3% of the global supply of the nutrient, although it has about 12% of the world's population.[52]

The part of the world that most urgently needs increased fertilizer use is sub-Saharan Africa, where an insufficient supply of nutrients results not just in low crop yields but in the continuing decline of soil fertility. So far the most comprehensive assessment of soil degradation concluded that less than 30% of nitrogen (as well as just 15% K and less than 40% P) needed by the region's crops is actually replaced by fertilizers.[53]

In order to stop this degradation the region should quintuple its currently negligible use of nitrogen, averaging a mere 10 kg/ha. In order to ensure the region's food security this should be done within one decade, and fertilizer use would have to grow by nearly 20% a year.[54] In reality, the region's nitrogen applications actually fell during the early 1990s.[55]

In contrast to sub-Saharan Africa, Asia's food production has increased considerably, and the rapidly increasing use of nitrogen fertilizers—from 18.6 Mt in 1975 to 45 Mt N in 1995—applied to high-yielding rice and wheat crops has accounted for most of that gain.[56] East Asian gains have been particularly impressive, and China's transition from traditional cropping done without any inorganic fertilizers to the world's largest user of inorganic nitrogen is the best illustration of this rapid transformation.

Intensive recycling of a large variety of organic wastes and cultivation of green manures remained the mainstays of China's nitrogen supplies during the first two decades after the establishment of the Communist regime in 1949. Historical reconstructions of nitrogen inputs into China's agriculture show that synthetic fertilizers provided less than 5% of nutrition during the early 1950s and that the share was still less than $1/3$ of the total by 1970 (appendix R).[57]

The contribution of inorganic nitrogen added up to about $3/5$ of all inputs by the mid-1980s, and a comprehensive account of nitrogen inputs into China's agriculture shows that the recycling of organic matter, biofixation, atmospheric deposition, and seeds contributed less than 9 Mt N in 1996 compared to 23.6 Mt N applied in synthetic fertilizers during the 1995/96 crop year (appendix R; fig. 7.5).[58] This means that ammonia-based compounds have been recently supplying about 75% of all nitrogen, and that the mean nationwide applications is now around 170 kg N/ha of

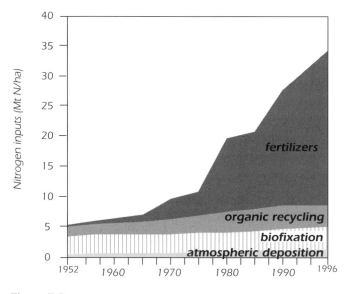

Figure 7.5
Nitrogen inputs into China's cropping, 1952–1996.

arable land, the rate higher than the declining Japanese average (about 160 kg N/ha in 1985, 120 kg N/ha in 1995/96).[59]

No other region in the world uses as much nitrogen as the four coastal provinces in East and South China: Jiangsu, Zhejiang, Fujian, and Guangdong. Using the official statistics of nitrogen consumption for 1995 and farmland areas derived from MEDEA's satellite study, their application rates ranged from about 275 kg N/ha in Jiangsu to 350 kg N/ha in Zhejiang, and the region's mean was almost exactly 300 kg N/ha (fig. 7.6).[60] Because the region's farmland is generally double-cropped, the average per crop would be around 150 kg N/ha. Elsewhere in Asia only South Korean and Taiwanese applications (also with general double-cropping) are now over 200 kg N/ha.

Several national accounts of nitrogen inputs in Western European agricultures show that in those countries the shares of synthetic fertilizers are somewhat lower than in East Asia. A very detailed Norwegian study, tracing every input of the nutrient into the country's food production, found that the applications of nitrogen fertilizers (averaging about 125 kg N/ha of farmland) supplied about 64% of the nutrient during the early 1990s.[61] Reconstructions of average nitrogen inputs into German cropping show the applications of inorganic nitrogen rising from about 2 kg N/ha

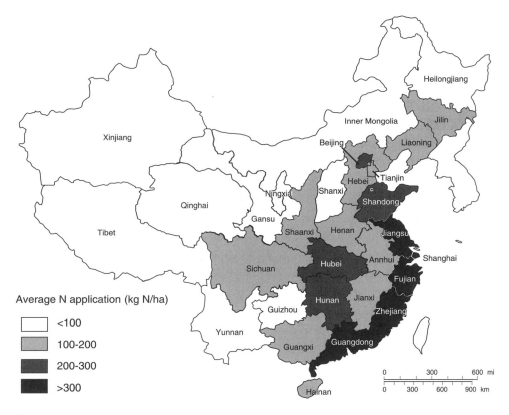

Figure 7.6
Average nitrogen fertilizer applications in China's provinces in 1995.

in 1900 to 125 kg N/ha by 1985, and its share in the total nutrient supply increasing from less than 5% in 1900 to ⅓ in the late 1950s, to about 60% by the year 1995 (appendix S; fig. 7.7).[62]

In the United Kingdom more than half of all nitrogen fertilizer has been applied to grasslands. A Royal Society study found that in the late 1970s average applications on pastures surpassed the inputs to arable land (172 vs. 135 kg N/ha), and that synthetic compounds accounted for 57–63% of all inputs.[63] The overall use of fertilizer nitrogen in the United Kingdom rose by almost 50% between the late 1970s and the mid-1980s, but it declined afterwards, and its average during the late 1990s has been only about 20% higher than a generation earlier, which means that the synthetic fertilizers supply between 65 and 70% of all nitrogen inputs.[64] But high-

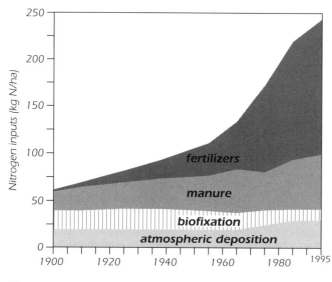

Figure 7.7
Nitrogen inputs into German cropping, 1900–1995.

yielding winter wheat—the 1998 mean was 7.97 t/ha—still receives more than 180 kg N/ha, double the amount applied in 1970 when the yield was around 4 t/ha, and the secular correlation between the rising applications of inorganic nitrogen and rising harvests is obvious (fig. 7.8).[65]

As already noted, Dutch farmers led the world in applications of inorganic nitrogen during the first decades of the twentieth century. Although the country lost this position, it remains a large consumer of nitrogen fertilizers. As in the United Kingdom, pastures receive high nitrogen applications. The nutrient balance of an average dairy farm in 1990 showed fertilizer inputs of 440 kg N/ha, amounting to ⅔ of the total supply.[66] The national mean per hectare of cropland and grassland rose from nearly 70 kg N in 1950 to about 230 kg N during the mid-1980s; since then it has declined to about 185 kg N by 1995.[67] Although other nitrogen inputs are relatively high—Dutch farmers practice intensive recycling of pig and cow manures, cultivation of legumes is widespread, and farmland receives one of the world's highest nitrogen inputs from atmospheric deposition—nitrogen from the Haber–Bosch fixation accounts for nearly 80% of the total input.[68]

When judged by its overall use of nitrogen, U.S. agriculture appears to be quite dependent on synthetic fertilizers: its annual consumption of more than 11 Mt N

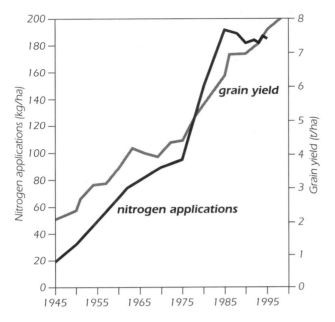

Figure 7.8
Nitrogen applications and yields of English winter wheat, 1945–1998.

in the late 1990s makes it the world's second largest user of nitrogen fertilizer, far behind China (more than 25 Mt N) but still ahead of India (more than 10 Mt N). But average applications remain fairly low, and a comprehensive account of nitrogen inputs received by U.S. farmland during the mid-1990s shows that, unlike in Western Europe or East Asia, nonfertilizer nitrogen still supplies at least half of the nutrient.

Average U.S. applications remained below 50 kg N/ha until the late 1970s, and they reached about 60 kg N/ha in 1980, when the overall consumption surpassed 10 Mt N. Since that time total applications have fluctuated mostly between 10 and 11 Mt N, so the average rate during the 1990s was still no more than 60 kg N/ha, far behind the typical European or East Asian means (fig. 7.9).[69] But, much more so than in the European case, the U.S. mean hides great disparities in fertilization of major crops. Corn has been the most widely fertilized crop ever since the widespread adoption of hybrid varieties during the 1950s, and it has also received the highest average applications and hence the largest share of the country's total nitrogen consumption.

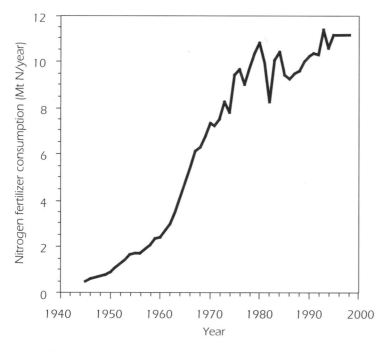

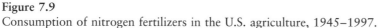

Figure 7.9
Consumption of nitrogen fertilizers in the U.S. agriculture, 1945–1997.

By the mid-1960s 90% of all corn fields received nitrogen fertilizer, and the rate was 98% during the late 1990s. Average annual applications rose from less than 50 kg N/ha in the early 1960s and have been above 140 kg N/ha since 1979. The crop received about 25% of all N consumption in the early post–World War II years, 40% by 1960, and about 50% since the mid-1980s.[70] In contrast, half of all U.S. wheat fields received some nitrogen fertilizer only in 1964, and although that share is now well over 80%, average applications are just around 60 kg N/ha, less than half the amount used on European wheats. And the country's second most important crop—soybeans, whose total production has been recently slightly higher than the combined winter and spring wheat harvest—receives only small amounts of supplementary nitrogen (about 20 kg N/ha) on less than 20% of its planted area.[71]

Several special characteristics add up to a relatively large nonfertilizer nitrogen input to U.S. cropland. Widespread cultivation of alfalfa, soybeans, and legume-grass mixtures means that symbiotic biofixation provides more than 5 Mt N every year;[72] although it is done primarily in order to limit soil erosion, the extensive recycling of crop residues returns about 2.5 Mt N, and animal manures produced by the country's enormous livestock sector add a further 2 Mt N. Atmospheric deposition recently has been the source of as much as 2 Mt N to America's cropland.[73] Inorganic fertilizers thus supply between 45 and 50% of all nitrogen received annually by America's farmland (appendix T).

The Most Productive Agroecosystems

Three of the world's most important cereal grains—rice, wheat, and corn—receive the largest share of available nitrogen fertilizers, and some of the highest applications are reserved for these crops, particularly in regions where soils and climate combine to provide optimal growing conditions. Comparisons of nitrogen inputs to rice grown in Jiangsu or Hunan (the two provinces with the highest yields in China), winter wheat grown in northwestern Europe (the highest yields are in the coastal region from the Loire to Denmark and in England), and corn grown in the U.S. Corn Belt (from Indiana in the east to eastern Nebraska in the west, and from northern Missouri to southern Minnesota) show similar shares of fertilizer nitrogen in the total supply of the nutrient.

Where Hunanese rice is still grown in a rotation including other food crops as well as green manures, the nonfertilizer inputs remain relatively high.[74] The three-year cropping sequence of rice, wheat, rice, green manure, rice, rapeseed would receive 300–320 kg N/ha as urea and another 150 kg N/ha from biofixation, organic recycling, and atmospheric deposition annually. Synthetic nitrogen would thus supply at least 67% of the total input.[75] In Jiangsu—where green manuring is virtually absent, biofixation may supply just 25 kg N/ha, and organic recycling, irrigation, and atmospheric deposition will add no more than 80 kg N/ha compared to 380 kg N/ha applied to rotations of rice and wheat—inorganic nitrogen will supply almost 80% of the total input.[76]

French, Dutch, and English winter wheats, whose yields average close to 8 t/ha, receive close to 200 kg N/ha from synthetic fertilizers, and 60–80 kg N/ha from recycled crop residues, biofixation and atmospheric deposition. Fertilizer nitrogen is thus the source of 70–75% of the total supply.[77] With corn monocropping there is only a small input of nitrogen from free-living diazotrophs. Atmospheric deposition adds much less than 10 kg N/ha in the driest parts of the Corn Belt, and where no manures are applied, crop residues are the only substantial nonfertilizer input of nitrogen. In such cases synthetic compounds may supply just over 70% of the nutrient for a crop yielding 8 t/ha.[78] Rotations of corn with soybeans may not reduce that rate too much as the harvests of high-yielding soybeans may leave no symbiotically fixed nitrogen for the subsequent corn crop.[79] Only rotations with alfalfa would lower the share of fertilizer nitrogen to well below 60%.

Other major crops receiving high nitrogen applications include potatoes and sugar beets. With no manure applications, but with all the tops (about 35 t of fresh weight/ha) plowed in, applications of about 300 kg N/ha will account for almost 80% of the total nutrient supply in Dutch beets.[80] The fertilizer share in Maine potatoes, again with no manure and with all tops recycled, would be around 70%.[81] Very high fertilizer inputs are also common in European farms combining intensively managed pastures and field cropping. In many of these cases supplementary feed brought from the outside (often actually imported from overseas) represent a major source of nitrogen deposited on the grassland in animal excrements.[82]

After subtracting the shares retained in milling residues, Jiangsu rice-wheat rotation can produce about 800 kg protein/ha, and Hunanese rotations, centered on the double-cropping of rice, as well as the most productive European wheat monocropping yield between 650 and 700 kg protein/ha.[83] This is 2.5–4 times more than the output of the most productive traditional agricultures relying on intensive recycling of organic matter and planting of legumes. With an overwhelmingly vegetarian diet this production could feed 35–45 people/ha, compared to 12–15 people/ha in the best premodern settings.[84] Rotations of corn and soybeans in Iowa can produce as much as 850–900 kg of protein/ha, and populations willing to subsist on a vegetarian diet based on these two foodstuffs could support fifty people from every hectare of farmland.

Actual maxima of large-scale means of modern carrying capacities in low-income countries are inevitably lower. They reflect diets that are overwhelmingly but not purely vegetarian, as well as a variety of environmental stresses limiting the frequency of cropping and its yields and the necessity to cultivate non-food crops. A hectare of farmland receiving more than 150 kg N and averaging 1.5 crops a year could support ten to eleven people on a vegetarian, but nutritionally adequate, diet. With China's current diet, still relatively low in animal products, the nationwide average for the mid-1990s was almost 9 people/ha.[85] Asia's highest apparent national densities—Japan's 25 people/ha, and South Korea's 20 people/ha—are reduced to rates similar to China's mean once the proper corrections for high food imports are made.[86]

Maximum population densities supportable by vegetarian diets would also be quite high in North America and Western Europe, but these highly carnivorous societies now use a great deal of fertilizer nitrogen in order to support that particular dietary habit. Naturally, this reality must be taken into account when trying to estimate the extent of the world's existential dependence on Haber–Bosch ammonia synthesis, the principal goal of the next chapter.

8

Our Dependence on Nitrogen
Agricultures and Populations

There is perhaps no better way to open this chapter than to recall once again Liebig's characterization that agriculture's principal objective is the production of digestible nitrogen.[1] The world's staple cereal grains contain between 7 and 14% of protein, that is, between 1.1 and 2.2% of nitrogen. Even when assuming that all crop residues would be recycled, harvested grain from a hectare of high-yielding winter wheat or corn (8 t/ha) would contain about 130 kg N, and even when the plant roots would recover as much as 60% of the nutrient available in the soil, the initial supply would have to amount to some 220 kg N/ha.[2] Rice has less protein than wheat or corn, but its double-cropping (6 t/ha for each crop) would also remove about 130 kg N/year, and, because nitrogen recovery in wet fields is commonly just 35–40%, more than 300 kg N/ha would be needed in order to reap the two harvests.[3]

As shown in chapter 2, intensification of inputs through organic recycling and planting of nitrogen-fixing legumes is limited by the availability of recyclable crop residues and animal wastes and by the extent of land that can be devoted to green manures. Only the most intensive, and hence spatially restricted, forms of traditional agriculture could have supplied more than 200 kg N/ha. No cropping based on residue and manure recycling, on rotations of cereals with legumes, and on planting of green manures could provide a regular supply of more than 200 kg N/ha over extensive areas.[4]

Today's high staple crop yields achievable on large expanses of intensively cultivated land could not be maintained without applications of synthetic nitrogen fertilizers. Nitrogen budgets presented in the previous chapter show that those compounds now provide anywhere between 60 and 80% of the nutrient available to the most commonly produced grain staples. We can also conclude with a high degree of confidence that the Haber–Bosch synthesis of ammonia has recently been supplying about half of the total nutrient input into the world's agriculture.

This means that, everything else being equal, today's global crop harvest would be cut in half without the applications of nitrogen fertilizers. But this does not mean that roughly half of all the food produced today could not be supplied without nitrogen fertilizer—or that half of the world's population would not be alive without the Haber–Bosch synthesis. Finding the share of global protein attributable to nitrogen fertilizers and calculating the fraction of mankind existentially dependent on fertilizer nitrogen are tasks that require further disentangling of nitrogen's complex pathways in the global agroecosystem.

How Many People Does Fertilizer Nitrogen Feed?

While most of the world's dietary protein is obviously derived from crops—directly by eating plant foods, and indirectly through feeding crops and their residues to domestic animals—a part of it comes from ocean and freshwater fish and other aquatic species, and a part of it is supplied by animals grazing on nonagricultural land that does not receive any fertilizer. The first task in finding the total amount of nitrogen in the global food harvest is to establish the mass of the nutrient that was actually available for human consumption in the harvested phytomass during the mid-1990s.

About 10 of the 60 Mt of nitrogen incorporated by annual and permanent crops were harvested in forages grown on arable land, and this phytomass is obviously inedible; 4 Mt N were in nonfood crops (above all fibers).[5] Nitrogen lost after the harvest and during crop storage and processing amounts to about 3 Mt, as does the nutrient returned annually to fields in seeds. Grain, legume, and tuber crops containing 20 Mt N were fed to domestic animals, and another 3 Mt N came from grain milling residues and oilseed cakes used as feed. Finally, distribution and storage losses of food at retail level amounted to about 1 Mt N.[6] Consequently, nearly 16 Mt N, or roughly ¼ of the nutrient incorporated in the world's crop harvest, ended up in directly consumed plant foods in 1996, enough to supply almost 47 g of plant protein daily to every one of the world's 5.75 billion people (fig. 8.1).[7]

In order to find the total amount of protein derived from crop nitrogen, appropriate shares of meat, eggs, milk, and fish protein that originate in harvested crop phytomass must be added. The global output of protein in animal foods was equiva-

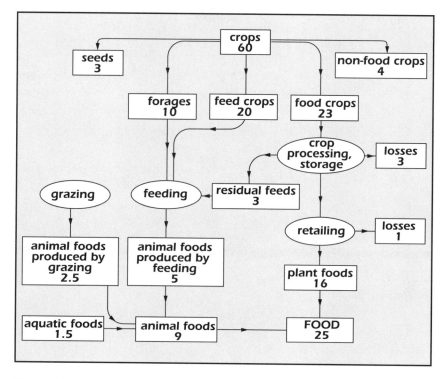

Figure 8.1
Nitrogen in the global food and feed harvest of the mid-1990s (all values are in Mt N/year).

lent to about 9 Mt N in 1996, providing almost 27 g of protein a day per capita, but 1.5 Mt N were in aquatic harvests, and about 2.5 Mt of nitrogen in livestock proteins were derived from grazing on nonarable land.[8] This means that about 5 Mt N in animal proteins originated in crop phytomass used as concentrate and roughage feeds.

About 85% of all nitrogen in food proteins available for human consumption (21 out of 24.5 Mt N) thus came—directly in plant foods or indirectly in animal products—from the world's cropland. And because synthetic nitrogen fertilizers provided about half (44–51%) of the nutrient in harvested crops, roughly 40% (37–43%) of the world's dietary protein supply in the mid-1990s originated in the Haber–Bosch synthesis of ammonia (fig. 8.1). But this global mean both overestimates and underestimates the degree of our dependence on the Haber-Bosch process because the applications of nitrogen in affluent nations have a very different role from the nutrient's use in low-income countries.

The global per capita mean of the dietary protein supply—about 73 g/day in 1996—is made up of two disparate parts: a huge excess in the high-income nations (1996 per capita mean of almost 100 g/day, including about 55 g from animal foods), and a much less comfortable rate in the low-income countries of Asia, Africa, and Latin America, where the 1996 per capita mean was about 66 g/day, with only some 18 g coming from animal foods (fig. 8.2).[9]

Clearly, protein supply has not been a concern in affluent nations, where higher use of nitrogen fertilizers during the latter half of the twentieth century merely added more meat and dairy products to diets that were already sufficient in animal protein.[10] But protein supply is still a challenge in most low-income countries, and although their average diet has improved during the past two generations, all but a very small fraction of their higher use of nitrogen fertilizers has gone into ensuring basic dietary protein requirements. Estimating the share of dietary protein derived

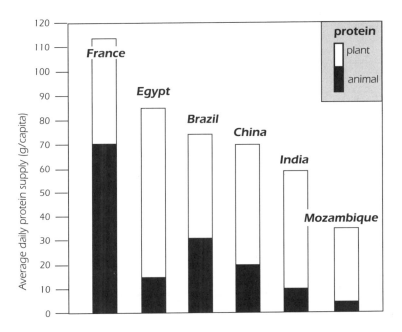

Figure 8.2
Average daily per capita supplies of dietary protein during the mid-1990s ranged from surfeit in affluent countries to inadequate rates in most of sub-Saharan Africa.

from nitrogen fertilizers that are applied in low-income countries would thus give us a good idea of the extent of a truly existential dependence on the Haber–Bosch synthesis.

In 1996 low-income countries consumed 64% of the world's nitrogen fertilizers, which provided about 55% of the total nutrient supply reaching their fields.[11] As no less than 92% of their food proteins were derived from crops, inorganic fertilizers supplied at least half of all nitrogen in their diets.[12] This would be an equivalent of feeding no fewer than 2.2 billion people, or roughly 40% of the world's 1996 total population. These people now depend on the Haber–Bosch synthesis for what is, as we shall see shortly, still an insufficient source of their basic food needs, that is, for their very survival. An excellent confirmation of this conclusion can be obtained by an entirely different approach.

In 1960 poor countries of Asia, Africa, and Latin America used just 2 Mt N in synthetic fertilizers, a negligible amount compared to biofixation in their fields and to recycling of their organic matter. By 1996 these countries counted about 4.5 billion people, 2.5 billion more than in 1960 when their agricultures were still overwhelmingly traditional, and they used 53 Mt N in synthetic fertilizers.[13] Food for about 400 million of these additional people, most of them in sub-Saharan Africa and Latin America, has come from extensive farming: from converting roughly 200 Mha of forests and grasslands to fields and farming them with no, or with merely marginal, inputs of synthetic fertilizers.[14]

On the other hand, food exports from affluent countries, produced overwhelmingly by intensive cropping relying on high fertilizer applications, now feed about 200 million people throughout the Third World (the total is about equally split between Africa and Asia).[15] Consequently, intensive farming—whose increasing harvests have depended almost entirely on additional nitrogen from synthetic fertilizers—now produces basic food needs for about 2.3 billion more people than the traditional, fertilizer-free, practices did, and 2.3 billion people is almost exactly 40% of the world's total population in 1996 (5.745 billion).

Estimates of our dependence on ammonia synthesis could be also approached by calculating the population totals supportable by specified diets. In 1900 the virtually fertilizer-free agriculture (less than 0.5 Mt N were applied to crops worldwide) was able to sustain 1.625 billion people by a combination of extensive cultivation and organic farming, on a total of about 850 Mha. The same combination of agronomic practices extended to today's 1.5 billion hectares would feed about 2.9 billion people, or about 3.2 billion when adding the food derived from grazing and fisheries.

This means that only about half of the population of the late 1990s could be fed at the generally inadequate per capita level of 1900 diets without nitrogen fertilizers.[16] And if we were to provide the average 1995 per capita food supply with the 1900 level of agricultural productivity, we could feed only about 2.4 billion people, or just 40% of today's total.[17]

The range of our dependence on the Haber–Bosch synthesis of ammonia thus is as follows: for about 40% of humanity it now provides the very means of survival; only half as many people as are alive today could be supplied by prefertilizer agriculture with very basic, overwhelmingly vegetarian, diets; and prefertilizer farming could provide today's average diets to only about 40% of the existing population.

The only way this global dependence on nitrogen fertilizer could be lowered would be to adopt an unprecedented degree of sharing and restraint. Because the world's mean daily protein supply of nearly 75 g/capita is well above the needed minimum, equitable distribution of available food among the planet's 6 billion people content to subsist on frugal, but adequate, diets would provide enough protein even if the global food harvests were to be some 10% lower than they are today.

In such a world the populations of affluent countries would have to reduce their meat consumption as hundreds of millions of people would have to revert to simpler diets containing more cereals and legumes.[18] Needless to say, the chances of this dietary transformation, running directly against the long-term trend of global nutritional transitions, are extremely slim. But even in such an altruistic and frugal world ammonia synthesis would still have to supply at least ⅓ of all nitrogen assimilated by the global food harvest![19]

A much more persuasive case regarding the future nitrogen needs would demand just the opposite: an effort should be made to increase the dependence on fertilizer nitrogen as fast as possible. The only way to eliminate the existing stunting and malnutrition caused by protein shortages among hundreds of millions of disadvantaged rural inhabitants and the poorest urban dwellers in low-income countries is, in the absence of any altruistic sharing, to increase global food production. On the other hand, the Western nations, using roughly ⅔ of their cereal and leguminous grain output to feed domestic animals, could easily reduce their dependence on synthetic fertilizers by significantly lowering their high consumption of meat and dairy foods without compromising the nutritional adequacy of their food supplies.[20]

In order to approximate the total people affected by malnutrition, it is necessary to look first at human protein requirements and at access to food. Contrasting the options and the needs of affluent and low-income nations can best be done by comparing the situation in the United States—a country that in spite of its exceedingly carnivorous ways has a relatively low dependence on fertilizer nitrogen—and in China, a nation that exemplifies the high, and increasing, existential dependence on the Haber–Bosch synthesis.

Human Protein Requirements

Dietary proteins are needed in infancy and childhood in relatively high amounts in order to construct growing body tissues. The relative need (per kg of body weight) declines fast with maturation, but it does not disappear in adulthood. Inefficiencies and oxidations entailed by the relatively rapid breakdown and reutilization of proteins in the human body lead to steady nitrogen losses. Nitrogen loss in urine and feces remains fairly constant throughout adult life, averaging about 53 (range of 41–69) mg/kg of body weight; sweating, shedding, and cutting of skin, hair, and nails, and other minor losses, add another 8 mg/kg to that total.[21]

An adequate rate of growth is the obvious sign of sufficient protein intake in infants and children. For older age groups most of the evidence for setting recommended levels of protein intake comes from direct studies of nitrogen balance. Unlike fat, protein cannot be stored in the body, and its excessive intake results in higher excretion of nitrogen-containing metabolites. Requirements can be set by feeding protein below and above the predicted adequate level and then interpolating to zero balance. Numerous studies confirm that healthy adults can maintain a nitrogen balance with daily protein intakes no higher than 0.6–0.7 g per kg of body weight—as long as those proteins are of high quality, combining the presence of adequate amounts of all the essential amino acids with easy digestibility.[22]

In practice, egg and cow milk proteins have been used to represent these ideal, or reference, proteins (scoring 100) employed by nutritionists to express the recommended rates of daily intakes; meat and fish proteins are almost as good. Proteins containing a suboptimal amount of one or more essential amino acids, or those that are not easy to digest, will not be able to support growth equally well. The essential amino acids that are likely to limit the protein quality in typical mixed diets include lysine (present in relatively low amounts in cereal grains), sulfur-containing methionine (relatively low in leguminous grains), threonine, and tryptophan (fig. 8.3). Consequently, the value of consumed proteins must be adjusted according to the least abundant essential amino acid.[23]

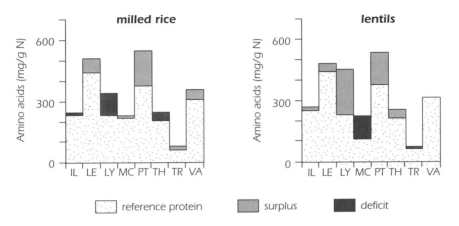

Figure 8.3
Comparison of the actual amino acid composition of rice and lentils with the ideal pattern illustrates the fact that all staple cereals are deficient in lysine while legumes have an inadequate amount of methionine.

Mixed diets with plenty of animal foods will get fairly high-quality scores; vegetarian diets including rice, beans, and corn may score almost as high; but those based on wheat, sorghum, and cassava will, mostly because of relatively low amounts of lysine in those staples, score only around 70, or even as low as 60. In order to ensure adequate amounts of all the essential amino acids, infants given such a low-quality diet would have to consume 40–70% more of the actual dietary protein than if they would have access to dairy products, meat, and fish.[24]

Proteins in eggs, milk, and cheeses have digestibility rates 95% and higher, and rates for meat and fish are nearly as high. Most proteins in staple grains have digestibilities less than 90%, but those in fine wheat flour average 96%, while proteins in beans have digestibilities below 80%. Rates for mixed diets obviously rise with higher shares of animal foods and highly milled grain. The U.S. mean has been around 95%, but the average for a vegetarian Indian diet may be just short of 80%.[25]

Net protein utilization (NPU) is a product of amino acid score and digestibility. Typical American diets have an NPU well over 80, but rural diets in Asia, dominated by staple grains and vegetables, may score below 50; a recent study of a traditional Chinese diet found that the net protein utilization was only 43.[26] Vegetarian wheat-based diets of low-income populations in Mediterranean countries would score similarly low. In order to get enough protein for their growth, children raised on such diets would have to consume daily 2.0–2.3 g of mixed proteins per kg of their body weight compared to just 1.0–1.1 g of dairy or meat protein per kg of body mass, sufficient to cover the protein needs of youngsters in more affluent countries.

A long-held standard conclusion that adult requirements for specific amino acids are much lower than those of infants—and hence could be met by all normal mixed diets, including the largely vegetarian ones in low-income countries—came under revision in the early 1990s. New studies recommended roughly tripling the total daily amino acid requirements (measured in mg/kg of body weight) for adults, and the proposed new amino acid pattern closely resembles new recommendations on protein quality for preschool (two- to five-year-old) children.[27]

This new amino acid pattern would tend to increase the demand for dietary protein among populations that are largely, or completely, vegetarian, but it would make no difference in mixed diets that contain a large share of animal proteins. In such cases a simple protein/energy (P/E) ratio (percentage of total food energy derived from proteins) provides a ready approximation of protein adequacy. For example, in the United States the recommended protein intakes translate to protein/energy ratios between 7 and 11%, while actual intakes range between 14 and 18%. In contrast, China's P/E ratio is adequate, and Africa's P/E ratio, with animal proteins supplying a mere 2% of all food energy, is significantly below the desirable intake.[28]

A good indicator of protein inadequacy is the number of stunted (low height-for-age), underweight (low weight-for-age), and wasted (low weight-for-height) children. The FAO estimates the global totals of these nutritional deficiencies at 215, 179, and 48 million, respectively, during the early 1990s.[29] The amount of additional food needed to redress these shortfalls is not stunningly high in the global aggregate. Bender calculated that if everyone's height and weight were increased to desirable levels, global food consumption in the early 1990s would have to rise by less than 8%.[30] But as these nutritional deficiencies are not evenly distributed, their elimination would require considerably higher gains in a number of low-income countries.

This need is particularly critical for children with their relatively high demand for essential amino acids needed to synthesize new tissues.[31] During the early 1990s one in three children under the age of five living in low-income countries was underweight, and the prevalence of stunting during the early 1990s was as high as 31% in China, 59% in Pakistan, and 61% in India.[32] Reducing, and eventually even eliminating, these deficiencies will inevitably require higher nitrogen applications even in those countries that are already highly dependent on nitrogen fertilizers. On the other hand, the relatively high consumption of synthetic fertilizers will continue to produce rich diets in affluent countries.

Comparing the world's two most important agricultures is perhaps the best way to contrast the two very different modes of our dependence on fertilizer nitrogen. U.S. agriculture receives relatively high contributions of nonfertilizer nitrogen, and its nitrogen fertilizer use stagnated during the 1980s—but it had increased once again during the 1990s (appendix J). In contrast, the already high dependence of Chinese farming on synthetic nitrogen compounds is rising rapidly (appendix R). While U.S. agriculture could greatly lower its nitrogen use without compromising the country's nutritional supply, China will have to greatly increase its reliance on nitrogen fertilizers during the coming two generations.

Nitrogen in U.S. Agriculture

According to the comprehensive account of nitrogen inputs in the preceding chapter, U.S. agriculture derives about 45% of its total nitrogen supply from ammonia synthesis. But the country's food balance sheets make it clear that adequate nutrition would be easily maintained with even a substantial cut in existing nitrogen applications. The two most important reasons for this conclusion are America's large food and feed exports, and its unusually high provision of animal foods.

The United States is the largest exporter of agricultural products in history. During the 1990s between 20 and 25% of the grain corn harvest, about 1/3 of all soybeans, and 40 and 50% of all wheat and rice were regularly sold abroad, as well as about 10% of the country's large meat production and millions of tonnes of oilseeds and protein meal.[33] This means that recent U.S. food exports have contained between 3 and 3.5 Mt N/year, or about 1/3 of all nitrogen in the U.S. crop harvest, and that in order to produce the country's rich diet dominated by animal foods, U.S. agriculture could use only 7–7.5 Mt of fertilizer nitrogen rather than 10–11 Mt N a year. Without exports the planted area would be proportionately reduced, and the share of fertilizer nitrogen in the overall supply would hardly change.[34]

Further substantial reductions of nitrogen applications—and in this case in both absolute and relative terms—would be possible if the only objective of U.S. agriculture would be to produce a healthy diet. About 70% of the country's cereal and legume harvest is fed to animals in order to make available almost 75 g of animal protein a day per capita.[35] Production of high-quality animal protein entails inevitable losses of plant energy and protein metabolized by animals reared for meat, eggs, and milk. Meat protein is inherently more costly to produce than fat; for example, the efficiency of metabolizable energy conversion to protein in pigs peaks at about 45%, while conversion to fat can be as much as 75% efficient.[36]

Feed conversions vary, but even for very efficiently produced broilers the rate is more than 4 kg of feed for 1 kg of meat, implying the transformation of at least 400 g of plant proteins into 100 g of meat protein. Protein conversion efficiencies for other domestic animals are substantially lower: 10–15% for pork, and just 5–8% for beef (fig. 8.4).[37] An equal mix of chicken, pork, and beef protein would have thus required about seven times as much plant protein to produce. Reducing America's per capita consumption of animal foods by 1/3 would result in cutting its nitrogen applications used to produce domestically consumed food by about 1/4.[38] The total use of fertilizer nitrogen could then decline from 7–7.5 Mt N to 5.3–5.6 Mt N.

This would cut the recent nitrogen applications in half, and if the reduced harvest would be produced with lower yields from roughly the same amount of farmland, then ammonia synthesis would supply between 1/3 and 2/5 of all nitrogen inputs. Even then the average annual meat consumption would be close to 80 kg/capita, as high as it is now in Switzerland and higher than in Sweden, and per capita supply of animal proteins would average almost 50 g/day, about as much as it is now in the United Kingdom.[39]

	Milk	Eggs	Chicken	Pork	Beef	Carp	Salmon
Feed (kg/kg of edible weight)	1.1	2.8	4.5	7.3	20	2.3	1.4
Protein content (% of edible weight)	3.5	13	20	14	15	18	22
Protein conversion efficiency (%)	30	30	20	10	5	20	40

Figure 8.4
Protein conversion efficiencies in animal food production.

There is also considerable scope for reducing the United States' overall food supply: its daily average is now 3,600 kcal/capita while food consumption surveys show that only about 2,000 kcal/capita are actually eaten.[40] Lowering the average food intake by 20% would mean that only 4.2–4.5 Mt N would have to be assimilated by crops—and even then the country would still waste ⅓ of its food! Finally, America could adopt a traditional Mediterranean diet, now considered just about the healthiest well-tested choice, and produce a more than adequate amount of plant foods and just enough dairy products and meat from cropland pastures and grasslands and from feeding of grain milling and oil processing residues to provide high-quality protein.[41]

This is, of course a hard-to-imagine proposition, and it is raised here merely in order to illustrate the country's low existential dependence on ammonia synthesis. Such a choice might need no more than 2.5–3.5 Mt N in plants, or 5–6 Mt N of initial supply when assuming an average 50% recovery rate by crops. This amount of nitrogen is well below even the most conservative account of nonfertilizer nitrogen inputs into the U.S. agroecosystem.[42] U.S. agriculture in the 1990s could have supplied a healthy diet for roughly 0.25 billion people without using any synthetic nitrogen compounds.

In 1977 White-Stevens noted correctly that nitrogen fertilizers supplied about half of all nitrogen in U.S. agriculture, but his conclusion that denying American farmers the use of fertilizers "would require the removal of 100 million Americans, some of whom, in a democracy it would seem, are likely to vote against the idea" was definitely wrong.[43] In the absence of synthetic fertilizers the United States would cease to be the world's largest exporter of food, and the average diets of its citizens would have to be profoundly adjusted—but the country's population might be actually healthier for the change.

U.S. cropping is clearly a prime example of a land-rich agroecosystem that has the luxury of being able to produce mostly animal feed rather than food and still provide enormous nutritional reserves. Its applications of synthetic nitrogen fertilizers—not very high per average harvested hectare, but substantial for certain crops—have made it possible to provide diets unusually high in animal foods supply and to expand food exports to the extent unmatched by any other country.

America's reliance on nitrogen fertilizer then is not a matter of existential, nutritional necessity; rather, it is tied to maintaining a very high (and, from a public health perspective, arguably excessive) consumption of animal foods and generating economic benefits through higher domestic sales of meat and dairy products and through large exports of feeds (primarily grain corn and soybeans), food cereals (wheat, rice), as well as oils, fruits, vegetables, cotton, and animal foodstuffs.

Nitrogen in Chinese Farming

In contrast to the United States, China has little room to reduce substantially its high dependence on nitrogen fertilizer. This dependence is overwhelmingly existential—to secure basic adequate nutrition for the country's more than 1.2 billion people—rather than a matter of economic choice or of a preferred diet as in the U.S. case. Origins of this recent dependence are closely connected with the most tragic period of China's modern history, and their recounting provides perhaps the most compelling illustration of the epochal difference made by the Haber-Bosch synthesis.

When the Communist regime established a new republic in 1949 the country had just two small fertilizer plants producing about 27,000 t of ammonium sulfate annually.[44] Construction of small, coal-based ammonia plants producing ammonium bicarbonate began in 1958, the year Mao Zedong launched the Great Leap Forward. Ignorant of economic and technical complexities but obsessed with the idea of making China a great power, Mao followed a primitive Stalinist model of development that equated economic modernization with the large output of steel and other commodities produced in small plants by mass mobilization of the country's huge population.[45]

At the same time, peasants were forced to abandon all means of private food production, and less land was planted to grain, the source of more than 80% of China's food energy, as fraudulent Lysenkoist precepts guided another disastrous "Great Leap" in China's farming and as fabricated reports announced record grain harvests.[46] The real harvest in 1959 was just 165 Mt, but a grossly inflated claim of 270 Mt was used as the basis to expropriate higher shares of produced grain for cities. Imagined plenty also justified the decision that all peasants should eat free meals in communal mess halls set up in the fall of 1958.[47]

This led both to a sharply higher demand for food and to enormous food waste. Given a very close fit between China's pre-1958 average per capita food supply (2,100–2,200 kcal/day) and demand (age- and sex-adjusted needs were about 2,200 kcal/day), the combination of such irrational decisions had a drastic effect on food availability: by the spring of 1959 there was famine in $\frac{1}{3}$ of China's provinces. Weather only exacerbated the suffering.[48]

During the three years from 1959 to 1961 at least 30 million Chinese died in the greatest famine in human history.[49] More pragmatic policies favored by Liu Shaoqi and Deng Xiaoping finally put the end to that tragedy. One of their results was the purchase of five midsized ammonia-urea plants from the United Kingdom and the

Netherlands between 1963 and 1965. By 1965 synthetic fertilizers supplied more than ¼ of all nitrogen. Then the more normal development was cut short again in 1966 with the launching of Mao's destructive time of ideological frenzy, political vendettas, and localized civil war that became known, most incongruously, as the Cultural Revolution.

The country, in a state of anarchy, let its population grow without any controls from 660 million people in 1961 to 870 million by 1972. Addition of more than 200 million people in a single decade represented the fastest population growth in China's long history and the highest absolute decadal increment ever recorded by any single nation (fig. 8.5).[50] At the same time, industrial and urban expansion and Maoist policies of agricultural mismanagement were shrinking the extent of China's arable land, and the traditional organic agriculture had reached its production limits set by the availability of recyclable nutrients.

Figure 8.5
China's annual population growth during the second half of the twentieth century plotted as consecutive five-year averages.

In just two to three years after the end of the great famine average yields of staple crops recovered to the pre-1959 level, but then they began to stagnate. By the end of the decade China's harvests could not keep up with even basic food needs of the country's growing population. By 1972 China's average per capita food supply was below the levels of the early 1950s, when the country had emerged from decades of instability and war (fig. 8.6). All food in cities was strictly rationed, meals had to be stir-fried with just teaspoonfuls of precious (and low-quality) cooking oil, and hundreds of millions of peasants subsisted on a monotonous and barely adequate vegetarian diet, eating meat and fish no more than a few times a year.[51]

Moreover, it was obvious that even an unprecedented degree of population controls—at that time contemplated by the ruling gerontocracy and soon put into effect in the world's most drastic and personally intrusive program—would not prevent the total population from rising to at least 1 billion by the year 1980 and to 1.2

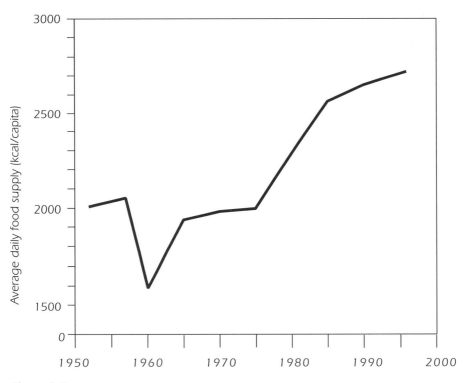

Figure 8.6
China's average per capita food supply, 1952–1996.

billion during the 1990s. This anticipated doubling would then result in cutting the per capita arable land by more than half in a single generation. But that land could not produce twice as much food with traditional ways of farming.

The only effective way out of this existential predicament was to intensify China's agriculture, to boost its average staple crop yields. In a country where nearly half of all farmland was already irrigated, this meant one change above all: a rapid increase in applications of nitrogen, the nutrient critical for raising yields. China's nitrogen fixation capacity had to increase quickly, and so shortly after President Nixon's visit reopened China to world trade, the country placed orders for thirteen of the world's largest and most modern ammonia-urea complexes, the biggest purchase of its kind ever. Such an enormous order was best filled by turning to M. W. Kellogg, the world's leader in ammonia synthesis. Eight of the thirteen large ammonia plants came from the company, and their product is fed directly to urea plants built by Kellogg's Dutch subsidiary.[52]

Output from these newly completed plants soon surpassed the country's pre-1972 fertilizer nitrogen production. More purchases of ammonia-urea complexes followed during the 1970s, as did more fertilizer imports. In 1979 China surpassed the United States to become the world's largest consumer of nitrogen fertilizers, and ten years later it surpassed the Soviet Union to become the world's largest producer of these compounds as well (fig. 8.7).[53] As shown in appendix R, during the late 1990s about 75% of all nitrogen reaching Chinese farmland originated in nitrogen fertilizers, and as no less than 90% of the country's protein supply is derived from domestically grown crops, $2/3$ of all nitrogen in China's food originates in the Haber–Bosch synthesis of ammonia.

A historical comparison confirms this estimate. During the early 1950s traditional organic farming produced 85–90 kg of plant protein per hectare of farmland; in 1996 China's plant protein output averaged about 240 kg/ha, or about 2.75 times above the early 1950s rate. As virtually all of the 1950–1996 increase in the supply of fixed nitrogen came from inorganic fertilizers, this implies roughly a $2/3$ contribution of nitrogen fixed by ammonia synthesis.[54]

The average Chinese diet in the late 1990s has been fairly adequate, but even if the Chinese were willing to forgo their recent, and much welcome, dietary improvements and reduce their protein intake to the barely adequate level of the mid-1950s, they would still have to rely on synthetic nitrogen fertilizer to produce at least half of their food protein.[55] Even with its greatly moderated fertility China will add an-

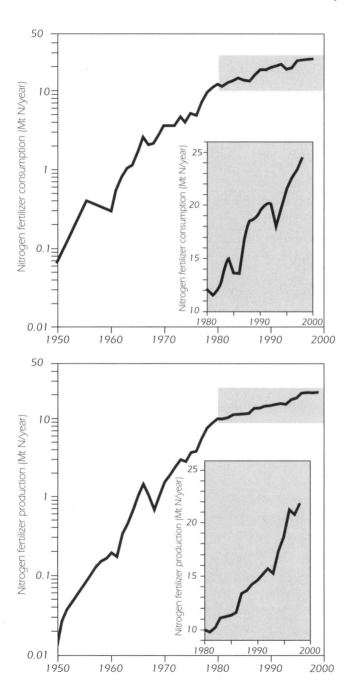

Figure 8.7
China's production and consumption of nitrogen fertilizers, 1950–1998.

other 300 million people by the middle of the twenty-first century before its total population may level off somewhere between 1.5 and 1.6 billion people.

Moreover, this much larger population will expect continuing improvements in the quality of its diet. Opportunities for expanding cultivated areas are very limited, and higher food imports can never play such a large role in China's case as they did in the development of its smaller Asian neighbors.[56] Consequently, all but a relatively small fraction of China's greatly increased demand for food will have to come from further intensification of cropping, which is unthinkable without further substantial increases of average nitrogen applications.

China is not alone in this predicament. Similar degrees of high existential dependence on repeated applications of synthetic nitrogen fertilizers have been evolving in other populous, land-scarce countries. Except for Nigeria, the average output of protein per hectare is now surpassing the maxima that can be produced by traditional farming in all of the world's ten most populous low-income nations, and fertilizer-free agricultures could not produce adequate food supply even if the populations of those countries were willing to subsist on purely vegetarian diets.

Growing Dependence during the Twenty-First Century

Instead of adding to the recently popular genre of detailed long-term quantitative forecasts (which turn out to be almost invariably wrong), I will review the factors and trends that will tend to increase the world's dependence on the Haber–Bosch synthesis during the twenty-first century. Foremost among them is the continuation of the very reality that was made possible by the synthesis: high absolute rates of global population growth.

There can be no doubt that the transition to low fertility—a shift completely accomplished in all the affluent countries and well advanced in much of Asia and Latin America—is now under way even in sub-Saharan Africa, the largest remaining region of very high fertilities. As a result, the world's population growth has recently began declining even in absolute terms.[57] Depending on the more distant, and as yet uncertain, course of the fertility decline, the global population may be anywhere between 7.3 and 10.7 billion people by the year 2050 (the UN's latest medium forecast being about 8.9 billion), and it may not increase very much afterwards.[58]

The conclusion that yet another doubling of world population is unlikely is most welcome news, but it is still a great reason for concern regarding the future food production. The UN's forecasts actually foresee that the rich nations' population in

2050 will be a bit below the 2000 total, which means that virtually all net population growth between 2000 and 2050 will be in today's low-income countries.[59] About 40% of the world's population growth during the next two generations will originate in only a handful of large countries, and at least $^1/_3$ of the total increment will be added in Africa.

According to the UN's medium forecast, this global growth would amount to about 3 billion people, surpassing the total of 2.8 billion people added throughout the poor world during the past 50 years. We may be moving toward a stabilized global population faster than anticipated, but the challenge of providing basic existential requirements for the world's poor will become greater, not smaller, in absolute terms. Among the five countries with the largest population increases, three are in Asia—India (with almost 550 million more people), China (about 220 million), and Pakistan (almost 200 million)—and two are in Africa: Nigeria will add nearly 140 million people, and Ethiopia about 110 million.[60]

While some African and Latin American countries may be able to expand their cultivated area—albeit at the price of the additional significant loss of tropical forests—worldwide opportunities for a more widespread practice of extensive farming are limited. A detailed appraisal of long-term agricultural prospects in ninety poor countries (excluding China) found that fifty-one of them had abundant or moderately abundant reserves of arable land, but their population was just $^1/_3$ of the assessed total. In contrast, eighteen countries with extreme land scarcity, already cultivating an average of 96% of their potentially arable land, supported half of the population.[61]

The need for further intensification of farming, a trend inevitably involving higher applications of nitrogen, will be particularly acute in Asia: with a single exception the continent contains low-income countries sharing, or approaching, China's high dependence on nitrogen fertilizer. These countries have populations larger than 50 million people, their arable land is limited to less than 0.2 ha/capita, and they are already producing more than 100 kg of edible protein/ha, the large-scale limit of intensive traditional agricultures relying solely on recycled nitrogen. Moreover, losses of farmland to nonagricultural uses (urbanization, industrialization, and expansion of transportation infrastructures) and the declining quality of arable land will add to the naturally decreasing per capita availability of cultivable soils.[62]

Besides China (currently with about 0.11 ha/capita and producing about 225 kg of food protein/ha) these countries include Bangladesh (0.09 ha and 190 kg), Indonesia

(0.12 and 170), Pakistan (0.19 and 145), the Philippines (0.13 and 140), and India (0.18 and 120); the non-Asian exception is Egypt, which has a mere 0.05 ha/capita and in spite of its very intensive cultivation relies on heavy food imports.[63] During the past generation these countries have been moving along remarkably similar trajectories of increasing dependence on synthetic nitrogen (fig. 8.8). This pattern will be followed by a large number of other densely inhabited countries.[64]

Inevitably, during the coming decades the dependence on nitrogen fertilizers would grow even if these countries would merely maintain their current average per capita food supply. That, of course, should not be enough. Nationwide means of per capita protein, and food energy, consumption are above the existential minima in most low-income countries, but in many of them there is hardly any safety margin, and in all of them common distribution inequalities mean that protein malnutrition, as already noted, continues to be widespread.[65]

Moreover, the future increase of food demand throughout the modernizing world will not be limited merely to satisfying the basic nutritional needs of every person. The wish to eat more animal foods is a universal one, and many studies of long-term trends in the rich countries demonstrate an unmistakable correlation between

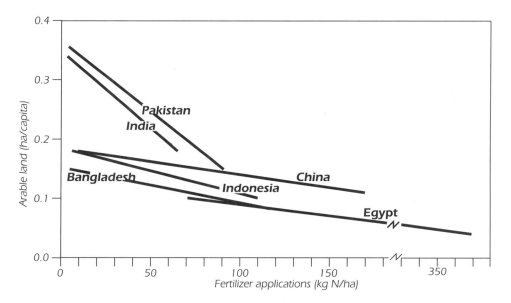

Figure 8.8
Increasing dependence on nitrogen fertilizers in land-scarce populous nations, 1960–1996.

rising incomes and dietary transitions from high levels of direct consumption of cereals, tubers, and legumes to a rising demand for meat and dairy products.[66] The initial phases of this transition are now clearly noticeable in many relatively better-off low-income countries. While the poor world's daily per capita availability of animal protein remains low, it had doubled between 1970 and 1995, and it has roughly tripled in China since the beginning of the country's modernization drive in the early 1980s.[67]

Another important consideration that will affect future demand for fertilizer nitrogen is the declining response of crop yields with increasing nutrient applications. Worldwide agronomic experience shows a wide range of responses.[68] In rice agriculture the returns range between 5 and 19 kg of unhusked grain/kg N (typical values are around 12 kg). The leveling appears even with applications as low as 30–40 kg N/ha, but with the best agronomic practices the diminishing response can be maintained with applications up to 200 kg N/ha. In wheat crops the returns range between 5 and 30 kg of unmilled grain per kg N (typically between 15 and 20 kg) for the first 50–70 kg N/ha; declining returns set in at around 100 kg N/ha, but, as with rice, good responses can continue up to 200 kg N/ha. Inevitably, these realities will be felt throughout the intensely multicropped regions of Asia, as well as in Mexican or Middle Eastern wheat cultivation.[69]

Finally, environmental change—ranging from higher concentrations of stratospheric ozone to possibly rapid global warming, and from excessive soil erosion to growing regional shortages of water—will further increase the need for intensive, high-yield cropping on the diminishing amount of prime farmland.[70] Quantifying the effects of various forms of environmental pollution and ecosystemic change is extremely uncertain. Although there will be some positive effects, it is most unlikely that the overall impact would not be substantially negative.

The minimum scenario of the world's future dependence on the Haber–Bosch synthesis is not difficult to construct. According to the medium version of the UN's long-term population projection, virtually the whole increase of 3 billion people to be added between 2000 and 2050 will be in today's low-income countries. Merely to maintain their average diets, these countries would have to use at least 85% more nitrogen fertilizer than they do today. Adding at least 10% for eliminating malnutrition and stunting, and another 10% for producing more animal foods, would double their 1995 nitrogen fertilizer consumption.[71] With unchanged consumption in affluent countries the world might be using some 140 Mt N in synthetic fertilizers by the year 2050.

This would mean that Haber–Bosch fixation would supply 60% of the nutrient reaching the world's cropland, and its product would be indispensable for ensuring basic nutrition for some 60% of the world's people. More meaty diets and the necessity of making up harvest losses due to environmental change could further raise the need, and hence the dependence, even higher, but more efficient use of fertilizers, lower food waste, and gradual dietary transformations can combine to lower it substantially. All we can say with certainty is that our dependence on the Haber–Bosch synthesis of ammonia will rise during the next two generations, but its actual level will be determined by a complex interplay of our increasing, and all too often more wasteful, needs and of our innovative and more efficient uses.

Consequences of the Dependence
Human Interference in Nitrogen's Biospheric Cycling

Haber–Bosch synthesis of ammonia is now the single largest cause of the human-driven intensification of the biospheric nitrogen cycle, and its annual output of fixed nitrogen is now almost as large as a conservatively estimated total rate of natural nitrogen fixation on the continents. In relative terms, human activities now perturb the global nitrogen cycle to a greater extent than they interfere in the cycles of the other two doubly mobile elements, carbon and sulfur. But synthesis of inorganic fertilizers is not just the largest cause of the roughly doubled rate of preindustrial nitrogen fixation: the complexity of nitrogen's biospheric cycling and the element's double mobility make it inevitable that large fractions of nitrogen fertilizers will not be assimilated by crops.

Higher rates of nitrogen fertilization will bring higher losses of the nutrient, both in water-borne forms, mainly as highly soluble nitrates, and in gaseous compounds, above all as ammonia and nitrogen oxides but also as nitrous oxide. Only a complete denitrification, converting NO_3-N to N_2, has no negative environmental impacts. All the other transfers and transformations of the element have some undesirable consequences, ranging from risks to human health to possibly major alterations of affected aquatic and terrestrial ecosystems.

While we are fairly certain that, as a global mean, about half of all nitrogen applied in inorganic fertilizers does not end up in crop tissues, we have much less confidence about apportioning this huge nitrogen loss among the major processes that remove the element from agroecosystems to the atmosphere and to both underground and surface waters. This chapter will review the best available evidence concerning the rates and the magnitudes of these losses, first in general terms and then in some quantitative detail for the world's agroecosystem. Then I will look at the possible harmful effects that the nitrogen that has escaped from inorganic fertilizers may have on human health and on the productivity, stability, and diversity of both aquatic and terrestrial ecosystems.

Intensifying the Global Cycling of Nitrogen

Although the actual rate for the preindustrial natural fixation cannot be pinpointed, it is most likely that in the absence of human activities terrestrial processes fixed between 150 and 200 Mt N annually.[1] Fixation by lightning is a minor component of the natural production of reactive nitrogen, which is dominated by symbiotic diazotrophs.[2] Of the three principal types of human interference in the nitrogen cycle—synthesis of ammonia, planting of leguminous crops, and combustion of fossil fuels—the first one is by far the largest.

With a recent annual production rate of about 100 Mt N, the Haber–Bosch synthesis introduces about four times as much reactive nitrogen into the biosphere as does the combustion of fossil fuels. Ammonia synthesis (for all its uses, not just for fertilizers) now fixes annually about three times as much nitrogen as does the biofixation associated with the cultivation of food and feed legumes.[3] This means that the annual nitrogen fixation by the fertilizer industry alone now approaches the lower estimates of natural terrestrial fixation, and that the combined anthropogenic inputs of some 150 Mt N of reactive nitrogen—100 Mt N from the Haber–Bosch process, about 30 Mt N from legume cultivation, and more than 20 Mt N from fossil fuel combustion—are as large as some estimates of the worldwide aggregate of the element fixed by natural terrestrial ecosystems during the preindustrial era.

Human interference in the global biogeochemical nitrogen cycle—amounting to anywhere between a third (when the comparison is restricted to fertilizers) to about a half (when all anthropogenic inputs are taken into account) the overall throughput of reactive nitrogen—has thus reached a level higher than in the case of carbon or sulfur, the two other doubly mobile elements involved in rapid biospheric cycling (fig. 9.1).

In 1957 Roger Revelle and Hans Suess pointed out in a now classic paper in *Tellus* that with our rising releases of CO_2 from fossil fuel combustion we are carrying out an unprecedented large-scale geophysical experiment.[4] Since that time human interference in global carbon cycle and the resulting possibility of a rather rapid planetary warming have received enormous research and public attention, and eventual controls of CO_2 emissions have become the subject of intense international negotiations.

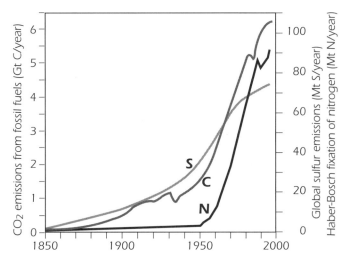

Figure 9.1
Trends in human interference in global carbon, sulfur, and nitrogen cycles, 1850–2000.

The anthropogenic emissions responsible for this concern amount to over 6 Gt of carbon from the combustion of fossil fuels annually, and 1–2 Gt C from land use changes, primarily from the deforestation in the tropics.[5] In contrast, terrestrial photosynthesis and respiration exchange annually about 100 Gt C, the marine primary production cycle involves about 50 Gt C assimilated and 45 Gt C respired (the rest is deposited in the deep ocean), and close to 100 Gt C are exchanged between the atmosphere and the ocean.

Although clearly important because of its role in affecting the Earth's thermal balance, the anthropogenic carbon is only a small fraction of the enormous natural fluxes of the element. Mass comparisons show that human interference in the sulfur cycle has been much more intense. Almost 80 Mt S are now released annually from the combustion of fossil fuels and from the smelting of color metals (overwhelmingly as SO_2), while the atmosphere receives annually at least 200 Mt, and possibly as much as 260 Mt, of sulfur in compounds released from terrestrial and marine bacteria and plants, from volcanoes, and in dust and (the single largest input) sea spray.[6] Because of their key role in acid precipitation, anthropogenic emissions of sulfur have also received a great deal of attention since the early 1970s.[7]

In comparison to the effort devoted to the understanding of carbon and sulfur cycles and to studies assessing the existing and potential impact of anthropogenic interferences in the two cycles, fertilizer-induced perturbances in nitrogen cycling have been a minor concern. Yet, as the preceding comparison clearly shows, they too constitute an unprecedented large-scale biogeochemical experiment whose eventual impacts we can predict only poorly. After all, while the tropospheric concentration of CO_2 has risen about 30% above the preindustrial level, the atmospheric deposition of NO_y and NH_x has roughly trebled. And while the human activities now account for about $\frac{1}{4}$ of all sulfur compounds released into the environment, they add up to at least $\frac{2}{5}$, and perhaps to $\frac{1}{2}$, of all terrestrial reactive nitrogen.

An argument can be made that the human interference in the global nitrogen cycle deserves special attention, because our existential dependence on the nutrient has no conceivable substitutes. High CO_2 emissions could be reduced very effectively by a combination of economic and technical solutions, and the transition to nonfossil energies will be inevitable even without any threat of global climatic change.[8] Compared to their peak during the late 1970s, emissions of SO_2 already have been reduced by about 20% in North America and by more than 40% in Europe, and there is no shortage of effective technical solutions to reduce them eventually to a fraction of their current rate.[9]

In contrast, there is no way to grow crops, and human bodies, without fixed nitrogen, and there are no bioengineered substitutes waiting to displace the Haber–Bosch synthesis and to supply the nutrient needed to grow food directly to crops. While we can certainly improve the typical efficiencies with which we use nitrogen fertilizers, there is no way to eliminate substantial losses accompanying field applications of nitrogen compounds, and hence to avoid all the undesirable environmental consequences of fertilization.

What Happens to Fertilizer Nitrogen

Not all nitrogen that is not taken up by crops leaves the soil. As explained in chapter 1, appreciable shares of the applied nutrient may be fixed by clay minerals, and the trapped NH_4^+ cations can no longer participate in the rapid cycling of the element in the soil (fig. 9.2). A large part of the trapped nitrogen may be rereleased during the growing season, but some of it remains unavailable to plants, resisting both inorganic breakdown and bacterial attack for extended periods of time. In some soils the mineral-bound NH_4^+ accounts for up to 8% of all nitrogen in the near-surface layer, and up to $\frac{2}{3}$ of the element in deeper soil.[10] The actual amount of mineral-bound nitrogen thus ranges from just a few kilograms to more than 1 tonne per hectare.

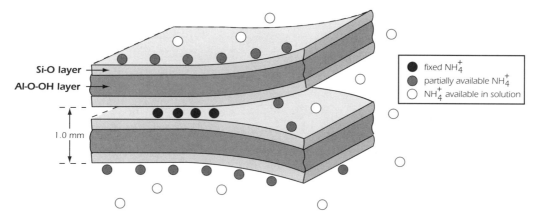

Figure 9.2
Fixation of ammonium by clay minerals.

Significant amounts of the nutrient can be also temporarily immobilized by soil biota or stored in long-lived organic compounds.[11] Complex chemical processes constantly sequester nitrogen in stable compounds: humus forms when polyphenols (aromatic alcohols) derived from decomposed lignin, or synthesized by microorganisms, react with amino acids and produce nitrogen-rich humic and fulvic acids of high molecular weight. These polymers lend yellow to black (mostly brown) colors to soil solutions, and they form complexes with clay minerals that allow humus to resist microbial breakdown. Humus may retain the incorporated nitrogen for several hundreds, or even thousands, of years.[12]

But the total amount of nitrogen subject to these kinds of redistribution within soils is small compared to losses of fertilizer nitrogen, which are an inevitable part of complex global fluxes of the nutrient. Nitrogen's easy double mobility—through the air and through the waters—ensures the intensive cycling that is the prerequisite of the element's perpetual biospheric availability but also the source of unwelcome economic losses and environmental concerns. An individual nitrogen atom can thus be sequentially incorporated in a number of compounds whose very presence, or whose excessive concentration, may be undesirable.

Only complete denitrification returning N_2 to the atmosphere has no unwelcome environmental consequences, but it can be a significant source of economic losses because of an excessive reduction of nitrates that would otherwise be available to soils. High denitrification rates are favored by high temperature, high soil water content, and plenty of available soil carbon, while the final conversion step from N_2O to N_2 is inhibited by high concentrations of soil NO_3^-.[13] Reliable direct measurements of complete denitrification in farmland are rare, and the fluxes, be they for a field or a region, are commonly estimated as residues after accounting for all other major nitrogen losses.

Total denitrification fluxes may be as high as 15% of applied nitrogen (inorganic and manure), particularly from heavily fertilized and irrigated farmland, but they may be only half as much, or less, in other fields.[14] Such a variability of fluxes from different soils and under different environmental conditions makes the choice of any typical or average value highly questionable. The same is true about the ratios of $N_2:N_2O$ produced by denitrification. N_2O, produced by incomplete denitrification, is a potent greenhouse gas (about 200 times more effective in absorbing the outgoing infrared radiation than CO_2), and its emissions can account for anywhere between less than 0.5% to about 5% of the initially applied nitrogen.[15] The gas is also involved in the depletion of stratospheric ozone.

Bacterial nitrification (oxidation of NH_4^+) and denitrification are the main sources of NO_x (mostly NO) emitted from soils. The key factors regulating nitrification are the availability of NH_4^+, soil acidity (the process is negligible below pH 4.5), moisture (waterlogging suppresses nitrification), and temperature (rates are low below 5° C).[16] Rates of fertilization, crop varieties, and tillage practices are the key management factors influencing the emissions, which are so variable that annual fluxes extrapolated from short-term measurements range over several orders of magnitude. Annual means of NO emissions from agricultural soils derived from long-term measurements range from mere traces to more than 20 kg N/ha.[17]

Reported volatilization losses of NH_3-N on the order of 10% of the initial nitrogen within one to three weeks of applications are common, and recorded maxima are above 60%, or even 70%.[18] Dry, calcareous soils, surface applications or shallow incorporation of fertilizers and manures, and high temperatures promote the loss in rain-fed fields. Volatilization losses are particularly high when ammoniacal fertilizers are broadcast directly onto flooded soils.[19] The principal factors promoting the process are shallow waters, their alkaline pH, high temperature and high NH_4^+-N concentration, and higher wind speeds. Consequently, the highest volatilization losses are in heavily fertilized paddy fields in the tropics.

Leaching rates depend on levels of fertilization, compounds used (NH_3 leaches very little in comparison to readily soluble NO_3), soil thickness and permeability, temperature, and precipitation. The single most important land management factor is the presence and quality of the ground cover, and long-term experiments with Oklahoma wheat also demonstrated that the buffering effect of soil and plants prevented accumulation of the leaching-prone inorganic nitrogen until the application rates exceeded those required for maximum yields.[20]

Everything else being equal, leaching from bare, fallow, or freshly ploughed soils is much higher than beneath row crops, which, in turn, is considerably higher than from soils under such dense cover crops as legume-grass mixtures whose roots can take up large amounts of added nitrogen. Annual leaching losses range from negligible amounts in arid fields to 20–30 kg N/ha in rainy temperate regions. Maxima of over 50 kg N/ha are not unusual in the most heavily fertilized crop fields of northwestern Europe and the American Midwest and Northeast.[21] Even higher leaching losses have been reported in some irrigated crops, but losses in many Asian paddy fields are low, just a few kg N/ha.

Recent concerns about soil erosion and its effects on crop productivity have not been matched by reliable information about the actual extent and intensity of the process or, more accurately, about soil losses in excess of natural denudation.[22] Quantification of these impacts is complicated by the fact that in some areas much of the eroded soil is not lost to food production: very little or no eroded sediment leaves the fields in areas with gentle relief without any major surface outlet, and when the sediment leaves the land most it may be deposited downstream as colluvium or alluvium.[23]

Short-term rates of cropland erosion have been measured and estimated with varying degrees of accuracy for many locales around the world. Their extremes range from negligible losses in rice paddies to more than 200 t/ha on steep tropical slopes and in the world's most erodible soils of China's Loess Plateau.[24] Average soil erosion estimates for larger areas are rarely available. Nationwide U.S. inventories, begun in 1977, are a major exception: recent annual means were just above 15 t/ha, with about 60% contributed by water erosion, and the preliminary value for 1997 is about 14 t/ha.[25] The first nationwide approximation for India indicates losses of at least 13 t/ha for water erosion alone, and regional estimates for China imply a national mean of about 30 t/ha.[26]

Although they are rarely discussed, losses from the tops of plants can account for a large share of the initially applied nitrogen. They occur mostly within two or three weeks after anthesis (full bloom), and at harvest time the crops commonly may have between 15 and 30% less nitrogen than was their peak content. Measured postanthesis losses in wheat ranged from less than 6 to 80 kg N/ha, or from less than 8% to almost 60% of all nitrogen present at anthesis. Postanthesis declines ranging mostly between 20 and 50 kg N/ha were measured in other cereal crops (rice, sorghum, and barley), with daily losses as high 2 kg N/ha. Rates as high as 45–81 kg N/ha were measured in corn fields, equaling 52–73% of all nitrogen unaccounted for by standard balance calculations.[27]

Neither translocation to roots nor root exudates explain the loss. A major part of the losses must be due to the shedding of various plant parts (pollen, flowers, leaves), leaching of nitrogen from aging leaves, and to herbivory: these losses (except for windborne pollen) are mostly internal redistributions of the nutrient, as the litter, leaching, and herbivory return the nutrient to soils. But most of the nitrogen lost from tops of plants is due to volatilization of NH_3.[28]

Nitrogen Losses in Modern Farming

Losses of fertilizer are easy to determine in indoor experimental settings (in phytotrons or greenhouses) by simply subtracting the recovery rates—the nutrient assimilated by harvested crops and their residues—from the total nitrogen inputs. Atmospheric deposition, mineralization of organic soil nitrogen, and contributions by free-living diazotrophs always complicate the situation in field studies. The most effective way to find the fate of fertilizer nitrogen in the open is to label it with an isotope, ^{15}N, and to trace its paths after the application.[29] But where nitrogen fertilizer applications dominate the supply of the nutrient, approximate recovery rates can be found as a quotient of the nitrogen content in the harvested phytomass (crops and their residues) and in the applied fertilizers.

Analyses of thirty agroecosystems located mostly in the temperate zone showed that $2/3$ of them had overall nitrogen recovery rates below 50%, while the most efficient cropping absorbed nearly 70% of applied nitrogen.[30] Where fertilization rates were below 150 kg N/ha uptake efficiencies were as high as high 60–65%, but with higher nitrogen applications recovery rates were scattered around 50%. An extensive European survey provided data on nitrogen uptake efficiencies both for the crops grown in wet climates of northwestern Europe and for less rainy climates of the southern part of the continent.[31] High-yielding winter wheats recovered 39–57% of nitrogen from the applied urea and 38–70% from ammonia in France, 52–65% in England, and 52–62% in the Netherlands; in contrast, nitrogen uptakes ranged between 27 and 40% in Portugal, and only between 18 and 37% in Greece.[32]

Long-term experiments with winter wheat on Broadbalk field at Rothamsted (fig. 9.3) show fairly steady recovery shares for more than a century of cropping, between

Figure 9.3
Aerial view of Broadbalk field at Rothamsted (Hertfordshire), the site of the world's longest continuous agronomic experiments. Courtesy of Rothamsted Experimental Station, Harpenden, Hertfordshire, England.

the early 1850s and the late 1960s. Regardless of the rate of application (48, 96, and 144 kg N/ha), wheat grain and straw recovered between 32 and 39% of applied fertilizer nitrogen.[33] A higher harvest index of new cultivars was the main reason for the post-1970s improvements in average rates of recovery, which rose as high as 69% for both the minimum and maximum applications (48 and 192 kg N/ha) and 83% for applications of 96 kg N/ha during the period 1979–1984.

Field experiments have shown that typical recovery rates in Asian flooded rice are only between 25 and 40% of all applied nitrogen; they may be considerably lower, and only rarely do they exceed 50% of the applied nutrient (fig. 9.4).[34] North American corn (grain and stover) recovers between 40 and 60%, but grain alone as little as 13–20%.[35] Multiplying conservative recovery rates for major crop categories—35% for rice and for rainfed crops grown in drier climates, 55% for crops in humid climates, 65% for legumes, and 75% for cropland forages—by areas sown to these crops results in a weighted worldwide recovery average of about 50%, an excellent confirmation of the mean calculated in chapter 7.

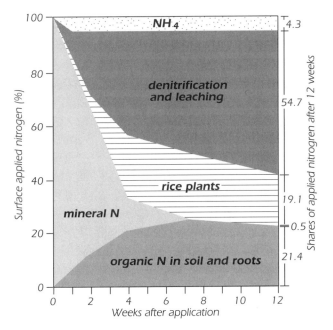

Figure 9.4
Partitioning of nitrogen fertilizer applied to a rice field.

We may thus conclude with a high degree of confidence that half of all the nitrogen added annually to the world's croplands is lost from the world's agroecosystems, most of it before it can get incorporated in the harvested biomass. Apportioning this loss among the handful of major causes cannot be done accurately, but even an imprecise attempt is useful as it reveals at least the relative magnitudes of the principal flows, and hence the potential for particular environmental impacts.

Complete denitrification is obviously the most desirable pathway that unassimilated nitrogen can take, and the process accounts almost certainly for the largest share of the nutrient's field loss. But estimating the annual return of N_2 to the atmosphere is by far the most elusive task in quantifying the postapplication fates of nitrogen fertilizers, be it for a small field or on a global scale. Taking 5 as the minimum ratio of $N_2:N_2O$ emissions would, with 2.5–3 Mt N_2O-N emitted annually from the world's agricultural soils, generate annually between 13 and 15 Mt N_2, or 10 and 14% of all nitrogen particularly susceptible to relatively rapid denitrification: that is, not only nitrogen in inorganic fertilizers, but also the nutrient introduced in atmospheric deposition, irrigation water, and animal manures.

The most likely total denitrification flux—the aggregate of N_2, N_2O, and NO—would then be around 20 Mt N/year, but uncertainties surrounding these emission estimates mean that the actual rate may be as low as 13–15 and as high as about 30 Mt N/year. Obviously, much more N_2 is eventually returned from agroecosystems to the atmosphere by denitrification of nitrogen compounds that were removed from fields by leaching and soil erosion and carried away in harvested feed and food crops.[36]

Uncertainties surrounding the quantification of leaching and erosion losses make it impossible to decide which route is more important in the global aggregate. Relatively abundant data on nitrate exports in streams and groundwaters are of little use in assessing the nitrogen lost from agricultural land as there is no reliable way to separate the contributions from atmospheric deposition, sewage, and industrial processes from the fluxes originating in inorganic fertilizers and animal manures.

A survey of forty agroecosystems on three continents indicated that with fertilization of less than 150 kg N/ha leaching equaled about 10% of applied nitrogen, while with additions of more than 150 kg N/ha about 20% of added nitrogen was lost.[37] Because leaching rates have been measured mostly over short periods of time (days to months after fertilizer application), prorating of these figures may underestimate long-term throughputs. A rather conservative estimate of the annual rate of nitrogen leaching from the world's agricultural land then would be about 17 (14–20) Mt N during the mid-1990s, averaging between 10 and 15 kg N/ha, and inorganic fertilizers should account for at least ⁴/₅ of this total.

Estimates of the nitrogen lost in eroded soil are even more uncertain. A global soil erosion mean of 20 t/ha implies annual removal of about 30 Gt of soil, a conservative total given the high erosion rates in parts of Africa, Latin America, and Asia. A disaggregated calculation using specific erosion means for continents ended up with about 35 Gt.[38] With at least ¼ of this soil redeposited on adjacent cropland or on more distant alluvia, the loss would be 22–25 Gt/year.

Soil nitrogen content is highly variable even within a single field, ranging over an order of magnitude (1.5–17 t N/ha) in cropped soils.[39] A value just short of 0.1% N is close to the current Chinese mean, while the U.S. average is around 0.125%.[40] A conservative range of 0.08–0.1% N would mean that 22–25 Gt of eroded soil would carry away about 20 (18–25) Mt N/year.

The last two major categories of nitrogen loss are ammonia emissions from soils and plant tops. The relatively large amounts that can be lost during the days and weeks following applications of ammoniacal fertilizers and animal manures have been a long-standing concern among agronomists.[41] In contrast, volatilization from non-NH_3 fertilizers is minimal, and large-scale (national or global) means used to calculate annual losses have been as low as 1% for the United Kingdom, and as high as 11% for the world.[42] Given the still increasing use of urea and the higher intensity of nitrogen applications in humid and warm environments in general, and in Asia's monsoonal rice fields in particular, the assumption of 8–12% (10%) of the nitrogen applied in inorganic fertilizers lost to volatilization is hardly excessive. It would produce annual fluxes of 6–10 (8) Mt NH_3-N.[43]

Even if the typical volatilization loss from the tops of plants would be just 10 kg N/ha, the world's croplands would lose no less than 15 Mt N annually, a flux similar to NO_3-N losses due to leaching. Because we do not know to what extent the gaseous nitrogen losses from the tops of plants may have been captured by previously outlined estimates of gaseous (NH_3 and NO) fluxes, I will assume that just 10 (5–15) Mt N/year are lost in this way.

After rounding these approximations, it can be concluded that almost 30% of all the nitrogen lost from the world's agroecosystems is directly denitrified, about 25% is taken away with the eroded soil, another 25% is lost to volatilization from fertilizing compounds, soils, and plant tops, and about 20% is removed by leaching (Appendix Q). Leached nitrates from synthetic fertilizers and manures have caused perhaps the greatest environmental concern because of their potential to affect human health and to transform aquatic ecosystems far downstream from farmed areas. Nitrous oxide produced by denitrification is a greenhouse gas, and volatilized ammonia and NO_x emitted from soils contribute to the atmospheric deposition of reactive nitrogen.

Excess Nitrogen and Human Health

Occurrences of excessive levels of nitrates in drinking water predate the use of inorganic nitrogen fertilizers. Although we lack any requisite measurement, it is certain that water wells on many traditional farms were contaminated with nitrates leached from barns, sties, and dung heaps improperly located near them. Indeed, contamination from these sources of organic nitrogen is still observed on many modern farms.[44] The increasing use of nitrogen fertilizers added to the pollution originating in organic sources and extended the problem of nitrate contamination far beyond individual farms and fields to whole watersheds and to waters in rivers far downstream from intensively farmed regions.

Health risks posed by this pollution were used by Barry Commoner during the late 1960s and the early 1970s to spearhead his widely publicized message about the degradation of the U.S. environment.[45] Commoner concluded that the country's nitrogen cycle was seriously out of balance, and he feared that the rising nitrate levels in the Corn Belt's rivers carried such health risks that limits on fertilization rates, economically devastating for many farmers, might soon be needed to avert further deterioration. Commoner's fears—based, as subsequent critiques have shown, on some major misinterpretations—were greatly exaggerated.[46]

Nitrate concentrations have not declined substantially since the early 1970s. In many streams and aquifers in North America, Europe, and Asia they have risen appreciably, but this rise has not been accompanied by any clearly discernible damage to human health. In the United States commercial fertilizers (whose overall use is now, as we have seen, much higher than in the late 1960s) are the primary nonpoint source of nitrogen in water.[47] Excessive nitrate levels have been present in water wells throughout the American Midwest for more than two decades. Concentrations above the maximum contaminant limit (MCL) are particularly common in the Corn Belt states, as well as in North Dakota and Kansas.[48]

The best available calculations show that 87% of nitrates accumulated in groundwaters of the countries of the European Union are leached from agricultural soils; in addition, a substantial share of leaching from nonagricultural lands is also caused indirectly through the nitrification of deposited ammonia that was volatilized from fertilized fields and from the wastes of farm animals.[49] European experience also shows that it may take not merely months, but many years or decades, before leached nitrates reach deep aquifers and before their concentrations, in the absence of denitrification, rise to potentially harmful concentrations. Parts of Western Europe where heavy nitrogen applications predate World War II provide excellent examples of this gradual process: nitrate concentrations began to rise quickly in both ground and surface waters only during the early 1970s.[50]

By the early 1980s nitrates were either near or above the European Union's MCL of 50 mg NO_3^-/L in a number of regions, especially in England and the Netherlands.[51] The average concentrations of nitrates in the most affected European rivers—the Thames, Rhine, Meuse, and Elbe—are now two orders of magnitude above the mean of unpolluted streams, and in the early 1990s more than $1/10$ of western Europe's rivers had NO_3^- levels above the MCL.[52] Nitrate levels in the Mississippi have risen nearly fourfold since 1900, and they are increasing in both of China's two largest rivers, the Huanghe and Yangzi (fig. 9.5).[53]

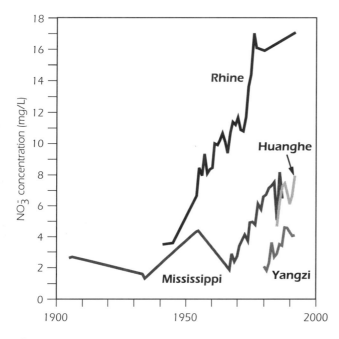

Figure 9.5
Rising nitrate concentrations in four of the world's major rivers.

Nitrate itself is not toxic to humans in concentrations encountered even in fairly polluted waters; only after its bacterial conversion to nitrite does it pose an acute threat to infants and a possible chronic risk to adults. Fortunately, there is no solid evidence that either of these risks has had any serious, or widespread, impact on human health. Methemoglobinemia (blue baby disease) is potentially deadly because infants, particularly those younger than three months, do not have enough acidity in their gastrointestinal tract to prevent its colonization with bacteria reducing nitrate to nitrite.[54] Toxic levels of nitrites may thus pass into the bloodstream where they change the ferrous iron (Fe^{2+}) of hemoglobin, the carrier and transmitter of oxygen, into the ferric iron (Fe^{3+}) of methemoglobin incapable of carrying out these vital functions.

Effects progress from cyanosis (blue baby syndrome) to shortness of breath and eventual asphyxiation (chemical suffocation) as the blood turns chocolate brown. Fortunately, an infant treated in a timely manner with ascorbic acid or methylene blue will experience a complete recovery, and adequate intake of vitamin C provides good prevention. Making the potentially affected rural populations aware of this risk has been a successful way to prevent this problem; reported incidence of methemoglobinemia has been increasingly rare in Western Europe and North America.[55] But methemoglobinemia remains a risk in low-income countries where many wells already contain nitrate levels well above 50 mg/L; so far, the improper handling of human and animal wastes, rather than the leaching of fertilizers, has been overwhelmingly responsible for this contamination, but a rapidly growing use of nitrogen applications will obviously increase the overall risk in the future.[56]

A suspected link between drinking nitrate-contaminated water and a higher incidence of some cancers in adults has been difficult to prove. The suggested etiology is as follows: a variable fraction of nitrates present in drinking water is reduced by bacteria in saliva to nitrites, adding to the quantity of nitrites taken directly, mostly from cured meats. Nitrites then react with amines to produce carcinogenic N-nitroso compounds.[57] An epidemiological association between high nitrate intake and oral, esophageal, gastric, and intestinal cancers was first reported from Chile in 1970; a number of studies done subsequently in Europe, North America, and Asia offered some confirmation of such links, as well as some evidence of an inverse relationship between nitrate intakes and various cancers.[58]

But even if nitrates were to be unequivocally implicated in the etiology of some cancers, their intake in drinking water would be only a part of the problem. Most of the nitrates that we take in (about four-fifths of the total daily intake in the United States) do not come from water but from vegetables, above all from beets, celery, spinach, lettuce, and radishes, and our bodies may synthesize daily an amount that matches that total, or even surpasses it.[59]

Moreover, a British study, which measured nitrate levels in saliva, found that their concentrations were significantly higher in areas where the risk of stomach cancer was low.[60] Perhaps most reassuringly, workers in fertilizer factories exposed to nitrate-containing dust do not have a higher incidence of stomach cancer, and the cancer has been declining throughout the Western world just as nitrogen applications, and nitrate concentrations in drinking water, have undergone unprecedented increases.[61]

The most recent review of the epidemiological evidence merely extends the inconclusive verdict: although no firm links have been found between dietary intakes of nitrate and stomach, brain, esophageal, and nasopharyngeal cancers, an association cannot be ruled out.[62] In contrast to these continuing epidemiological uncertainties, the undesirable impacts of water-borne reactive nitrogen on aquatic ecosystems have been much easier to demonstrate, and the 1990s have also seen new evidence concerning the effect of excess nitrogen on terrestrial ecosystems.

Nitrogen and Natural Ecosystems

The availability of reactive nitrogen limits photosynthesis in a wide variety of terrestrial and marine ecosystems. Nitrogen limitation is particularly common in temperate and boreal forests, temperate grasslands, and Arctic and subalpine tundra, and in estuaries and coastal marine ecosystems in temperate zones.[63] Introduction of relatively large external inputs of the nutrient must be expected to change the productivity of affected ecosystems, as well as the modes of their nitrogen storage and composition of their species.

Eutrophication of shallow waters is the oldest concern arising from this change: it was actually one of the prominent factors in the genesis of modern environmental awareness during the late 1960s and early 1970s.[64] Enriching streams, lakes, ponds, bays, and estuaries with what is normally a photosynthesis-limiting nutrient promotes the growth of algae. Their decomposition can deoxygenate water and hence seriously affect or kill aquatic species, particularly the bottom dwellers (shellfish, molluscs). Algal blooms may also cause problems with water filtration and produce harmful toxins. Nitrogen-induced eutrophication threatens above all shallow lakes and coastal waters that receive high inflows of the nutrient from the land.

This nitrogen does not originate only in inorganic fertilizers; it also comes from animal manures, human wastes, industrial processes, and atmospheric deposition. But there is a clear correlation between a watershed's average rate of nitrogen fertilization and the riverine transport of the nutrient (fig. 9.6).[65] As the streams, or groundwaters, gather these inputs from large areas, nitrogen additions to coastal ecosystems may be ten times, and even more than 100 times, higher per unit of their area than they are in fairly heavily fertilized fields.[66] Most estuaries in North America—including the Long Island Sound, San Francisco Bay, and the mid–Chesapeake Bay—have the productivity of their plant and algae species limited by nitrogen, rather than by phosphorus, and hence they are particularly susceptible to nitrate-induced eutrophication.[67]

The worst affected offshore zone in North America is a large region of the Gulf of Mexico where the nitrogen load brought by the Mississippi and Atchafalya rivers has doubled since 1965, and where every spring eutrophication creates a large hypoxic zone that kills many bottom-dwelling species and drives away fish.[68] Worrisome

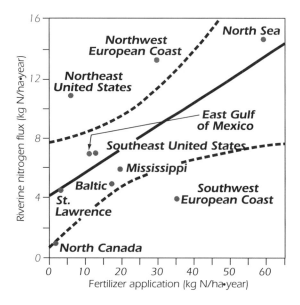

Figure 9.6
Correlation between nitrogen fertilizer applications and riverine flux of nitrogen.

reports have also come from the lagoon of the Great Barrier Reef, the world's largest coral formation, in Australia.[69] Since the early 1950s applications of nitrogen on the adjoining farmland have risen nearly tenfold, and those of phosphorus have quadrupled. The resulting eutrophication of shallow waters threatens coral by smothering them with algae as well as by promoting the survival and growth of the larvae of *Acanthaster planci,* the crown-of-thorns starfish, that has recently destroyed large areas of the reef.

In Europe the Baltic Sea, which receives almost 1.5 Mt N annually from agricultural runoff and atmospheric deposition, has been one of the most noticeably eutrophied marine ecosystems. In many places the hypoxic, or outright anoxic, conditions of the sea's bottom sediments are attested to by the presence of extensive mats of *Beggiatoa,* bacteria that reduce CO_2 and produce sulfates in the absence of oxygen.[70] Appreciable eutrophication has also been observed in the Black, Mediterranean, and North Seas.

While the emissions of SO_2 dominated the concerns about acid precipitation and its effect on aquatic ecosystems during the 1970s, it was later realized that nitrogen compounds contribute significantly to the process in both Western Europe and North

America. They do so directly in the form of nitrates generated from oxidation of NO_x emitted mostly from combustion of fossil fuels but also from nitrogen fertilizers. Their indirect contribution to acidification begins with ammonia volatilized from animal wastes and nitrogen fertilizers and released during the combustion of fossil fuels: NH_4^+ deposited from the atmosphere is either taken up by plants or, more likely, nitrified, and both of these processes produce hydrogen ions.

After about three decades of extensive research on acid deposition some of the effects of acidity in regions devoid of any buffering capacity are well known. Leaching of calcium and mobilization of toxic aluminum from soils and often profound changes of plant and heterotrophic species composition in lakes, including the demise of the most sensitive fishes, amphibians and insects, are the most deleterious and indisputably established impacts.[71] In contrast, the exact role of acid deposition in the reduced productivity and dieback of some forests remains uncertain.[72]

Atmospheric deposition of ammonia and nitrates is now taking place on such unprecedented scales that nitrogen inputs to natural ecosystems have become significant even by agricultural standards: in parts of eastern North America, northwestern Europe, and East Asia they are between 20 and 60 kg N/ha a year, and the peaks, in the Netherlands, are over 80 kg N/ha a year.[73] The rates around 60 kg N/ha are as high as average fertilizer applications to North American spring wheat, and they are much higher than the fertilization means in most African countries. For most forests such rates are an order of magnitude higher than the means of the preindustrial world, and they equal or surpass the quantity of the element made available through net mineralization of organic matter in the forest floor.

Given the widespread nitrogen shortages in natural ecosystems, especially in temperate and boreal forests that have developed under constant nitrogen stress, such an enrichment should initially stimulate photosynthesis and appreciably increase the carbon and nitrogen stored in plants, litter, and soil organic matter. But this response is self-limiting: total nutrient inputs eventually surpass the combined plant and microbial demand, and nitrogen saturation of an ecosystem leads to the element's transfer to waters as well as its emission to the atmosphere. The production (and storage) phase is followed by a destabilization phase and then by a decline (nutrient release) phase (fig. 9.7).

This expected sequence of events has been confirmed by recent research in temperate forest ecosystems in Europe and in the United States.[74] Initial nitrogen retention has been relatively high (particularly on previously disturbed, or deforested, sites), as

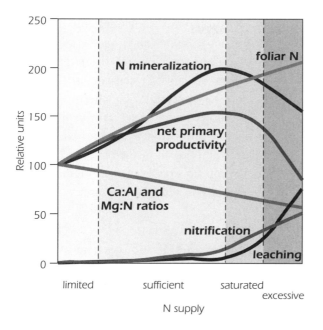

Figure 9.7
Stages of forest response to nitrogen enrichment.

has been the increase in foliar nitrogen concentration.[75] But in spite of this, nitrogen fertilization experiments in both New England and Europe have provided preliminary evidence of declining forest productivity and even some greater tree mortality.[76] Where the critical load for nitrogen—the deposition rate that would not result in excess release of the nutrient—has been exceeded, ground or surface waters are receiving much higher amounts of nitrates. While pristine Nordic forests often release less than 0.5 kg N/ha to their streams, some areas of dense forest in southwest Sweden now have leaching rates exceeding 8 kg N/ha a year.[77]

Nitrogen saturation can lead to significant changes in ecosystem composition and diversity. Nitrogen-fixing species will become less competitive, and lichens associated with nitrogen-fixing cyanobacteria may retreat. In ecosystems adapted to low nitrogen supply, individual species or whole plant communities may become extinct because a few nitrophilic varieties, able to grow faster under high inputs of the nutrient, will thrive. Declines of plant diversity with rising nitrogen loading have been already observed in grasslands in both North America and Europe and in European heathlands and forests.[78]

In view of these complexities it is difficult to estimate what effect the nitrogen enrichment already had, and what future impact it might have, on the additional storage of carbon in the terrestrial phytomass. The variety of growth patterns, carbon reservoir turnovers, and phytomass C/N ratios make large-scale estimates of vegetation's response to higher nitrogen inputs very uncertain. Northern wetlands, storing most of their production in peat, should become a larger carbon sink with nitrogen enrichment. But most of the additional productivity and carbon storage should take place in nitrogen-limited temperate and boreal forests: high C/N ratios of wood (often above 200) guarantee high carbon uptakes even with modest additions of nitrogen, and the longevity of tree trunks ensures sizable long-term carbon storage.

Published estimates have already credited annual increments as low as 0.2 Gt C and as high as 2.3 Gt C to stimulation by the releases of anthropogenically fixed nitrogen.[79] If values close to the upper limit of this range were correct, then the nitrogen-stimulated uptake of carbon would provide a sink large enough to balance global carbon losses from land use changes and phytomass combustion—and it would also solve the puzzle of the missing carbon.[80] At this time we do not know how long these enrichment benefits will last, how individual species will respond, how the gains will be partitioned among tree tissues, and what the eventual consequences of competition among species will be.

We must have a reasonably accurate account of all relevant nitrogen inputs in order to estimate the share of the overall nitrogen flux that is caused by the losses of the nutrient from nitrogen fertilizers. This is generally easiest for water-borne flows: in some cases the dominant role of fertilizers will be obvious (a shallow lake surrounded by heavily fertilized fields), in other instances we have enough data to apportion the causes with a fair degree of certainty. For example, nitrogen fertilizers are estimated to account for 56% of the nutrient's presence in the Mississippi's discharge to the Gulf of Mexico, manures for 25%, and municipal waste water for only 6%.

But in the case of atmospheric deposition it is possible to offer only rough estimates. Total anthropogenic gaseous emissions of reactive nitrogen compounds recently have been about 80 Mt N/year, with between 10 and 15 Mt N coming from inorganic fertilizers and at least 30 Mt N (as NH_3) from domestic animals. With about ¼ of the latter flux released by grazing animals, emissions originating in synthetic fertilizers would account for 40–50% of recent atmospheric deposition generated by human activities. Estimating the share of fertilizer-derived N_2O is even more uncertain; it may be anywhere between 20 and 50% of the total anthropogenic flux.[81]

Future impacts of nitrogen losses from fertilizers will not be simply proportional to the rising nutritional needs of larger global populations. There are many effective ways to improve the efficiency of fertilizer use and to reduce undesirable losses through better agronomic practices—and no smaller opportunities for reducing the biospheric throughput of the anthropogenic nitrogen are latent in the composition of our diets.

10

Nitrogen and Civilization
Managing the Nitrogen Cycle

Great social, political, and technical changes of the twentieth century have a common root in the most remarkable development that has affected humans since 1900: the unprecedented multiplication of our species. The world's population reached 10 million people by about 3000 B.C., the era just preceding the Old Kingdom's pyramid builders. Then it took it some 2,500 years to surpass 100 million around the time of the Athenian splendor, and the 1 billion mark was reached just a few years after the beginning of the nineteenth century, during the time of the Napoleonic expansion (fig. 10.1).[1] Growth rates for the last two intervals translate, to a mere 0.09% and 0.18% per year, respectively.

In 1900 the global count stood at 1.625 billion after a century of growth averaging 0.5% a year. Twentieth-century growth averaged about 1.3%, and as the year 2000 began the global population just surpassed 6 billion people. This expansion—currently equivalent to adding the population of Germany every year—has been accompanied by an equally unprecedented rise of food production. While the twentieth-century population total has grown 3.7-fold, the global harvest of staple cereal crops, as well as the total production of crop-derived food energy, has expanded sevenfold (fig. 10.2).[2] As a result, never before have so many people—be it in absolute or in relative terms—enjoyed such an adequate to abundant supply of food. Continuing malnutrition and stunting are caused by unequal access to food rather than by absolute supply shortages.

This enormous accomplishment has been made possible only by a combination of many advances, including rapidly rising energy subsidies in farming, the introduction of new crop varieties, and the availability of better pest and weed controls. But neither the mechanization of farming nor better cultivars or new pesticides would have produced today's harvests without a massive increase in the supply of fixed

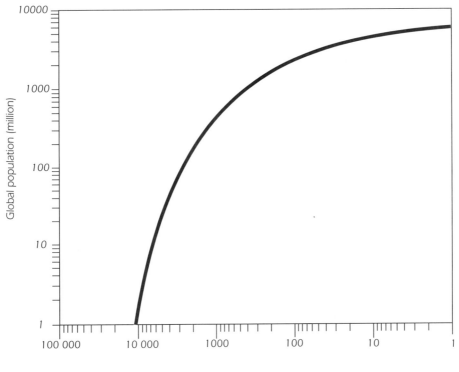

Figure 10.1
Growth of the world's population during the past 10,000 years.

nitrogen, the nutrient that is required in the largest quantity by agricultural crops and whose shortages are the most common reason for low yields.

Haber–Bosch synthesis of ammonia removed the limits on nitrogen supply that had constrained all traditional agricultures. By doing so it opened the way to universal adoption of high-yielding cultivars responsive to high applications of synthetic fertilizers. Recounting the accomplishments of Fritz Haber, Carl Bosch, and their colleagues and retracing the subsequent advances of the nitrogen fertilizer industry, agricultural productivity, and global population growth makes clear the revolutionary nature of the invention and its increasing importance for the survival of a growing share of humanity.

But, as with any fundamental change, the abundance of fixed nitrogen has, as reviewed in the preceding chapter, many undesirable consequences. That makes it

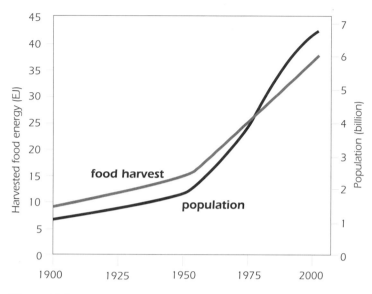

Figure 10.2
Growth of global population and food harvests during the twentieth century.

essential to improve the efficiency with which we use the nutrient in agriculture. These improvements will be critical in order to reconcile the anticipated rise in fertilizer demand during the twenty-first century with the necessity to minimize the human impact on the environment. But better agronomic practices may not suffice to bring about this highly desirable compromise; rational diets and stabilized populations should contribute to the quest for minimizing the nitrogen burden on the biosphere.

What Has Been Accomplished

At the beginning of the twentieth century the world faced the prospect of a steeply rising demand for fixed nitrogen and of limited, insecure, or expensive opportunities for its supply. Modernizing agricultures could break the yield limits imposed on traditional farming only through the access to inexpensive and abundant supplies of the nutrient. Between 1901 and 1907 laboratory work by Le Châtelier, Ostwald, Nernst (with Jost), and Haber (with Oordt and Le Rossignol) laid the theoretical foundations of ammonia synthesis.

Any one of these chemists could have been the actual inventor of technically viable ammonia synthesis, but unique circumstances—the explosion in Le Châtelier's laboratory, Ostwald's erroneous interpretation of experimental results, and Nernst's opinion that reaction yields were too low to justify further exploration—led the three scientists to abandon their ammonia-related work. Only Haber was not discouraged by his early setbacks, and he and Le Rossignol persevered in their search for a commercially acceptable solution. In two years, between May 1907 and July 1909, they advanced the project to the level of a satisfactory working bench-top apparatus. This was a truly revolutionary accomplishment: biospheric evolution has not endowed any eukaryotic organisms with the capacity to fix nitrogen, and none of the surprisingly small number of prokaryotes able to perform this reduction is symbiotic with any staple cereal or tuber crop.[3]

Bosch's conviction that unprecedented difficulties could be overcome in order to commercialize the process proved to be correct. He lead a successful effort, much aided by Mittasch and Lappe, to transform a laboratory device into a continuously operating high-pressure, high-temperature synthesis process, a prototype for a new kind of chemical syntheses, in just four years. Oppau, the first commercial plant, built in just fifteen months, was intended to fix 10 t NH_3/day, but eventually it ended up producing three times as much.

Knowledgeable individuals immediately appreciated the importance of the discovery and marveled at the rapidity of its commercialization. W. J. Landis, the director of the American Cyanamide Company, wrote in 1915 that "too much honour cannot be shown to the courageous chemists who have succeeded in placing this process on a commercial working basis," and two years later, during a debate on nitrogen in the British House of Commons, Sir William Pierce said that the Haber–Bosch process is "one of the greatest achievements of the German intellect during the war. . . . It is really a wonderful achievement."[4] Yet just a few generations later surprisingly few people know of the discovery, and even fewer are aware of its fundamental, and steadily increasing, importance for modern civilization.

Until the 1930s the most important technical improvements necessary for the further expansion of ammonia synthesis—including the first designs for the reforming of natural gas—continued to come from Germany. Gradual post–World War II advances ended in the early 1960s with an innovative U.S. design of single-train ammonia synthesis plants using centrifugal compressors. Since that time the highest capacities of these plants have risen from less than 600 to nearly 2,000 t/day. This growth was accompanied by a declining energy intensity of ammonia synthesis, from around 120 GJ/t N in the first German plants to about 55 GJ/t N in the best plants using reciprocating compressors and to just over 30 GJ/t N for today's most efficient plants built by such leading producers as Kellogg Brown & Root, Haldor Topsøe, and Uhde.

The combination of technical innovations (including lower pressures in reforming and synthesis loops, better catalysts, and integrated management of the process's energy demands) has made it possible to satisfy the rising demand for fertilizer (as well as for industrial ammonia) with products sold at progressively lower prices (fig. 10.3).[5] The global capacity for ammonia synthesis has surpassed 120 Mt N a year, and actual production during the closing years of the twentieth century was over 100 Mt N. After subtracting losses and industrial uses, more than 80 Mt N are now left to be used either directly as fertilizer or, a much likely course, as the feedstock for synthesizing more complex nitrogen compounds, above all urea, currently the most important solid nitrogen fertilizer.

Annual applications of about 80 Mt of nitrogen are ten times as large as the application of inorganic fertilizers in 1956 and 100 times as large as the annual rate in

Figure 10.3
Average price of U.S. ammonia, 1960–1998.

1928. Applications of nitrogen fertilizers prevented depletion of natural nitrogen stores in intensively cultivated soils and made it possible to plant more staple food crops (often, unfortunately, in continuous monocultures) instead of forage legumes that were previously needed to maintain soil fertility. Ready availability of the nutrient has fundamentally transformed the global agriculture: synthetic fertilizers have become the dominant source of nitrogen in modern, intensive cropping, supplying at least 60%, and commonly 70–80%, of the nutrient reaching intensively farmed fields.

Without the resulting rise in average yields—globally for cereal grains from about 0.75 t/ha in 1900 to 2.7 t/ha in 2000—it would not have been possible to support the unprecedented numbers of people with diets whose adequacy is unmatched in history. Even very conservative calculations indicate that during the closing years of the twentieth century at least 2.4 billion people—that is, some 40% of the world's population of 6 billion—could not be fed in their countries even by the most intensive, but always nitrogen-limited, traditional farming. These people are alive because the proteins in their bodies were synthesized from dietary amino acids whose nitrogen originated in the Haber–Bosch synthesis of ammonia.

By sustaining higher yields, nitrogen fertilizers have also made it possible to reduce the total area of land that has to be devoted to farming. Without intensive cropping we would not have had the post–World War II regrowth of forests in Europe and North America, and much larger areas of forests, grasslands, and wetlands throughout the poor world would have to be converted to fields (fig. 10.4).[6]

But, as with any fundamental change, the abundance of fixed nitrogen also has a number of troublesome consequences. Haber–Bosch synthesis has made it possible to sever the traditionally tight link between cropping and animal husbandry, and to move increasing amounts of fixed nitrogen not only within individual countries but also among nations and continents. This has given rise to disjointed, one might say dysfunctional, nitrogen cycling. Individual farms, even whole agricultural regions, have ceased to be functional units within which the bulk of crop nutrients used to keep cycling during centuries, or even millennia, of traditional farming.[7]

Synthetic compounds eliminated the necessity of returning the nutrient in animal wastes to fields in order to sustain new harvests, and they allowed the emergence of highly specialized cropping that is largely, or completely, separated from no less specialized and highly concentrated animal production. Long-distance transfers of fixed nitrogen have replaced this traditional pattern. In plant production the element moves in with purchases (imports) of fertilizers and goes out with sales (exports) of

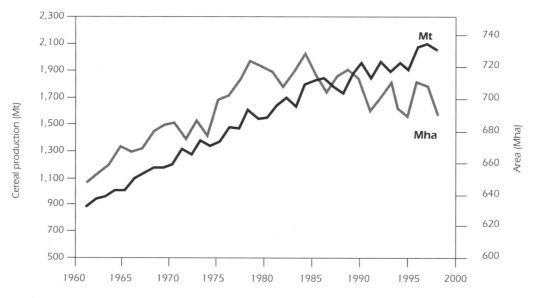

Figure 10.4
Higher yields are sparing the land as larger harvests come from an unchanged, or even declining, area of farmland as shown by this comparison of the global cereal harvest and planted area.

food and feed crops; in animal husbandry it comes in concentrate feeds (often imported from overseas) and goes out in meat, dairy, and aquacultural products. These movements have become significant even on the international scale: about 30% of the world's nitrogen fertilizer production is exported, as is about 15% of all staple crops and 10% of all meat.

These arrangements bring some impressive economies of scale, but they create environmental problems at both ends. Synthetic fertilizers can provide any conceivable nutrient needs for specialized monocropping, but reduced, or completely absent, recycling of organic wastes gradually lowers the concentrations of the soil's organic matter. This change has many undesirable consequences for the soil quality, above all greater soil compaction resulting in worsened tilth, easier erodibility, lowered water-holding capacity, and weakened ability to support diverse soil biota and to buffer acid deposition.

At the other end, it is increasingly difficult to dispose of nitrogen accumulation resulting from highly concentrated meat production; large beef feedlots, piggeries, and broiler houses now contain many thousands of animals. And animals use nitrogen rather inefficiently: about 30% of the nutrient consumed by dairy cows in feed is transferred to milk, and the analogical rates are only 20% for chicken, 10% for pork, and 5% for beef. Fields located within a radius allowing a profitable recycling of manures (generally no more than 10 km) can absorb only a small fraction of generated wastes, and even this limited recycling is questionable because of the relatively high concentrations of heavy metals present in animal and poultry excreta.[8]

Higher applications of nitrogen fertilizers have also increased the opportunities for losing the nutrient from fields and transferring it to fresh and coastal waters, to soils, and into the atmosphere. Because nitrogen is commonly the most important limiting nutrient in natural ecosystems, such enrichment must have a variety of consequences for aquatic and terrestrial biota that are now subjected to this steady, and often increasing, eutrophication. Some nitrogen losses from the agricultural land are an inescapable part of the nutrient's complex cycle, but most leakages result from inappropriate agronomic practices.

On the whole, the record is unsatisfactory because no more than about half of the globally applied inorganic nitrogen finds its way to plant tissues; mean uptakes are higher in intensive Western agricultures, but considerably lower (even below 30%) in Asia's rice fields, now the leading consumers of fixed nitrogen. Producing food for larger populations that aspire to a higher standard of living will require substantial increases in the Haber–Bosch synthesis during the first half of the twenty-first century, but increasing the supply through expanded synthesis of ammonia should be only one part of meeting that challenge. More efficient fertilizing and more rational diets of stabilized populations can obviate a great number of nitrogen applications and lower the burden on the biosphere.

More Efficient Fertilizing

There is no shortage of means to improve the efficiency of fertilization, but when using them we should be mindful of the lessons that we have learned about more efficient use of energy after OPEC did the world a great favor and quintupled crude oil prices in 1973–1974.[9] Even if it were to be universally adopted and assiduously practised, no single approach or technique can make a decisive difference. Substantial efficiency gains are always a matter of gradual, aggregate advances resulting from the widespread embrace of various improvements.

These improvements range from such simple, relatively inexpensive, and widely available practices as repeated soil testing to the most sophisticated methods of variable fertilizer applications using advanced satellite and computer techniques. Fortunately, the efficacy of these measures is not directly proportional to their cost. Soil testing, selection of the most appropriate fertilizing compounds, maintenance of proper nutrient ratios, and optimized timing and placement of fertilizers are the most important direct, and low-cost, measures, whose common adoption has the potential to cut the world's fertilizer use. Indirect approaches rely primarily on good agronomic practices embracing crop rotations, conservation tillage, and weed control.[10]

Periodic soil testing does not just provide recommendations for appropriate macronutrient applications. It can also uncover growth-limiting micronutrient deficiencies. At the same time, the inherent complexities of N flows in agroecosystems mean that accurate predictions of actual fertilizer demand will be always difficult.[11] Soil testing can make the greatest difference in many low-income countries where too many farmers simply follow general recommendations for fertilizer applications, but the best results might be achieved only when farmers adjust fertilizer application rates to the highly variable nitrogen supply capacity of soils. This adjustment can push nitrogen recovery by rice over 50% even with the high applications needed to support yields in excess of 9 t/ha.[12]

Many experiments have shown that correct applications generally make a much greater difference than the choice of a particular synthetic compound, and a balanced nutrient supply is a key prerequisite for achieving good recovery rates. Liebig's law of minimum asserted this need at the beginning of modern agricultural science, and numerous trials have demonstrated the effectiveness of this balancing act.[13] In spite of this, excessive use of nitrogen, a practice promoting high losses of the nutrient, is still common. Unfortunately, China, the world's largest consumer of nitrogen, is its leading improper user. While the worldwide N:P:K mean is now about 100:44: 28, Chinese applications have been chronically deficient in both P and K, with a nationwide ratio of 100:33:10.[14]

Proper timing of fertilizer applications is another well-proven way of saving nitrogen: applications should coincide, as much as possible, with periods of the highest nutrient need, which comes generally during the rapid vegetative growth stage before crops shift to reproductive growth.[15] Applications many weeks ahead of planting will inevitably increase the risk of substantial losses of nitrogen. Split applications—the initial one before or at the time of planting, the second, or the third one, nearer the time of maximum uptake—are the best choice.

Correct placement of fertilizers is no less important. They should be put as close to the position of future, or existing, plant roots as is practical, but they should not cause injury to plants. Surface applications (broadcasting or sidedressing) are almost always less efficient than subsurface incorporation (broadcasting followed by plowing or disking, row application with seed, deep banding apart from seed).[16] But too deep a placement of nutrients may also be wasteful; studies with ^{15}N show that the use of nitrogen decreases appreciably with depth. A counterintuitive result regarding fertilizer placement emerged from a study in Missouri that demonstrated that with adequate water supply applications of nitrogen, and water, in alternate furrows produced a harvest at least equivalent or even higher than those resulting from fertilizing every furrow.[17]

Cheaper microcomputers and the commercial availability of global positioning systems (GPS) opened up the possibility of site-specific applications of fertilizers. Rather than choosing the fertilizer amount according to average conditions within a field, precision farming adjusts the applied amount according not only to field-to-field variations in soil quality but also to often substantial differences of nutrient content within a single field.[18] Maximized yields and profits and minimized nitrogen applications, and hence reduced nutrient losses to the environment, should be the ideal outcome. Evaluations of recent experiments with precision farming have ranged from highly encouraging results to no significant difference when compared with standard practices to unprofitable experiences, but there is little doubt that the technique will be pursued and perfected in the future.[19]

A totally different approach to increasing the efficiency of fertilizing is to slow down the rate of nitrification, and hence to reduce the water-borne losses of the nutrient. This has been possible since the early 1960s when nitrapyrin, the first nitrification inhibitor, became commercially available. Many inhibitors are now on the market—nitrapyrin and dicyandiamide are the two most popular choices—but their use is rather expensive, their performance is highly variable, results from field tests are often contradictory, and their applications have been restricted to just a few field crops (above all corn and root crops) and special climatic conditions.[20]

Many agronomic practices, some of them not directly connected with fertilization, can help boost the efficiency of fertilizers.[21] Crop rotations including leguminous cover species are particularly desirable, but it must be kept in mind that some legumes will actually remove rather than add fixed nitrogen. Reduced tillage, adequate (but not excessive) water supply, and recycling of residues are perhaps the most generally applicable ways of rational farming that help maintain good soil structure (by replenishing soil organic matter and enhancing water absorption capacity), enrich soil fauna, enhance mycorrhizal associations, and minimize nutrient losses due to erosion and leaching. Adequate pest and weed control also prevents excessive losses of nitrogen.

The Codes of Best Agricultural Practices—mandated by a European Union directive published in December 1991 (commonly known as the Nitrate Directive) and designed to protect fresh, coastal, and marine waters against nitrate pollution from diffuse sources—promote an integrated approach to more efficient fertilization.[22] Nutrient applications should not exceed crop needs (after the contributions from organic sources are taken properly into account), soils should not be left bare during rainy periods, and nitrates present in soil between crops should be limited through planting of trap crops.[23] Several countries have also introduced maximum fertilizer levels by crop and per applications in the most vulnerable areas.

Gains in fertilizer efficiency can already be seen in the records of many Western countries, as well as in Japan. Nitrogen applications to English winter wheat peaked in 1985, and since that time they have stagnated at slightly lower levels, while the amount of nitrogen assimilated by the crop rose by about 20% between 1985 and 1995 (fig. 10.5). Average nitrogen applications to Japanese rice fell more

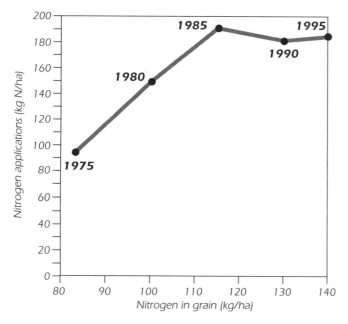

Figure 10.5
Increasing efficiency of nitrogen fertilizer use in English wheat cultivation.

substantially since their peak in the early 1980s; in combination with a slightly higher yield this has translated to about a 30% gain in the apparent uptake of the nutrient between 1980 and 1995 (fig. 10.6).[24] In 1996 1 kg of nitrogen applied to U.S. corn helped to produce, on average, nearly 56 kg of grain compared to 46 kg in 1975, roughly a 20% gain in a generation.[25]

Potential future gains remain impressive. Good agronomic practices should raise the worldwide mean of nitrogen use efficiency by at least 20%, and possibly 25%, during the first quarter of the twenty-first century. This would push the average uptakes to no less than 50% in low-income countries and to around 65% in affluent nations. These higher fertilizer efficiencies would use on the order of 10 Mt of the currently wasted nitrogen applications. At the same time, the effective supply of nitrogen from organic wastes, biofixation, and atmospheric deposition should also increase because of the combination of reduced nutrient losses from soil erosion, more frequent rotations, and more vigorous recycling.

Reducing erosion losses by about 20% would save roughly 5 Mt N from nonfertil-izer sources, and expanding biofixation (largely through proper rotations with le-

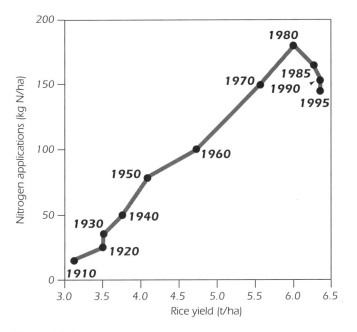

Figure 10.6
Increasing efficiency of nitrogen fertilizer use in Japanese rice cultivation.

gumes) and waste recycling by just 10% could add another 5 Mt N. The cumulative effect of adopting well-proven, and mostly low-cost, agronomic measures aimed at increasing the efficiency of nutrient uptake could then be equal to expanding the effective nitrogen supply to crops by about 20 Mt a year, the mass equivalent to $1/4$ of total nitrogen fertilizer applications during the late 1990s.

As such a large share of harvest is now fed to animals, further substantial gains in fertilizer efficiency could be achieved by integrating or reintegrating crop and animal farming and reversing the spatial separation of the two segments of the trophic pyramid. Such a reconnection would allow for more extensive, and more efficient, recycling of manures, but it would go against the entrenched trend of the increasing separation of the two activities as traditional ways of mixed farming are being replaced by specialized cropping and centralized animal feeding.[26]

Reversing another widespread trend—eating more food from more distant places rather than consuming locally produced food (even in places where environmental conditions provide no obstacle to doing so)—would also be extremely helpful in improving the efficiency with which nitrogen is used in modern agriculture. But a realistic assessment must conclude that this shift, too, has only a low probability.[27] In contrast, chances of an early stabilization of the global population appear to be much greater, and the twentieth century has already seen some desirable long-term changes in the composition of average diets.

Stabilized Populations

Some milestones in history are eagerly anticipated, much celebrated when they come and repeatedly recalled afterwards. Other pass unnoticed and elicit surprisingly few comments even when their importance eventually becomes so obvious; global population milestones belong in this second category. The first key turning point for the world's population in the twentieth century was reached, virtually unnoticed at the time, during the late 1960s, when its relative rate of growth peaked at just over 2% a year (the 1965–1970 mean was 2.04%; fig. 10.7).[28]

At that time the world's population was increasing by between 70 and 75 million people a year. The global growth rate fell to 1.72% a year a decade later, but because of the higher population totals absolute annual additions kept increasing; they reached 85 million people a year during the late 1980s. In 1991 the medium variant of the United Nations forecast put the average annual growth at 1.73% during the first half of the 1990s and at 1.63% for the years 1995–2000.[29]

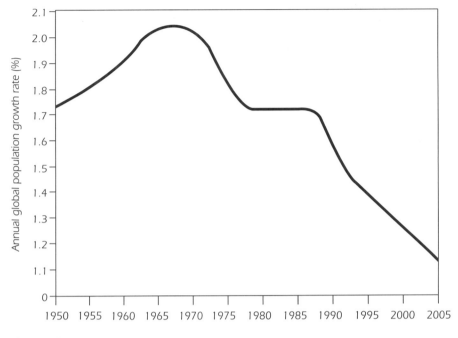

Figure 10.7
Declining growth rates of global population, 1950–2005.

In absolute terms this would have meant additions of almost exactly 100 million people a year by 1995 and over 100 million people a year during the late 1990s. But then an unexpectedly rapid decline of fertilities reversed the trend. In the 1994 revision of their forecasts the UN demographers lowered the expected growth rate for the first half of the 1990s from 1.73% down to 1.57%, a 10% cut, and in 1996 they put the actual rate at just 1.48%.[30] In the latest edition of the UN's *World Population Prospects* they revised the rate to 1.46% (fig. 10.7).[31]

This means that annual global population growth has been declining even in absolute terms. It fell to 80 million by the year 1995, and it will keep falling slowly. Again, this milestone event received surprisingly little public attention. There is a fairly high probability that the absolute rate may never rise again—and that the growth curve of our species is now approaching its asymptote faster than was anticipated by demographers even just a few years ago.[32] After millennia of a very slow but steady increase (the initial flat phase of the growth curve), and after a few centuries of rapid additions, reflected by a steeply rising segment of the growth curve, we are now finally beyond the midpoint of the S-curve with the asymptote discernible only a few generations away (fig. 10.8).

While there can be no doubt that the transition to low fertility—a shift already accomplished in all the affluent countries and well advanced in much of Asia and Latin America—is now under way even in sub-Saharan Africa, the largest remaining region of very high fertilities, the eventual outcome of this global shift cannot be forecast with high accuracy.[33] Depending on the future course of fertility declines, by the mid–twenty-first century the global population may total anywhere between 7.3 and 10.7 billion people, with 8.9 billion the UN's latest most likely forecast. This outcome is predicated on the assumption that by 2050 virtually all countries will have a total fertility rate at, or below, the replacement level of 2.1.

Although this would require steep fertility declines in a number of populous countries with high fertilities—most notably in Nigeria and Pakistan (whose current fertilities are 5.9 and 5.0, respectively)—such reductions would not be unprecedented. China cut its fertility from about six in the mid-1960s to less than two by the mid-1990s, and a number of smaller Asian countries accomplished their demographic

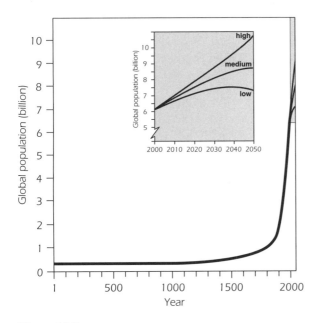

Figure 10.8
Possible growth curves of global population, 1–2050.

transition as fast or even faster.[34] Of course, it is easy to argue that similar performances are much less likely throughout sub-Saharan Africa or in parts of the Middle East. Consequently, the possibility that the global population will surpass 10 billion people by the year 2050 cannot be excluded, but even the UN's high forecast variant of 10.7 billion means that the year 2000 population of 6 billion people would not double.

The conclusion that yet another doubling of world population is unlikely has recently received independent confirmation from a probabilistic appraisal using expert opinions on fertility, mortality and migration trends.[35] Simulations used to derive probability distributions of population sizes and age structures for thirteen regions of the world indicate that there is about a 66% probability that the world's population is not going to double during the twenty-first century.

We may be thus glimpsing the beginning of the end of the world's population growth. Everything else being equal, this development would obviously limit the future need for nitrogen fertilizers. The faster the transition would be accomplished and the lower the eventually stabilized total, the less nitrogen would be needed to satisfy basic food needs. On the other hand, even an unexpectedly rapid decline of fertilities and an early leveling off at what would have seemed an incredibly low total just a generation ago could still lead to a much higher demand for fixed nitrogen if typical diets will be including higher food intakes in general, and more animal foods in particular.

Rational Diets

The eventual impacts of this nearly universal dietary transition are highly uncertain because the process will not conform to a single pattern.[36] All of the world's population growth between now and 2050 will occur in today's low-income countries, where the average per capita food energy supply is just marginally above the average per capita food intake and where meat availability stands now at less than 20 kg/year.[37] Should we put the mean of the ultimate future global per capita food energy demand near the current North American and European rates, currently the world's highest? Should we assume that the whole world will want eventually to eat as much meat as Americans or Dutch do?

If that is clearly improbable, where will the most likely rate be: will the newly rich Indians and Chinese generally prefer average meat intakes at, or below, the current Italian or Swedish levels, or will the Japanese pattern, characterized by a relatively low meat intake combined with a high demand for fish, be a more appropriate example?[38] As explained in chapter 8, different supply levels made up of different shares of animal foods will have far-reaching consequences for the overall demand for animal feedstuffs and hence for nitrogen applications. The extremes of possible future demand are rather easy to calculate.

Merely extending the prevailing global means of food and fertilizer use to 10 billion people would call for the supply of 130–135 Mt N/year by 2050. With a modest 10% increase in the efficiency of fertilizer use this total could be reduced to 120–125 Mt N/year, 50% above the late 1990s rate.[39] As the underlying assumptions are clearly too conservative, the total of 120–130 Mt N/year is almost certainly too low. On the other hand, securing the rich world's current mean intake of animal protein for 10 billion people would call, with unchanged fertilizer use efficiency, for applications of about 230 Mt N by 2050.[40] This total is obviously unrealistically high: Bangladesh will not emulate Texan diets, and the efficiencies of fertilizer use will increase.

I will not offer any actual detailed quantitative forecasts for 2050: our rich experience with long-range forecasting proves that the complex assumptions underlying such exercises guarantee their failures.[41] But I want to stress that in order to keep the total nitrogen applications as low as possible—preferably below 150 Mt N/year by the mid–twenty-first century—we should not rely merely on higher rates of nitrogen utilization but also engage in judicious demand management, that is, in influencing food consumption.[42]

I believe that a rational position is to advocate a moderate consumption of those animal foods that require the largest amount of feed (feedlot-raised beef raised on a mixture of corn and soybeans is the most obvious case in point) and to promote the intake of those foods that can be produced most efficiently. The latter category includes above all aquacultured herbivorous fish, all dairy products, and chicken.[43]

At the same time, I have to point out that the promotion and widespread adoption of diets considered optimal by modern nutritional science may still entail high, and highly wasteful, applications of nitrogen fertilizers. These diets conform broadly to traditional Mediterranean patterns of relatively frugal consumption dominated by cereals, pulses, vegetables, fruits, nuts, olive oil, cheese, and milk, with the occasional eating of fish and meat, and with moderate drinking of wine.[44] But such a dietary shift may not lower the demand for nitrogen fertilizers. Because of large differences between the prevailing and recommended intakes, adjustments in supply of fruits, vegetables, and nuts would have to be large,[45] and modern cultivation of these crops uses generally over 100 kg N/ha, and often in excess of 150 kg N/ha. A recent California study shows that these crops commonly take up much less nitrogen than cereal grains, with losses frequently amounting to 66–75%.[46]

Of course, the argument for moderated intakes of food energy in general and of animal foods in particular does not require evoking any risks arising from our interference in the biospheric nitrogen cycle: fundamental health concerns are obvious enough. Obesity—the condition of having at least a 35% excess over ideal body weight—already has a truly alarming incidence in many rich countries. The problem is particularly serious in the United States where, after being stable during the 1960s and 1970s (at about 25% of the adult population), the prevalence of obesity increased during the 1980s by about 30%. By the early 1990s every third U.S. adult was overweight, and the ratio still appears to be advancing.[47]

A growing incidence of obesity is not limited either to Western or to affluent countries. A recent survey documents surprisingly high shares of obesity in societies as different as Mexico (50% of adult males, and 58% of females obese), Thailand (with about ¼ of all adults obese), and Kuwait (more than ⅓ of all adults obese).[48] In China's cities the proportion of overweight adults rose from 9.7% in 1982 to 14.9% in the early 1990s, and in Beijing the rate is now over 30%, similar to the North American incidence.[49]

The dietary shift from coarser to more refined grains and to higher intakes of lipids, both as plant oils and in meat, is only a part of this undesirable trend. The changing employment structure brought by the rapid mechanization of agriculture, urbanization, and industrialization is reducing physical activity and increasing passive leisure time.[50] While it would be naive to suggest that exhortations to eat less will suffice to reverse this worrisome trend, I believe that we should use every approach—from nutritional education starting in childhood to bioengineered foodstuffs providing the feeling of satiety with ingestion of less energy—to reduce the extent of obesity.

Unless we do so, an increasing share of nitrogen fertilizers will be used to produce food whose consumption will seriously degrade the health of affected populations, as obesity will become one of the world's leading health concerns. A wealth of medical evidence associates obesity with reduced longevity and with a variety of health problems ranging from such metabolic disorders as insulin resistance, hyperglycemia, and hypercholesterolemia to hypertension, gallbladder diseases, orthopedic impairment, pulmonary difficulties, and surgical risk.[51]

A Long View

No eukarya are able to fix nitrogen; surprisingly few free-living prokaryotes that can do so fix it at rather low rates; and *Rhizobium* bacteria, which can fix considerable amounts, are symbiotic only with legumes, and they exact a fairly high price for the service from its host plants. Agricultures have always had to live with these peculiar biospheric limitations, and there is little to indicate that these limits can be eliminated during the next one or two generations.[52]

Three solutions would suffice: direct incorporation of nitrogen-fixing genes into the plant genome independent of bacterial infection; development of artificial symbiotic systems between nitrogen-fixing prokaryotes and various staple crops; and transgenic crops able to use the available nitrogen more efficiently. The first two options do not appear to be on even the distant horizon. In spite of enormous progress in our theoretical understanding of biofixation achieved during the 1990s—including identification of the complete nucleotide sequence and gene complement of the *Rhizobium* plasmid that endows the bacterium with its ability to associate symbiotically with legumes[53]—it is appropriate to repeat the conclusion reached at the end of a major nitrogen fixation symposium in the mid-1980s: "little of that progress has yet been applied in practical sense to improve crop production."[54]

But the third possibility has been under intensive, and promising, development.[55] In the presence of high concentrations of nitrogen, the green alga *Chlorella sorokiniana* competes with bacteria by producing high levels of glutamatae dehydrogenase (GDH); the enzyme converts ammonium directly into glutamate, which is then used in a variety of metabolic processes. Transgenic wheat producing high levels of GDH yielded up to 29% more with the same amount of nitrogen fertilizer than did the normal crop. Other transgenic crops with a high GDH capacity are under development, but it still unclear how and when Monsanto will release any commercial cultivars, and, of course, how they will perform in field conditions in the long run.

Nor are there any clear commercial prospects for radically new ways to synthesize ammonia. In January 1998 a group of Japanese researchers reported on obtaining a moderate NH_3 yield after a 24-hour treatment of the tungsten dinitrogen with dihydrogen at atmospheric pressure and 55 °C, and in October 1998 two Greek chemical engineers described ammonia synthesis from its elements at atmospheric pressure in a solid state proton(H^+)-conducting cell reactor.[56] Combining dinitrogen and dihydrogen to synthesize ammonia under mild conditions would be superior to the Haber–Bosch process, but commercial applications of these laboratory tests seem remote.[57]

The Haber–Bosch process is not going to be displaced any time soon, and increased global dependence on this synthesis is inevitable even with an early stabilization of the world population and a widespread adoption of rational diets. Intensive cultivation cannot be sustained without applications of nitrogen fertilizers. The yield increases required to produce larger harvests—needed to provide food for additional 2–4 billion people by 2050, as well as to improve the diets of some 3 billion people— are unthinkable without a substantial overall increase of nitrogen applications.

We know that harvests in completely unfertilized fields would eventually decline to a small fraction of today's yields. How fast this would happen depends, obviously, on the initial stores of the nutrient. The Exhaustion Land Experiment at Rothamsted demonstrated what happens in the absence of nutrient replenishment in a field rather well-endowed with nitrogen when different fertilizer treatments between 1856 and 1901 were followed by decades of continuous cereal cropping without any fertilization.[58] Soil nitrogen declined exponentially, with half of the starting level gone after about fifty years, and the total soil residue of the nutrient built up by previous recycling was forecast to disappear after 250–300 years.

Because most of the world's agricultural soils have lower natural stores of the nutrient than present in the Rothamsted field, it is inevitable that continuous cropping without any fertilization would reduce the annual supply of fixed nitrogen to minuscule rates in less than two centuries.[59] Total nitrogen inputs to such fields— a combination of atmospheric deposition, biofixation by scarce free-living bacteria, and minimal mineralization—could then be as low as 15–20 kg N/ha, enough to produce no more than about 45–60 kg of protein per hectare. This would return us to the levels that prevailed during the millennia of traditional extensive agriculture when staple grains yielded often no more than 0.5 t of grain per hectare and when fewer than 3 people could be supported by 1 hectare of farmland.[60]

As I demonstrated in chapter 2, we also know that even the most assiduous recycling of inorganic wastes combined with crop rotations including leguminous crops and green manures cannot supply more than 120–150 kg N/ha in highly intensive traditional cropping. Such agroecosystems can produce around 200 kg of protein per hectare and feed at least ten to eleven people on largely vegetarian diets. In contrast, today's most productive agroecosystems yield around 800 kg of protein per hectare in multicropped fields receiving high applications of inorganic nitrogen.[61]

The only proven option that could increase nitrogen inputs without the Haber–Bosch synthesis would be a large expansion of biofixation. There are no biophysical reasons why the current global rate of managed biofixation—most likely between 30 and 35 Mt N/year produced by rhizobia associated with leguminous grains and cover crops—could not be doubled, or even tripled.[62] But nutritional preferences, agronomic realities, and carrying-capacity imperatives will greatly limit any future expansion.

Per capita consumption of pulses has been declining everywhere except for India; moreover, the most commonly cultivated pulses (lentils, beans, peas) often fix only small amounts of nitrogen.[63] Soybeans, the legume whose cultivation has expanded enormously since the 1950s, are a major foodstuff only in East Asia, and there does not seem to be a high probability of making other cultures into eager beancurd consumers.[64] As already noted (in chapters 2 and 7), good soybean harvests may actually use soil nitrogen rather than add to the nutrient's soil stores.[65]

Only leguminous cover crops can leave behind appreciable amounts of fixed nitrogen, but, as I also stressed before (in chapter 2), a large-scale increase in planting such crops in land-scarce countries would inevitably limit their cultivation of staple crops.[66] Consequently, I see only a limited scope for reintegrating green manures into Asian cultivation; given Africa's greater land availability, prospects eventually may be better there. Still, I think it unlikely that managed biofixation could supply twice as much nitrogen as it does today.[67]

Improved efficiency of nitrogen fertilization must become a major new source of nitrogen. As shown earlier in this chapter, such gains already have been appreciable in affluent countries. There is no doubt that it will be much more difficult to adopt the highly managed, information-rich farming that is required to achieve such savings in low-income countries, whose food requirements will generate nearly all of the world's additional needs for nitrogen fertilizers. But this, of course, is precisely the reason for engaging the farmers of these countries in the quest for gradual savings.[68]

Asia will be the continent where these changes will be needed most: its nitrogen consumption already accounts for almost 60% of the global total, and it may double by 2020.[69] At the same time, the continent's rice agriculture uses the nutrient with relatively low efficiency, and as average applications now surpass 100, or even 200 kg N/ha, in many of Asia's regions the environmental impacts of excess nitrogen will be felt more acutely during the decades ahead.

Regardless of our achievements in more efficient fertilizing, during the next generation we should get a much clearer understanding of the real impact of this excess nitrogen in the biosphere. As noted previously (in chapter 9), not all of the nutrient not assimilated by crops moves into waters or into the atmosphere. High fertilizer inputs mean that in spite of large nitrogen removals in high-yielding crops, nutrient balances for individual fields or for entire intensive agroecosystems are either steady or are slightly positive.

A study of nitrogen balances of some forty agroecosystems on three continents found gains in soil organic nitrogen (the pool that usually contains more than 90% of the nutrient present in soil) in almost 60% of cases, with a mean annual increase of 35 kg N/ha.[70] However, without synthetic fertilizers more than half of those agroecosystems would have had net nutrient losses. Two more recent national agricultural nitrogen balance sheets, for Germany and the Netherlands, found average annual accumulations of 47 and 38 kg N/ha, respectively.[71] A global balance that I calculated for the mid-1990s indicates a moderate gain even on the global level (appendix Q). Many properly farmed soils now may be now gaining annually anywhere between 10 and 35 kg N/ha (equivalents of 0.15–0.5% of total soil N stores), and in other soils inputs and outputs are in equilibrium. This combination could outweigh nitrogen losses from deteriorating soils, particularly in Africa where nutrient mining is widespread.[72]

The extent of our future concerns about our interference in nitrogen's biospheric cycle will not depend only on the eventual demand and our success in more efficient utilization of the nutrient. The fact that scientific attention to the excess nitrogen in the biosphere grew considerably during the 1990s is not a reliable indicator of eventual public recognition.[73] And even if it were so, how long would that last? Societies shift their worries: after all, concerns about nitrates in water helped to launch the Western environmental awareness of the late 1960s and the early 1970s, but they soon yielded to other worries.[74]

Acid rain caused by SO_2 emissions gained prominence during the 1970s, the 1980s brought worries about the loss of stratospheric ozone, and the 1990s were dominated by concerns about the potentially destabilizing effects of rapid climate change.[75] So far, concerns about fertilizer-induced alterations of nitrogen cycling have not ranked anywhere near the attention paid to these three environmental changes, and particularly to the possible risks of rapid global warming. Yet they too constitute an unprecedented large-scale experiment whose eventual impacts we cannot predict.

A long-term perspective supports a readjustment of our interest. High CO_2 emissions could be reduced very effectively by a combination of economic and technical solutions, and the transition to nonfossil energies will be inevitable even without any threat of global climatic change.[76] In contrast, there is no way to grow crops and human bodies without nitrogen, and there are no imminent substitutes for the Haber–Bosch synthesis. Consequently, we will soon enter the second century of our dependence on the Haber–Bosch process.

When the Swedish Academy of Sciences awarded the Nobel Prize for chemistry to Fritz Haber in 1920, it noted that he created "an exceedingly important means of improving the standards of agriculture and the well-being of mankind."[77] Even such an effusive description appears inadequate in the year 2000. In 1920, when commercial ammonia synthesis began to take off, the world's population had reached 2 billion. Now nearly 2.5 billion people are here because proteins in their bodies are built of amino acids whose nitrogen came—via plant and animal foods— from the Haber–Bosch synthesis. Virtually all the protein needed for the growth of 2 to 4 billion children to be born during the next two generations will have to come from the same source, from the synthesis of ammonia from its elements.

Postscript

My earliest concept was to write this book basically as a biography of Fritz Haber. But although Haber's life, fulfilled and tragic, is a fascinating subject, I soon realized that in order to convey the epochal nature of his invention I must take a much broader view, spanning the realities of preindustrial agricultures, detailing Carl Bosch's fundamental contributions, and explaining the enormous consequences of ammonia synthesis. Once this decision was made, the book's contents fell logically in place, and the lives of the two protagonists exited the story shortly after the commercialization of ammonia synthesis. I feel a postscript is needed to describe the eventful lives of Carl Bosch and Fritz Haber after the two scientists turned to other matters.[1]

Carl Bosch

My narrative (in chapter 5) followed Bosch almost to the end of the World War I, when he organized the expansion of ammonia synthesis and nitrate production at Oppau and the construction of a new ammonia plant at Leuna. Shortly after the war's end, in December 1918, Bosch was delegated to Spa, and in March 1919 to Versailles, as the representative of German industry in the armistice and peace treaty delegations. After his return from France he became the chairman of the BASF board, and in that capacity he had to deal with the worst disaster in the company's history, an explosion at Oppau that caused more than 500 fatalities and destroyed the homes of more than 7,000 people in September 1921. In May 1923 Ludwigshafen was occupied by French troops, and Bosch, who fled across the Rhine, was sentenced in absentia for his refusal to cooperate with French authorities.

In December 1925 Duisberg's initiative led to the establishment of the consortium of six largest German chemical companies, I. G. Farben (*Interessengemeinschaft der deutschen Farbenfabriken*). The BASF, Bayer, and Hoechst each contributed a 27.4% share, Agfa 9%, CFGE 6.9%, and Weiler-ter Meer 1.9%.[2] Bosch was named the first chairman of the I. G. Farben board of directors in 1926. Under his leadership the consortium greatly expanded its production line. Most notably, as he had done once with the synthesis of ammonia, Bosch took an enormous risk by promoting the development and commercialization of coal hydrogenation producing liquid fuels from Germany's inferior lignites (brown coals) at Leuna, the site of the world's largest ammonia plant.

During the 1920s Bosch traveled widely and received a large number of honors and awards in the business world, from German academic institutions, and from governments (fig. P.1).[3] The high point of his professional career was reached in 1932 when, most deservedly, he accepted his Nobel Prize for the synthesis of ammonia in a double ceremony with Friedrich Bergius, who was rewarded for his invention of coal hydrogenation. As for so many Germans, Hitler's seizure of power was a shock-

Figure P.1
Photograph of Carl Bosch from the late 1920s. Courtesy of BASF Unternehmensarchiv, Ludwigshafen, Germany.

ing surprise to Bosch, who favored the policies of Gustav Stresemann, the pragmatic and moderate foreign minister of the Weimar Republic, and who used I. G. Farben funds to support three centrist parties during elections of the late 1920s and early 1930s.[4]

After the Nazi takeover Bosch tried to protect Jewish scientists and businessmen, argued for freedom of research and for international economic cooperation, and refused to join the NSDAP.[5] Pain killers and alcohol were his way of dealing with deepening depression and fear about the future. On April 20, 1940, with Bosch critically ill, his son was summoned from Berlin. Next day Bosch told him what he foresaw:

To begin with, it will go well. . . . But then he will bring the greatest calamity by attacking Russia. Even that will go well for a while. But then I see terrible things. Everything will be totally black. The sky is full of airplanes. They will destroy whole Germany, its cities, its factories, and also the I.G.[6]

Bosch died on April 26, 1940, before he could see his premonitions turning into realities—and before he could witness how his achievements helped the regime he despised, and how his successors enmeshed the company in Nazi crimes against humanity.[7]

I. G. Farben lost money on developing the synthetic fuel process. In 1932 the company had to ask for government price and purchase guarantees, and it eventually licensed the process to a new consortium and concentrated instead on the production of synthetic rubber.[8] After 1935 the consortium built twelve large coal hydrogenation plants, the last one between 1942 and 1944 in the North Bohemian Lignite Basin near Brüx in the German-occupied Sudetenland. The Third Reich could thus rely on hydrogenation of coal to produce synthetic gasoline during the World War II.

After the Romanian oilfields were lost on August 30, 1944, to the advancing Red Army, hydrogenation of coal was the only source of liquid fuels for the Wehrmacht and the Luftwaffe for the remaining eight months of the war. Bosch's technical acumen thus helped, posthumously, to prolong yet another war. In the year of Bosch's death it was his protégé Carl Krauch, at that time the chairman of the I. G. Farben *Aufsichtsrat* (Supervisory Board of Directors), who reversed the company's earlier decision against building a third synthetic rubber plant, by doing so "not only impelling the selection of a site near Auschwitz, but also instigating the use of concentration camp inmates as construction workers."[9]

Fritz Haber

In contrast to Bosch's life, Fritz Haber's fate already turned tragic during World War I. He left Karslruhe two years after his demonstration of ammonia synthesis. The newly established Kaiser-Wilhelm-Gesellschaft was setting up three research institutes in Berlin-Dahlem, and a rich industrialist, Leopold Koppel, offered to build a separate Institute for Physical Chemistry and Electrochemistry, providing that Fritz Haber would be its director.[10] Haber moved to Berlin in 1911 and became a member of an emerging elite of science managers, a class that was to grow in all the industrial countries during the twentieth century.

Within a few months after August 1914 the Dahlem institutes became integrated into the German war effort, and Haber, along with many other eminent German chemists and physicists, was almost immediately drawn into the practical tasks of strengthening the country's military power and improving its precarious resource foundations. He first acted as a consultant in expanding Germany's nitrate supplies, and he solved the problem of producing gasoline with a low freezing point. But even before the end of 1914 he began working on gas warfare by taking over, and expanding, the experiments began by the *Auergesellschaft* in October 1914.

The Hague Conventions of 1899 and 1907 forbade the use of poisonous gases in war, but Haber was not concerned about the legality of his work. The German government made the decision to use gases in war, and Haber, ever a German patriot, made sure that their deployment was technically possible.[11] Moreover, Haber believed that the use of chlorine gas against entrenched troops would bring a speedier victory to Germany and thus end protracted trench warfare suffering. During the first gas attack, on April 22, 1915, German troops released about 168 t of chlorine from more than 5,000 steel cylinders along a 6 km-long front at Ypres. The total French casualties were 15,000 people, including 5,000 fatalities.[12] A week later Haber was again in Berlin, getting ready to travel to the Eastern Front to launch the first gas attacks there.

In the early hours of May 2, 1915, Clara Haber shot herself through the heart with her husband's army revolver; thirteen-year old Hermann discovered his mother's body. To what, if any, extent was her death connected with Haber's war work in general, and with the Ypres gassing in particular? How deeply was she, a dedicated scientist, opposed to using scientific knowledge in the way her husband was doing?[13] We will never know all the reasons that drove her to suicide, but that she had been unhappy in the marriage almost from the very beginning is certain (fig. P.2). In 1909

Figure P.2
Clara Immerwahr (1870–1915). Courtesy of Bibliothek und Archiv zur Geschichte der Max Planck Gesselschaft, Berlin, Germany.

she wrote to Richard Abegg that "What Fritz has won during these eight years (since our marriage), that—and still more—I have lost, and what remains ahead of me fills me with the deepest dissatisfaction.[14]

After Clara's suicide Haber left Berlin as planned for the Russian Front. A year later he became the head of the newly formed Chemical Warfare Service. In February 1916 he received the Iron Cross II Class, in 1917 one of the I Class. By the war's end the casualties of gas warfare amounted to about 1.3 million. If Haber was tormented by these facts, he never let anybody know.[15] In his postwar writings and speeches he never apologized for his wartime work, and he not only maintained that poisonous gases are not at all more cruel than flying steel projectiles, but also used U.S. Army findings to argue that they belong among the most humane weapons.[16]

Haber met Charlotte Nathan, according to her memoirs, in March 1917 at the Deutsche Gesellschaft, a club for German power brokers and celebrities established at the beginning of World War I, where she worked as a secretary. His poems quoted in her book of reminiscences (Haber was an accomplished and frequent versifier) and his letters to her show a man almost boyishly in love with a woman nearly twenty years younger.[17] They were married in the Kaiser-Wilhelm-Gedächtniskirche in late October 1917. Haber insisted on a church ceremony, and Charlotte had to convert to Christianity.

The new family—Eva Charlotte, was born in July 1918, Ludwig Fritz two years later—did little to lift Haber's deepening postwar depression. There are no indications that he was ever close to his small children, and his relations with Charlotte began changing just a few years after the marriage. The collapse of imperial Germany, the state that provided him with opportunities for his triumphs and patriotic devotion, changed him profoundly. The farewell letter from War Minister Heinrich Scheüch extolled Haber's wartime contributions, but it provided no consolation. Haber was a completely disillusioned and broken man who, according to a friend, was 75% dead.[18] His name appeared on the first list of war criminals sought for extradition by the Allies, and he hid for a short time in Switzerland.

When the announcement of his Nobel Prize for 1918 came in November 1919, the Swedish Academy's choice of "the inventor of gas war" (as he was often labeled abroad) was attacked in France, England, and the United States. The award ceremony was postponed to 1920, and it took place not in the fall, but exceptionally on Nobel's birthday on June 1. Three generations later Haber's acceptance speech still reads as a masterful overview of the challenge and the solution to one of the key limiting factors on the growth of modern civilization.

Before his trip to Stockholm Haber took the first steps toward an ambitious, but ill-fated, goal of helping his fatherland in peace. The Versailles treaty required Germany to pay 20 billion gold marks by May 1921, and 132 billion (reduced from the original 269 billion) subsequently in forty-two installments.[19] This amounted to 50,000 t of gold, equivalent to some $2/3$ of the world's total reserves at that time. Haber looked for a way to help the country out of this desperate predicament, to prevent the frightening spiral of inflation and attendant political polarization and radicalization.

He knew that Svante Arrhenius had calculated in 1903 that with an average concentration of several milligrams of gold per tonne of seawater, oceans contain billions of tonnes of gold. If this conclusion could be confirmed and a commercial extraction of the metal found, Germany's hopeless position would be erased by yet another scientific triumph. "Unrest consumes me," declared Haber, as his coworkers began collecting and analyzing seawater samples. In July 1923 Haber and four colleagues boarded *Hansa,* a steamship of the Hamburg-America Line equipped with an onboard laboratory, for a journey to New York. In October of that year they took another steamer to Buenos Aires to sample equatorial waters. Other water samples were obtained from an extended cruise of the research ship *Meteor*; others were sent by friends and colleagues from locales ranging from Chile to Greenland.

In 1923 Haber was Germany's delegate to the First World Power Conference in London, where he met again with Le Rossignol. A year later he and his wife took a grand world tour, setting off on September 2, 1924, to New York. After a month in the United States they sailed from San Francisco to Honolulu and Yokohama and traveled extensively in Japan.[20] Before they returned to Europe in March 1925 they visited Manchuria, Shanghai, Hong Kong, Singapore, Java, Colombo, and Egypt.

Meanwhile the small gold-from-seawater team, working in secrecy, was analyzing thousands of seawater samples, and by perfecting microanalytical techniques, it was able to exclude possible interferences, most notably those arising from trace contamination of chemical glass. By the spring of 1926 the verdict was clear: initial reports, converging on values of several milligrams of gold per tonne of seawater, were badly off. Real concentrations were two orders of magnitude lower, with the average for the Atlantic being mere a 0.008 mg/t. There was no hope of commercial extraction.[21]

Haber's near-chronic depression deepened; he suffered from sleeplessness, constant thirst, breathing difficulties, and increasing cardiac pains (fig. P.3). Nitroglycerin helped to relieve his angina, but nothing restored his sleep. "It is so bad with this disease, one dies of it so slowly," he complained.[22] During the fall of 1927 came the final break with Charlotte. He made sure that she and the children were provided for, but they never lived together again.[23] Hermann, forced by his father to study chemistry rather than law, was unable to hold a steady job, drifting among employers and countries (Czechoslovakia, United States, and France), depressed and alcoholic.[24] And as the Weimar Republic weakened and the German society sank into anarchy after the crash of 1929, anti-Semitism was on the rise.

Haber's Institute, which he had now led for two decades, seemed to be the only source of his remaining strength. That changed rapidly after Hitler's rise to power.

Figure P.3
Fritz Haber in 1927 in a faithful drawing by Emil Orlik. Courtesy of the E. F. Smith Collection, Rare Book and Manuscript Library, University of Pennsylvania.

On April 7, 1933, a new law—*das Gesetz zur Wiederherstellung des Berufsbeamtentums* (the Law for the Restoration of Professional Civil Service)—made it impossible to employ anyone of "non-Aryan descent." World War I participants were the exceptions, and Haber, when filling out the requisite questionnaire, noted both his non-Aryan descent and his wartime service, as he was initially determined to stay and help many of the Institute's employees. Max Planck, as president of the Kaiser-Wilhelm-Gesellschaft, personally intervened with Hitler on Haber's behalf on May 16, 1933. Hitler assured him that he had nothing against Jews themselves, but as all Jews are Communists, they were his enemies and he must fight them.[25]

Haber had no choice but to resign. His patriotism, his wartime work for the army, his service "beyond all measure," his early conversion to Protestantism, his devotion to his country—all made no difference: in 1933 he was just another Jew in Nazi Germany. A linden tree planted on the grounds of the Institute to commemorate his sixtieth birthday in 1928 was uprooted. Haber wrote to Willstätter: "I am bitter as never before. I was German to an extent that I feel fully only now and I find it odious in the extreme that I can no longer work enough to begin confidently a new post in a different country."[26] Writing from exile, Einstein captured Haber's predicament precisely in a scientific analogy: "I can imagine your inner conflict. It is as if one must give up a theory which one has worked on all one's life. It is not the same for me because I never believed in it the least."[27]

On October 1, 1933, the date he was officially pensioned, Otto Hahn, who five years later with Fritz Strassmann demonstrated the possibility of nuclear fission, posted Haber's parting words to the Institute's staff: "With these words I take the leave of the Kaiser-Wilhelm-Institute which was established for you according to my proposals by the late *Oberbaurat* Leopold Koppel, and which has tried for 22 years under my leadership to serve the mankind in peace and the fatherland in war."[28]

By the end of October Haber went into exile; after a few days in Paris he traveled to Cambridge where Sir William Pope, who had developed a new synthesis of mustard gas during his wartime work for the British Chemical Warfare Service, invited him to work in his laboratory. Chaim Weizmann asked him to join the new Daniel Sieff Institute in Palestine. Haber accepted, but he never saw Jerusalem: he died on January 29, 1934, in Basel while on a short visit to Switzerland. His last will, which he drew up in Cambridge, asked for cremation and return of his ashes to Dahlem. That has never happened; instead, Hermann moved his mother's ashes to his father's grave in Basel.

Haber's name is now virtually unknown beyond academic circles, particularly those of industrial chemists and World War I historians. His role in that conflict will always remain controversial, an epitome of a moral dilemma that so many other scientists had to face during subsequent, and even more destructive, wars.[29] But there can be no doubt that Haber's invention of the industrial ammonia synthesis—unchanged today in its fundamentals but greatly improved in its performance—has been one of the cornerstones of modern civilization. And it will remain so until we find an effective substitute. Haber himself anticipated such a development. In concluding his Nobel Prize acceptance speech he noted:

It may be that this solution is not the final one. Nitrogen bacteria teach us that Nature, with her sophisticated forms of the chemistry of living matter, still understands and utilizes methods which we do not as yet know how to imitate.[30]

Appendixes

Appendix A
Physical and Chemical Properties of Ammonia

Color	Colorless gas at ambient temperatures
Odor	Pungent, suffocating
Gas Density	0.7714 g/L at 0 °C and 101.3 kPa
Liquid Density	0.6386 g/cm^3 at 0 °C and 101.3 kPa
pH Water Solution (1%)	11.7
Melting Point	−77.7 °C
Boiling Point	−33.4 °C at 101.3 kPa
Explosive Limits	16–27% NH$_3$ by volume in air at 0 °C
Vapor Pressure	1,013 kPa at 25 °C
Relative Vapor Density	0.6 air
Solubility in Water	529 g/L at 20 °C

Sources: Shah, K. D. 1997. *Safety, Health and Environmental Aspects of Ammonia.* York: Fertiliser Society; EFMA. 1995. *Production of Ammonia.* Brussels: EFMA, p. 40.

Appendix B
Nitrogen Content of Organic Wastes (All Values Are in % N)

Organic Matter	Boussingault's 1841 Analyses[1]	Ranges and Means of Modern Analyses[2]
Cereal Straws		0.6
Barley	0.26	0.6–0.7
Buckwheat	0.54	
Oats	0.36	0.5–0.7
Rice		0.3–0.7 (0.6)
Rye	0.20–0.50	
Wheat	0.30–0.53	0.2–0.9 (0.5)
Stover		
Corn		0.4–0.9 (0.6)
Sorghum		0.6–0.8
Legume Straws		
Beans		1.0–1.7
Chickpeas		1.2
Peanuts		1.4
Peas		1.4
Soybeans		0.8–1.3
Plant Tops		
Beets	4.50	2.4
Potatoes	2.30	1.6

Appendix B (continued)

Organic Matter	Boussingault's 1841 Analyses[1]	Ranges and Means of Modern Analyses[2]
Sugarcane		0.3
Sweet potatoes		3.1
Seed Cakes		
Cotton		3.4–6.5
Linseed	6.00	5.8–6.1
Peanut	8.89	4.5–7.8
Rapeseed	5.50	4.6–5.1
Sesame		5.8–6.2
Soybean		6.7–7.0
Green Manure		0.20–0.55
Alfalfa		0.6
Astragalus		0.4
Barley		0.3
Broad Beans		0.6
Clover		0.4
Field Peas		0.5
Gliricidia		0.6
Sesbania		0.6
Turnips		0.2
Aquatic Plants		0.3
Urine		
Cattle	3.8	0.8
Horses	12.5	
Human		1.0–1.2
Night Soil	2.6–4.4	1.2–1.5
Manures		
Mixed	1.95	0.8–2.3*
Cattle	2.59	1.1–3.0*
Goats	3.93	
Horses	3.02	
Pigs	3.37	1.9–4.0*
Poultry		1.3–5.1*
Sheep	2.99	1.2–2.7*
Manure + Bedding		0.2–0.6
Cattle		0.2–0.7
Goats		0.5–0.6

Appendix B (continued) ·

Organic Matter	Boussingault's 1841 Analyses[1]	Ranges and Means of Modern Analyses[2]
Horses		0.4–0.6
Pigs		0.2–0.6
Poultry		1.0–1.8
Sheep		0.5–0.6
Fish Waste		4–10
Horns and Hooves		11–13
Hair and Wool	20.2	10–12
Bones	7.6	1–4
Dried Blood	15.5	10–12
Compost		0.3–3.9
City Garbage		0.8–1.0
Mud and Silt		0.1–0.3
Ashes	0.71	

Notes: Only values with an * are for dry matter.

1. Boussingault, J. B., and A. Payen. 1841. Mémoire sur les engrais et leurs valeurs comparées. *Comptes Rendus de l'Academie des Sciences* 13:323–332. All values are for dry matter.
2. Buck, J. L., O. L. Dawson, and Y. Wu. 1966. *Food and Agriculture in Communist China.* New York: Praeger, p. 138; Dawson, O. L. 1970. *Communist China's Agriculture.* New York: Praeger, p. 118; Chao, K. 1985. *Agricultural Production in Communist China.* Madison: University of Wisconsin Press, p. 145; Parnes, R. 1986. *Organic and Inorganic Fertilizers.* Mt. Vernon, Me.: Woods End Agricultural Institute; Parr, J. F., and D. Colacicco. 1987. Organic materials as alternative nutrient sources. In Z. Helsel, ed., *Energy in Plant Nutrition and Pest Control.* Amsterdam: Elsevier, pp. 81–99; Singh, A. C. G., and G. Singh. 1995. Organic and biological plant nutrient sources: potential, methods for reducing the bulk and improving the availability of nutrients. In R. Dudal and R. N. Roy, eds., *Integrated Plant Nutrition Systems.* Rome: FAO, p. 88; Bath, D., et al. 1997. Byproducts and unusual feedstuffs. *Feedstuffs* 69 (30):32–38.

Appendix C
Extremes of Published Biofixation Estimates and the Ranges of Typical Biofixation Rates of Major Leguminous Species

Crops	Extremes of Published Estimates (kg N/ha)	Ranges of Typical Biofixation Rates (kg N/ha)
Seed Legumes		
Beans	3–160	30–40–50
Broad Beans	45–300	80–100–120
Chickpeas	3–141	40–50–60
Lentils	10–192	30–40–50
Peanuts	37–206	60–80–100
Peas	10–244	30–40–50
Soybeans	15–450	60–80–100
Other Pulses	7–235	40–60–80
Forages		
Alfalfa	65–600	150–200–250
Clovers	28–300	130–150–170
Other Forages	9–180	80–100–120

Sources: Extreme ranges according to LaRue, T. A., and T. G. Patterson. 1981. How much nitrogen do legumes fix? *Advances in Agronomy* 34:15–38; Heichel, G. H. 1987. Legume nitrogen: symbiotic fixation and recovery by subsequent crops. In Z. Helsel, ed., *Energy in Plant Nutrition and Pest Control.* Amsterdam: Elsevier, pp. 62–80; Giller, K. E., and K. J. Wilson. 1991. *Nitrogen Fixation in Tropical Cropping Systems.* Wallingford: CAB International; Peoples, M. B., et al. 1995. Biological nitrogen fixation: an efficient source of nitrogen for sustainable agricultural production? *Plant and Soil* 174:3–28.

Typical biofixation ranges according to Smil, V. 1997. Some unorthodox perspectives on agricultural biodiversity: the case of legume cultivation. *Agriculture Ecosystems Environment* 62:135–144; Smil, V. 1999. Nitrogen in crop production: an account of global flows. *Global Biogeochemical Cycles* 13:647–662.

Appendix D
Average Annual Production of Animal Waste Solids, Their Typical Nitrogen Content, and Average NH_3-N Losses

Animals	Waste Solids (kg/head)*	N	N Output (kg/head)	Manure Produced in Confinement	NH_3-N Losses
Dairy Cattle					
Modern	2,000	4%	80	60%	36%
Traditional	1,200	3	45	65	36
Nondairy Cattle					
Modern	1,200	4	50	35	36
Traditional	900	3	30	25	36
Water Buffaloes	900	3	30	35	28
Horses, Mules	1,200	3	35	50	28
Pigs	300	3	10	100	36
Sheep, Goats	150	3	5	10	28
Poultry	8	4	0.3	100	36

* I assume the following average live weights: modern dairy cattle, 500 kg; modern nondairy cattle and horses, 400 kg; traditional dairy and nondairy cattle and water buffaloes, 300 kg; pigs, 100 kg; sheep and goats, 40 kg; and poultry, 1.5 kg.

Sources: Misra, R. V., and P. R. Hesse. 1983. *Comparative Analyses of Organic Manures.* Rome: FAO; Smil, V. 1983. *Biomass Energies.* New York: Plenum; Nordstedt, R. A. 1992. Selecting a waste management system. In H. H. Van Horn and C. J. Wilcox, eds., *Large Dairy Herd Management.* Champaign, Ill.: American Dairy Science Association, pp. 630–639; Bouwman, A. F., et al. 1997. A global high-resolution emission inventory for ammonia. *Global Biogeochemical Cycles* 11:561–587.

Appendix E
Composition of Various Guanos

Location	Moisture	Nitrogen	Phosphorus
Chincha Islands	16%	14%	4.4%
Pabellon de Pica	17	9.0–9.8	3.8
Punta de Lobos	16	4.1–6.2	5.8
Angamos	11	14	3.8
Ichaboe Island		6	6.1
Sombrero Island	5	0	6.5
Bolivia	13	3.0–3.5	3.8
California		1.2–7.0	6.0
Chile		1.0–1.2	
Ecuador		0.5–0.7	
Mexico		0.3–0.5	
Patagonia		1.5–2.0	8.9
Peru	16	12.5–14.2	4.8
Queensland		0.6–0.8	

Source: Church, A. H. 1890. Guano. In *Encyclopedia Britannica,* 9th edition. Chicago: R. S. Peale & Co., vol. XI, p. 235.

Appendix F

Extraction of Guano and Chilean Nitrate, and Production of Ammonium Sulfate, Cyanamide, and Calcium Nitrate, 1850–2000 (All Values Are in kt N)

	NaNO$_3$	Guano	Coke-Oven Gas (NH$_4$)$_2$SO$_4$	Cyanamide Synthesis CaCN$_2$	Electric Arc Ca(NO$_3$)$_2$	Total N
1850	5	—	0	0	0	5
1860	10	70	0	0	0	80
1870	30	70	0	0	0	100
1880	50	30	0	0	0	80
1890	130	20	—	0	0	150
1900	220	20	120	0	0	360
1905	250	10	130	0	0	390
1910	360	10	230	10	—	610
1913	410	10	270	30	10	730
1920	410	10	290	70	20	800
1929	510	10	425	255	20	1,220
1935	180	10	370	230	—	790
1940	200	10	450	290	—	950
1950	270	—	500	310	—	1,080
1960	200	—	950	300	—	1,450
1970	120	—	950	300	—	1,370
1980	90	—	970	250	—	1,310
1990	120	—	550	110	—	780
2000	120	—	370	80	—	570

Note: — signifies a negligible amount.

Sources: Appendix F; Timm, B. 1963. 50 Jahre Ammoniak-Synthese. *Chemie-Ingenieur-Technik* 35:823; Rintelen, P. 1961. Betriebswirtschaftliche Betrachtungen zum Stickstoffverbrauch. In O. Balz, et al., eds., *Der Stickstoff.* Düsseldorf: Fachverband Stickstoffindustrie, p. 430; Honti, G. D., ed. 1976. *The Nitrogen Industry.* Budapest: Akadémiai Kiado, vol. I, p. 52; Food and Agriculture Organization. 1999. *Fertilizers.* Rome: FAO (http://apps.fao.org); International Fertilizer Industry Association. 1998. *World Nitrogen Fertilizer Consumption.* Paris: IFA (http://www.fertilizer.org).

The following conversion rates were used: NaNO$_3$, 15% N; guano, 12% N in 1860 and 1870, 7% in 1880, 4% in 1890, 3% in 1900, 2% in 1910–1920, 1% in 1930 and 1940; (NH$_4$)$_2$SO$_4$, 25% N; CaCN$_2$, 20% N. All values are rounded to the nearest 5%.

Appendix G
Exports of Chilean Nitrate, 1835–1932 (All Figures Are in kt/Year)

Year	Exports	Year	Exports	Year	Exports
1835	3	1895	1,051	1917	2,776
1840	13	1900	1,339	1918	2,919
1850	27	1905	1,543	1919	803
1860	59	1910	2,251	1920	2,794
1870	181	1913	2,465	1925	2,518
1880	224	1914	1,847	1929	2,896
1885	436	1915	2,023	1930	1,785
1890	884	1916	2,980	1932	243

Sources: Munroe, C. E. 1910. The nitrogen question from the military standpoint. *Annual Report of the Board of Regents of the Smithsonian Institution,* Washington, D.C.: Smithsonian Institution, p. 230; Nitro de Chile. 1964. *Enciclopedia Universal Ilustrada.* Madrid: Espasa-Calpe, p. 832; Collier, S., and W. F. Sater. 1996. *A History of Chile, 1808–1994.* Cambridge: Cambridge University Press, pp. 163–164, 218. There are slight discrepancies among the cited sources.

Appendix H
Evolution of the Earliest BASF Ammonia Converters, 1910–1915

	1910	1911	1912	1913	1913	1914	1915
Length (m)	1.8	4.0	4.0	8.0	8.0	8.0	12.0
Outside Diameter (cm)	14.6	14.6	23.0	28.5	45.0	67.5	108.0
Weight (t)	0.3	0.5	1.0	3.5	8.5	25	75
Pressure (atm)	100	100	100	200	200	200	300
NH_3 output (t/day)	0.01	0.03	1.0	3.6	4.8	20	85

Sources: Appl, M. 1986. *Ammonia Synthesis and the Development of Catalytic and High-Pressure Processes in the Chemical Industry.* H. L. Roy Memorial Lecture, Hyderabad, December 19, 1986, p. 8; Nagel, A. von, et al. 1991. *Stickstoff.* Ludwigshafen: BASF, p. 56.

Appendix I
Annual Totals of Ammonia Produced by Haber–Bosch Synthesis in Germany, 1913–1943
(All Values in kt N/Year)

Year	Production	Year	Production	Year	Production
1913	0.8	1924	245	1934	514
1914	6	1925	360	1935	540
1915	12	1926	615	1936	588
1916	51	1927	711	1937	777
1917	75	1928	845	1938	914
1918	96	1929	837	1939	1,020
1919	63	1930	556	1940	1,016
1920	122	1931	448	1941	1,021
1921	140	1932	452	1942	954
1922	210	1933	526	1943	920
1923	190				

Sources: Schmidt, A. 1934. *Die industrielle Chemie.* Berlin: Verlag Walter de Gruyter & Co., pp. 330–332; Balz, O., et al. 1961. *Der Stickstoff.* Düsseldorf: Fachverband Stickstoffindustrie, pp. 431–435; Timm, B. 1963. 50 Jahre Ammoniak-Synthese. *Chemie-Ingenieur-Technik* 35:823. Different sources give slightly different annual values. I always chose the highest available figure.

Appendix J
Annual Production of Ammonia and Consumption of All Inorganic Nitrogen Fertilizers in the United States, 1900–1997

Year	Ammonia Production (kt NH$_3$)	Nitrogen Fertilizer Consumption (kt N)
1900		56
1905		82
1910	15	132
1915	25	187
1920	25	207
1925	50	253
1930	90	343
1935	70	283
1940	300	380
1945	520	571
1950	1,450	912
1955	2,990	1,778
1960	4,410	2,483
1965	8,120	4,207
1970	12,340	6,765
1975	14,320	7,801
1980	14,740	10,346
1985	14,260	10,423
1990	15,430	10,046
1995	15,800	10,629
1997	16,250	11,182

Sources: U.S. Bureau of the Census. 1975. *Historical Statistics of the United States: Colonial Times to 1970.* Washington, D.C.: USGPO, pp. 691–692; U.S. Department of Agriculture. 1971. *Commercial Fertilizers.* Washington, D.C.: USDA, pp. 2–3; U.S. Bureau of the Census. 2000. *Inorganic Fertilizer Materials and Related Products.* Washington, D.C.: USBC (http://www.census.gov/pub/cir/www/mq28b/html); FAO (http://apps.fao.org) and IFA (http://www.fertilizer.org).

Appendix K
Energy Cost of the Most Efficient Ammonia Synthesis (All Values Are in GJ/t NH_3)

Year	Feedstock	Energy Cost
1920	Coal	100
1930	Coal	95
1940	Coal	88
1950	Coal	85
1955	Natural gas	55
1958	Natural gas	50
1962	Natural gas	45
1966	Natural gas	40
1973	Natural gas	38
1977	Natural gas	34
1980	Natural gas	30
1982	Natural gas	29
1991	Natural gas	28
1996	Natural gas	27
2000	Natural gas	26
Stoichiometric Energy Requirement for NH_3 Synthesis		20.9
NH_3 Lower Heating Value		18.6

Sources: Appl, M. 1993. Modern ammonia technology: where have we got to, where are we going? *Nitrogen* 202:44–53; Radgen, P. 1996. Pinch and exergy analysis of a fertilizer complex. *Nitrogen* 224:39–48; Strait, R. 1999. Grassroots success with KAAP. *Nitrogen & Methanol* 238:37–43. Slightly different values can be found in other sources cited in chapter 6, note 62.

Appendix L

Global Consumption of Nitrogen Fertilizers and the Shares of Haber–Bosch Synthesis, 1910–2000 (Values, except for the last column, Are in kt N/Year)

	Haber–Bosch	Other Fixed N	Haber–Bosch Synthesis	
			Total	Share (%)
1910	0	610	610	0
1913	—	730	730	<1
1920	150	800	950	16
1925	600	1,090	1,690	36
1929	930	1,220	2,150	43
1935	1,300	790	2,090	62
1940	2,150	950	3,100	69
1950	3,700	1,080	4,780	77
1955	6,800	1,450	8,250	82
1960	9,540	1,750	11,290	84
1965	17,900	1,200	19,100	94
1970	30,230	1,370	31,600	96
1975	43,050	1,350	44,400	97
1980	59,290	1,310	60,600	98
1990	76,320	780	77,100	99
1995	78,240	560	78,800	>99
2000	85,130	570	85,700	>99

Note: — signifies a negligible amount.

Sources: Haber–Bosch fixation from: Honti, G. D., ed. 1976. *The Nitrogen Industry.* Budapest: Akadémiai Kiado, vol. I, p. 52; International Fertilizer Industry Association. 1998. *World Nitrogen Fertilizer Consumption.* Paris: IFA (http://www.fertilizer.org); other fixed nitrogen from Appendix F. All output values are rounded to the nearest 10 kt N/year.

Appendix M
Consumption of Nitrogen Fertilizers, 1960–2000 (All Figures Are in Mt N/Year)

Year	High-Income Countries	Low-Income Countries	World
1960	8.5	2.3	10.8
1965	14.7	4.4	19.1
1970	22.9	8.8	31.7
1975	30.7	13.7	44.4
1980	35.7	25.0	60.7
1985	38.5	31.3	69.8
1988	39.8	39.8	79.6
1990	35.2	42.0	77.2
1995	28.9	49.9	78.8
1996	30.0	52.9	82.9

Sources: Food and Agriculture Organization. 1999. *Fertilizers.* Rome: FAO (http://apps.fao.org). Slightly different totals are available from International Fertilizer Industry Association. 1998. *World Nitrogen Fertilizer Consumption.* Paris: IFA (http://www.fertilizer.org).

Appendix N
Highest Average Annual Nitrogen Fertilizer Applications to Major Crops during the Mid-1990s

Crop	Country	Average Applications (kg N/ha)
Corn	Greece	250
	Chile	250
	Egypt	250
	Portugal	160
	France	160
	United States	140
Rice	Spain	178
	South Korea	170
	Colombia	170
	Venezuela	152
	United States	147
	China	145
Wheat	United Kingdom	183
	Egypt	178
	Mexico	170
	Netherlands	170
	France	155
	Denmark	149
Soybeans	Iran	66
	China	59
	Mexico	29
	Romania	25
	Zimbabwe	20
	Canada	10

Source: Harris, G. 1998. *An Analysis of Global Fertilizer Application Rates for Major Crops.* Paris: IFA (http://www.fertilizer.org).

Appendix O
Annual Global Harvest of Crops and Crop Residues in the Mid-1990s (the First Column Is in Mt of Fresh Weight per Year; All Other Figures Are in Mt of Dry Matter per Year)

Crops	Harvested Crops		Crop Residues	Total Harvest
Cereals	1,900	1,670	2,500	4,170
Sugar Crops	1,450	450	350	800
Roots, Tubers	650	130	200	330
Vegetables	600	60	100	160
Fruits	400	60	100	160
Legumes	200	190	200	390
Oil Crops	150	110	100	210
Other Crops	100	80	200	280
Forages	2,500	500		500
Total	7,950	3,250	3,750	7,000

Sources: Fresh weight of harvested crops according to FAO (http://apps.fao.org); dry weight calculated by using standard moisture contents in Watt, B. K., and A. L. Merrill. 1975. *Handbook of the Nutritional Contents of Foods.* New York: Dover Publications; and Bath, D., et al. 1997. Byproducts and unusual feedstuffs, *Feedstuffs* 69(30):32–38. Derivation of crop residue totals is described in Smil, V. 1999. Nitrogen in crop production: an account of global flows. *Global Biogeochemical Cycles* 13:647–662. All figures are rounded to avoid the appearance of unwarranted precision.

Appendix P
Nitrogen Incorporated in the Global Crop Harvest of the Mid-1990s (All Figures Are in Mt N/Year)

Crops	Nitrogen Incorporated in		Total Harvest
	Harvested Crops	Crop Residues	
Cereals	30	15	45
Legumes	10	5	15
Sugar Crops	2	1	3
Roots, Tubers	2	1	3
Vegetables, Fruits	2	1	3
Other Crops	4	2	6
Forages	10		10
Total	60	25	85

Sources: Fresh weights of crops and dry weights of crop residues from appendix O multiplied by average N contents listed in Watt and Merrill and Bath et al. (see appendix O).

Appendix Q

Nitrogen Flows in the Global Agroecosystem of the Mid-1990s (All Values Are in Mt N/Year)

N Flows	Minimum	Mean	Maximum
Inputs	151	169	186
Seeds	2	2	2
Atmospheric Deposition	18	20	22
Irrigation Water	3	4	5
Crop Residues	12	14	16
Animal Manures	16	18	20
Biofixation	25	33	41
Inorganic Fertilizers	75	78	186
Outputs	143	165	190
Harvested Plants	85	85	85
Losses			
NO Emissions	1	4	6
N_2O Emissions	1	4	7
N_2 Emissions	11	14	18
NH_3 Volatilization	8	11	14
NO_3^- Leaching	14	17	20
Soil Erosion	18	20	25
Losses from Plant Tops	5	10	15
Balance	+8	+4	−4

Source: Smil, V. 1999. Nitrogen in crop production: an account of global flows. *Global Biogeochemical Cycles* 13:647–662. All figures are rounded to avoid the appearance of unwarranted precision.

Appendix R
Reconstruction of Nitrogen Inputs into China's Agroecosystem, 1952–1996

	1952	1957	1965	1970	1975	1980	1985	1990	1995	1996
Fertilizers	0.1	0.3	1.3	2.9	4.8	11.8	13.6	19.2	23.4	25.4
Symbiotic Biofixation	0.7	0.8	0.8	1.1	1.2	1.2	1.2	1.2	1.2	1.2
Nonsymbiotic Biofixation	2	2	2	2	2.1	2.1	2.1	2.2	2.2	2.2
Green Manures	0.2	0.2	0.2	0.3	0.4	0.5	0.4	0.3	0.2	0.2
Crop Residues	0.1	0.1	0.1	0.2	0.2	0.2	0.3	0.3	0.3	0.3
Animal Manures	0.5	0.6	0.6	0.8	1	1.2	1.6	1.9	2	2
Human Wastes	0.7	0.8	0.8	1	1	1.2	1.2	1.2	1.1	1.1
Oilseed Cakes	0.1	0.1	0.1	0.2	0.2	0.2	0.3	0.3	0.2	0.2
Seed	0.1	0.1	0.1	0.2	0.2	0.2	0.3	0.3	0.3	0.3
Atmospheric Deposits	0.6	0.7	0.7	0.7	0.8	0.8	1	1	1.1	1.1
Irrigation Water	0.1	0.1	0.1	0.1	0.1	0.2	0.2	0.3	0.3	0.3
Total	5.2	5.8	6.8	9.6	12	19.6	22.3	28.2	32	34.3
Fertilizer Share	2	5	19	30	40	60	61	68	73	74

Appendix S
Average Nitrogen Inputs in German Agriculture, 1900–1995 (All Values, Except for the Last Line, Are in kg N/ha)

Nitrogen Source	1900	1910	1925	1938	1955	1965	1975	1985	1995
Atmospheric Deposition	20	20	20	20	20	20	25	30	30
Biofixation	20	20	20	20	20	20	20	15	15
Manure	20	25	30	30	40	45	45	50	55
Fertilizers	2	5	10	20	35	50	90	125	145
Total	62	70	80	90	115	135	180	220	245
Fertilizer Share	3%	7%	13%	22%	30%	37%	50%	57%	59%

Sources: See chapter 7, note 60. All input rates (except for fertilizer in 1900) are rounded to the nearest 5 kg N/ha to avoid the appearance of unwarranted precision.

Appendix T
Nitrogen Flows in U.S. Crop Production of the Mid-1990s

Flows	Annual Rates
Inputs (Mt N)	23.0
Atmospheric Deposition	2.0
Biofixation	5.5
Organic Recycling	4.5
Crop Residues	2.5
Organic Wastes	2.0
Seeds	0.2
Irrigation Water	0.3
Synthetic Fertilizers	10.5
Removals (Mt N)	15.0
Crops	10.5
Crop Residues	4.5
Losses (Mt N)	8.0
Input Shares	
Biofixation	25%
Organic Recycling	20%
Synthetic Fertilizers	45%
Uptake Efficiency	65%
Losses	35%

Note: Because of considerable annual fluctuations in planted areas and yields, the rates are five-year averages for about 190 Mha of cropland and improved pastures. Crop areas and yields are from the USDA's *Agricultural Yearbook* (available at http://www.usda.gov/nass/pubs). All shares are rounded to the nearest 5.

Notes

Transforming the World

1. Needham, A. E. 1965. *The Uniqueness of Biological Materials.* Oxford: Pergamon Press, p. 149.

Chapter 1

1. By far the best source detailing the history of chemistry is Partington, J. R. 1962. *A History of Chemistry.* London: Macmillan. A good concise review is Brock, W. H. 1992. *The Norton History of Chemistry.* New York: W. W. Norton.

2. An excellent, systematic review of the early research on photosynthesis is Conant, J. B., et al., eds. 1957. *Harvard Case Histories in Experimental Science.* Vol. 2. Cambridge, Mass.: Harvard University Press, pp. 325–436.

3. Now we know that most plants, including such staple grains as wheat and rice, also respire during the day. In contrast to these so-called C_3 plants whose reduction of CO_2 first produces phosphoglyceric acid, a compound containing three carbons, C_4 plants, first producing a four-carbon acid, do not have any photosrespiration. This is a major reason why corn, sorghum, and sugarcane, the principal C_4 crops, have higher photosynthetic efficiency and inherently higher yields than C_3 species; see Edwards, G., and D. Walker. 1983. C_3, C_4: *Mechanisms and Cellular and Environmental Regulation of Photosynthesis.* Berkeley: University of California Press.

4. De Saussure also noted the contribution of atmospheric deposition, adding that nitrogen assimilated by plants also comes from "vegetable and animal materials held in suspension in the atmosphere, which deposit upon the plant" (Conant et al. (2), p. 431).

5. Thaer, A. D. 1809–1812. *Grundsätze der rationallen Landwirtschaft.* 2 vols. Berlin: Realschulbuchhandlung.

6. For a review of the humus theory and its demise see Waksman, S. A. 1942. Liebig: the humus theory and the role of humus in plant nutrition. In F. R. Moulton, ed., *Liebig and after Liebig: A Century of Progress in Agricultural Chemistry.* Washington, D.C.: American Association for the Advancement of Science, pp. 56–63.

7. Liebig preferred to call the process *eremacausis*. The term, derived from Greek for "quietly burning," implied that decomposition is essentially just a slow combustion.

8. Schwann, T. 1837. Vorläufige Mittheilung, betreffend Versuche über die Weingährung und Fäulnis. *Annalen der Physik und Chemie* 41:184–193; Cagniard-Latour, C. 1838. Mémoire sur la fermenation vineuse. *Annales de Chimie et de Physique* 68:206–222.

9. For details on the theory of fermentation and for Pasteur's work see Partington (1), vol. 4, pp. 301–310; "Pasteur's study of fermentation," in Conant et al. (2), pp. 439–485. By being wrong Liebig was ultimately closer to the truth than Pasteur was. When he eventually admitted that yeasts are living organisms, he concluded that fermentation does not depend on the yeast cells being alive, but rather on their ability to produce a substance that decomposes organic molecules. The discovery of enzymes confirmed that conclusion. A good historical review is Dressler, D., and H. Potter. 1991. *Discovering Enzymes*. New York: Scientific American Library.

10. When they began their research, chemistry was still beholden to Georg Stahl's (1660–1734) idea that all flammable bodies contained phlogiston, an ingredient that was transferred to the atmosphere during combustion, or that could combine with oxides. See Brock (1), pp. 78–84.

11. Partington (1), vol. 3, p. 222.

12. Ibid., p. 264.

13. Ibid., p. 249.

14. Lavoisier's two books—*Opuscules Physiques et Chimiques* (1774) and *Traité Élémentaire de Chimie* (1789)—as well as his other writings and memoirs are best accessible in *Oeuvres de Lavoisier* published by Imprimerie Impériale, Paris, in 1864 and reprinted by Johnson Reprint Corporation, New York, in 1965. Expectedly, there is a huge amount of biographical and scientific literature on Lavoisier in French (see Partington (1), vol. 3, pp. 363–364). More recent English-language works include Donovan, A. 1996. *Antoine Lavoisier: Science, Administration, and Revolution*. New York: Cambridge University Press; Guerlac, H. 1990. *Lavoisier—The Crucial Year: The Background and Origin of His First Experiments on Combustion in 1772*. New York: Gordon and Breach; McKie, D. 1980. *Antoine Lavoisier: Scientist, Economist, Social Reformer*. New York: Da Capo Press.

15. Lavoisier, A. L. 1789. *Traité Élementaire de Chimie*. Paris, p. 48.

16. Wallis Budge, E. A. 1920. *An Egyptian Hieroglyphic Dictionary*. London: John Murray, vol. 1, p. 407.

17. Partington (1), vol. 3, p. 514.

18. For details on Boussingault's life and research see Aulie, R. P. 1970. Boussingault and the nitrogen cycle. *Proceedings of the American Philosophical Society* 114:435–479; Aulie, R. P. 1974. The mineral theory. *Agricultural History* 48:369–382; McCosh, F. W. J. 1984. *Boussingault: Chemist and Agriculturist*. Dordrecht: D. Reidel.

19. Boussingault, J. B. 1836. Recherches sur l'quantité d'azote contenue dans les fourrages, et sur leur équivalents. *Annales de Chimie et de Physique* 63(2):225–244.

20. Boussingault, J. B., and A. Payen. 1841. Mémoire sur les engrais et leurs valeurs comparées. *Comptes Rendus de l'Academie des Sciences* 13:323–332.

21. Lightning is the largest natural source of NO_x, but even today the published rates of its N fixation remain highly uncertain. The highest published estimates based on field measurements (up to 220 Mt N/year) would make it the single largest producer of fixed nitrogen in the biosphere; see Liaw, Y. P., Sisterson, D. L., and N. L. Miller. 1990. Comparison of field, laboratory, and theoretical estimates of global nitrogen fixation by lighting. *Journal of Geophysical Research* 95:22,489–22,494. Such figures are impossibly high because there is no corresponding amount of N deposition as nitrates. The most likely annual totals of N fixed by lightnings are less than 10 Mt N/year; see Galloway, J. N., et al. 1995. Nitrogen fixation: anthropogenic enhancement–environmental response. *Global Biogeochemical Cycles* 9:235–252.

22. Dumas, J. B. A., and J. B. Boussingault. 1841. *Leçon sur la Statique Chimique des êtres organisés.* Paris: Sorbonne. A century later Hutchinson noted that "although in the absence of any bacteriological information, they inevitably made certain errors in interpretation, the cyclical character of the migration of the common non-metallic elements had never been demonstrated so lucidly and with such rhetorical skill." Hutchinson, G. E. 1944. A century of atmospheric biogeochemistry. *American Scientist* 32:129.

23. Liebig's best biography is Brock, W. H. 1997. *Justus von Liebig: The Chemical Gatekeeper.* Cambridge: Cambridge University Press. The book contains detailed discussions and assessments of his scientific contributions. Critical extracts from Liebig's work concerning agricultural sciences are presented in Liebig, J. 1989. *Boden Ernährung Leben.* Stuttgart: Paul Pietsch Verlag. Perhaps the most important of several volumes of Liebig's letters are his exchanges with Wöhler published in 1982; see *Wöhler und Liebig: Briefe von 1829–1873 aus Justus Liebig's und Friedrich Wöhler's Briefwechsel in den Jahren 1829–1873.* Göttingen: J. Cromm.

24. Before Wöhler and Liebig, the chemistry of carbon compounds lagged well behind inorganic analysis and synthesis. "Between them, Liebig and Wöhler had revealed the source of richness (and fascination) of organic chemistry, that the simple elements of carbon, oxygen, hydrogen and nitrogen could combine together in myriads of different ways to produce millions of different compounds" (Brock (23), p. 201). Liebig perfected the analytical method for determining organic compounds, found correct formulae for many known materials (including quinine and codeine), and discovered a number of new ones; his main theoretical contributions included the experimental establishment of the radical theory and the hydrogen theory of acids.

25. For a critical retrospective of the law see Browne, C. A. 1942. Liebig and the law of the minimum. In Moulton (6), pp. 71–82. Liebig's idea was too simple; we have come to realize that crop yields do not correlate with the amount of nutrient supplied to soil, but rather with the amount actually assimilated by plants. Efficiency of this transfer depends on a number of environmental conditions, including the availability of other nutrients.

26. The book whose insights, as well as errors, set the direction of agronomic research for several generations was actually written as an extended introduction to Liebig's *Traité de chimie organique,* which was published in Paris in April 1840. The German original came out as a separate volume in August 1840, and it was immediately translated from galley sheets into English by Lyon Playfair, who was studying during the summer semester at Giessen. It was released in September of the same year as *Chemistry in Its Application to Agriculture and Physiology* (London: Taylor & Walton). The adjective "organic," dropped in the English translation, was also dropped from subsequent German editions.

27. Liebig, *Chemistry* (1st edition), p. 89.

28. Ibid., p. 85.

29. Ibid. (2nd edition), p. 5.

30. Ibid., p. 70.

31. Ibid.

32. Ibid., p. 72.

33. Ibid. (1st edition), p. 85.

34. Ibid. (3rd edition), p. 213.

35. Ibid. (2nd edition), p. 74.

36. Almost 100 measurements of ammonia in precipitation on all continents, taken between the 1950s and the 1990s, can be found in Berner, E. K., and R. A. Berner. 1996. *Global Environment.* Upper Saddle River, N.J.: Prentice Hall, pp. 69–75.

37. The current two largest, and most concentrated, sources of volatilized ammonia—enormous feedlots containing tens of thousand heads of animals, and fields heavily fertilized with ammoniacal fertilizers—were, of course, entirely absent at the time of Liebig's writing.

38. For example, annual NH_4^+ deposition remains negligible (<1 kg N/ha) nearly everywhere west of the Rocky Mountains, and its peaks around and downwind from the highest concentrations of animals and the most intensive fertilizer applications in the Midwest (Iowa, Illinois, Nebraska) range between 4 and 5 kg N/ha. Annual isopleth maps of U.S. atmospheric deposition can be seen at National Atmospheric Deposition Program, Colorado State University, http://nadp.nrel.colostate.edu/NADP.

39. Liebig (20, 1st edition), p. 201.

40. Several clay minerals, above all vermiculite and illite, have crystal lattices whose individual sheets are expandable. When NH_4^+ is substituted for Ca^{2+} and Mg^{2+}, the two dominant interlayer cations, it becomes mineral-bound, resists both inorganic hydrolysis and bacterial attack, and becomes unavailable for rapid recycling. Little binding takes place where kaolinite is dominant, but in nonkaolinite soils the unavailable NH_4^+ accounts for up to 8% of all N in the near-surface layer, and to 45% of subsoil N. Actual amount of mineral-bound N thus ranges from just a few kilograms to more than 1 t/ha; see Stevenson, F. J. 1986. *Cycles of Soil.* New York: Wiley-Interscience, pp. 199–202.

41. Liebig, *Chemistry* (3rd edition), pp. 202–210. Liebig's basic conclusion that different crops require different amounts of minerals was correct, and today it is the basis of a thriving science and industry of micronutrient applications; see Mortvedt, J. J., et al., eds. 1991. *Micronutrients in Agriculture.* Madison, Wisc.: Soil Science Society of America.

42. Liebig (20, 1st edition), pp. 187–188.

43. Kargon, R. 1977. *Science in Victorian Manchester: Enterprise and Expertise.* Baltimore: Johns Hopkins University Press, p. 104.

44. Brock (23), pp. 122–129.

45. Lawes, J. B., and J. H. Gilbert. 1895. *The Rothamsted Experiments over 50 Years.* Edinburgh and London: Blackwood & Sons.

46. Ibid.

47. Quoted in Nutman, P. S. 1987. Centenary Lecture. *Philosophical Transactions of the Royal Society London* B 317:85.

48. Lawes, J. B. 1847. On agricultural chemistry. *Journal of the Royal Agricultural Society* 8:245.

49. Ville, G. 1853. *Recherches expérimentales sur la végétation*. Paris: V. Masson; Boussingault, J. B. 1853. Mémoire sur le dosage de l'ammoniaque contenus dans les eaux. *Annales de chimie* 39:257– 291; Lawes, J. B., and J. H. Gilbert. 1855. Reply to Baron Liebig. *Journal of the Royal Agricultural Society of England* 16:411–498.

50. Boussingault, J. B. 1838. Recherches chimiques sur la végétation, entreprises dans le but d'examiner si les plantes prennent de l'azote à l'atmosphère. *Comptes Rendus de l'Academie des Sciences* 7:889–891; Boussingault, J. B. 1838. Recherches chimiques sur la végétation. Troisième Mémoire. De la discussion de la valuer relative des assolements par l'analyse élémentaire. *Comptes Rendus de l'Academie des Sciences* 7:1,149–1,155.

51. Boussingault, J. B. 1860. *Agronomie, chimie, agricole et physiologie*. Paris: Gauthier-Villars.

52. Lawes, J. B., J. H. Gilbert, and E. Pugh. 1862. On the sources of nitrogen of vegetation; with special reference to the question whether plants assimilate free or uncombined nitrogen. *Philosophical Transaction of the Royal Society London* 151:431–577.

53. Nutman (47), p. 87.

54. The oldest published picture of leguminous nodules dates to 1542; during the sixteenth century they were considered to be just normal outgrowths of root tissues. Later reports saw in them anything from galls caused by insect larvae to abortive roots: Fred, E. B., et al. 1932. *Root Nodule Bacteria and Leguminous Plants*. Madison: University of Wisconsin, pp. 19–21.

55. An excellent recounting of this discovery and of the related studies before and after 1886 is Nutman's Centenary Lecture (see note 47). The original report is Hellriegel, H., and H. Wilfarth. 1888. Untersuchungen über die Stickstoffernährung der Gramineen und Leguminosen. *Beiläge der Zeitschrift des Vereins für die Rübenzuckerindustrie*. Berlin: Kayssler.

56. The translation is from Fred et al. (54), p. 11.

57. Beijerinck, M. W. 1888. Die Bakterien der Papilionaceen-Knölchen. *Botanische Zeitschrift* 46:725–804. *Bradyrhizobium* and *Azorhizobium* are the two other genera of less common symbiotic diazotrophs.

58. Phycocyanin, a bluish pigment, gives the cyanobacteria their characteristic color. Although most species live in water and many form long filaments, the only functional attribute they have in common with algae are chloroplasts needed for photosynthesis. The importance of these photosynthesizing bacteria goes far beyond their role in the nitrogen cycle: they are the oldest surviving organisms (3.5 billion years), and it was their photosynthesis during the Archaean and Proterozoic periods that enriched the atmosphere with oxygen.

59. Schloesing, T., and A. Muntz. 1877. Sur la nitrification par les ferments organisés. *Comptes Rendus de l'Academie des Sciences* 84:301–303.

60. Winogradsky, S. 1890. Sur les organismes de la nitrification. *Comptes Rendus de l'Academie des Sciences* 110:1,013–1,016.

61. Gayon, U., and G. Dupetit. 1885. Recherches sur la réduction des nitrates par les organismes microscopiques. *Annales de la Science Agronomique Francaise et Étrangere* 2(1):226–325.

62. Descriptions of individual bacterial species can be found in Laskin, A. I., ed. 1977. *CRC Handbook of Microbiology.* Vol. 1. Boca Raton, Fla.: CRC Press.

63. Our understanding of the nitrogen cycle, and its interactions with other biospheric circulations, has been summarized in Smil, V. 1985. *Carbon Nitrogen Sulfur.* New York: Plenum Press; Harrison, A. F., et al., eds. 1990. *Nutrient Cycling in Terrestrial Ecosystems.* Amsterdam: Elsevier; Schlesinger, W. H. 1991. *Biogeochemistry: An Analysis of Global Change.* San Diego: Academic Press; Tamm, C. O. 1991. *Nitrogen in Terrestrial Ecosystems.* Berlin: Springer; Butcher, S. S., R. J. Charlson, G. H. Orians, and G. V. Wolfe, eds. 1992. *Global Biogeochemical Cycles.* London: Academic Press; Dobrovol'skii, V. V. 1994. *Biogeochemistry of the World's Land.* Moscow: Mir Publishers; Mackenzie, F. T., and J. A. Mackenzie. 1995. *Our Changing Planet.* Englewood Cliffs, N.J.: Prentice Hall; Smil, V. 1997. *Cycles of Life.* New York: Scientific American Library.

64. Stevenson (40), pp. 111–112, 199–201.

65. Degens, E. T., et al. 1991. *Biogeochemistry of Major World Rivers.* New York: Wiley.

66. Even the most conservative estimates put nitrogen in soils at no less than 40 Gt; in contrast, N assimilated annually by all of the world's crops amounts to less than 100 Mt, a difference of two orders of magnitude.

67. Nitrogenases belong to a rather uniform group of metalloenzymes made up of two protein molecules. The larger one contains Mo and Fe, and its molecular weight is between 215,000–240,000. The smaller one (dinitrogenase reductase) has two equivalent subunits (with total molecular weight of 57,000–72,000), and it contains just nonheme iron. *Azotobacter* genus has an alternative nitrogenase with V-Fe protein. No nitrogenase tolerates more than just a fleeting exposure to O_2. The total mass of nitrogenases in the biosphere may be less than a dozen kilograms—and yet the enzyme is clearly not an evolutionary latecomer, as cyanobacteria are among the oldest autotrophs, reliably dated to more than three billion years ago.

68. Lumpkin, Thomas A. 1982. *Azolla as a Green Manure: Use and Management in Crop Production.* Boulder: Westview Press. Theoretical potential is as much as 140 kg N/ha for one *Azolla* standing crop; actual field rates are generally less than 1/4 of that amount.

69. Boddey, R. M., et al. 1995. Biological nitrogen fixation associated with sugar cane and rice: contributions and prospects for improvement. *Plant and Soil* 174:195–209.

70. Whitton, B. A., and M. Potts, eds. 1999. *Ecology of Cyanobacteria: Their Diversity in Time and Space.* Amsterdam: Kluwer Academic.

71. In the late 1960s a group of Brazilian researchers, led by Johanna Döbereiner, found that *Acetobacter,* as well as *Azospirillum* and *Herbaspirillum,* form associations with the roots of some tropical grasses. Unlike symbiotic rhizobia, which form visible root nodules and are engaged in a direct exchange of metabolic products with their hosts, these bacteria participate in more subtle interactions with roots. They live dispersed on and near plant roots (and in plant stems and leaves), competing for their exudates with other microbes and indirectly transferring some of their fixed nitrogen to them, but with a much lower efficiency than rhizobia. Similar associations were later discovered in corn and sugarcane; see Boddey, R. M., and J. Döbereiner. 1995. Nitrogen fixation associated with grasses and cereals: Recent progress and perspectives for the future. *Fertilizer Research* 42:241–250.

72. Twenty amino acids whose various combinations make proteins are all variations on a single theme: a central carbon atom bonded to four distinctive groups, an amino group (NH_2),

a carboxylic acid group (COOH), a hydrogen atom (H), and a variable group, the so-called side chain, which accounts for their differences. In glycine, the smallest amino acid, the side chain is just a single hydrogen atom; in contrast, arginine and lysine have long linear side chains, and the chains of phenylalanine, tryptophan, and tyrosine include hexagonal rings. Ultimate analysis of proteins shows 50–55% C, 21–24% O, 15–18% N (16% is used as the mean), 6.5–7.5% H, and 0.5–2% S.

Liebig's work also laid the foundations of amino acid chemistry. By treating milk solids with acid he released, purified, and analyzed the molecules of leucine and tyrosine, but it was only in 1935, nearly seventy-five years after Liebig's pioneering work, that the present list of twenty amino acids was completed; see Dressler, D., and H. Potter. 1991. *Discovering Enzymes.* New York: Scientific American Library, pp. 74–78.

73. The essential amino acids are arginine, histidine, isoleucine, leucine, lysine, methionine, phenylalanine, threonine, tryptophan, and valine. Nonessential amino acids can be produced in our bodies by transamination, a DNA-directed process of placing amino acids in proper sequences to form specific proteins.

74. FAO/WHO/UNU. 1985. *Energy and Protein Requirements.* Geneva: WHO; Pellett, P. L. 1990. Protein requirements in humans. *American Journal of Clinical Nutrition* 51:723–737. Human protein requirements will be detailed in chapter 8.

75. A nested definition is necessary: nucleic acids are polymers of nucleotides; nucleotides are nucleosides whose sugar carries one or more phosphate groups; and nucleosides are compounds made up of a purine or pyrimidine base linked to a pentose sugar, usually ribose or 2-deoxyribose. DNA (deoxyrobonucleic acid) stores and RNA (ribonucleic acid) processes all genetic information. Nitric oxide (NO) serves as a neuotransmitter or neuromodulator, a chemical signal used for communication between the brain's nerve cells.

76. In contrast to relatively widespread monitoring of wet deposition, we have neither extensive nor reliable measurements of dry deposition. Minimum estimates can be derived by assuming that dry NH_x deposition is equal to at least 30% of the wet flux; see Warneck, P. 1988. *Chemistry of the Natural Atmosphere.* London: Academic Press, p. 440.

77. The amount of nitrogen deposited in dissolved nitrates is now much larger than NH_x-N deposition everywhere in Western Europe and eastern North America, where combustion of fossil fuels in stationary and mobile sources releases large amounts of NO_x. North American maxima, in Pennsylvania and the neighboring northeastern states, are over 20 kg NO_3^-/ha. Dry NO_3 is equal to at least $^3/_5$ of the wet input; see Warneck (76), pp. 476–483.

78. For details on the role of N_2O in the global climate change see Houghton, J. J., et al., eds. 1996. *Climate Change 1995: The Science of Climate Change.* New York: Cambridge University Press.

Chapter 2

1. Practices and productivities of shifting agricultures are reviewed in Smil, V. 1991. *General Energetics.* New York: John Wiley & Sons, pp. 108–111.

2. For comparisons of population densities in foraging (hunting and gathering) societies with those of shifting sedentary farmers see Smil, V. 1994. *Energy in World History.* Boulder: Westview, pp. 27, 57–78.

3. This is a phenomenon documented by many measurements of nitrogen decline in cultivated soils. Losses of the nutrient amount commonly to 50% of the original mass after fifty to one hundred years of cultivation; see Campbel, C. A., et al. 1986. Land quality, trends and wheat production in Western Canada. In A. E. Slinkard and D. B. Fowler, eds., *Wheat Production in Canada: A Review*. Saskatoon: University of Saskatchewan, pp. 318–353; Buyanovsky, G. A., et al. 1997. Sanborn Field: Effect of 100 years of cropping on soil parameters influencing productivity. In E. A. Paul, et al., eds., *Soil Organic Matter in Temperate Agroecosystems*. Boca Raton, Fla.: CRC Press, pp. 205–225.

4. Cato, M. P. 1934. *De Agri Cultura (On Agriculture)*. Translated by W. D. Hooper. Cambridge, Mass.: Harvard University Press, p. 53.

5. Sung Ying-hsing. 1966 (1637). *T'ien-kung k'ai-wu (The Creations of Nature and Man)*. Translated by E. Z. Sun and S. C. Sun. University Park: Pennsylvania State University Press, p. 6.

6. The first extensive list of nitrogen content of crop residues and other organic wastes is Boussingault, J. B., and A. Payen. 1841. Mémoire sur les engrais et leurs valeurs comparées. *Comptes Rendus de l'Academie des Sciences* 13:323–332; extensive modern listings are Parnes, R. 1986. *Organic & Inorganic Fertilizers*. Mt. Vernon, Me.: Woods End Agricultural Institute; Parr, J. F., and D. Colacicco. 1987. Organic materials as alternative nutrient sources. In Z. Helsel, ed., *Energy in Plant Nutrition and Pest Control*. Amsterdam: Elsevier, pp. 81–99; Bath, D., et al. 1997. Byproducts and unusual feedstuffs. *Feedstuffs* 69(30):32–38.

7. For example, traditional wheat varieties cultivated at the beginning of the twentieth century were around 1 meter tall, producing 1.8–3 times as much straw as grain; see Singh, I. D., and N. C. Stoskopf. 1971. Harvest index in cereals. *Agronomy Journal* 63:224–226. In contrast, today's many short-stalked wheat cultivars have straw/grain ration just slightly above, or even slightly below, one, producing as much grain as straw.

8. Smil, V. 1999. Crop residues: agriculture's largest harvest. *BioScience* 49:299–308.

9. Lists of C/N ratios for many kinds of materials are published in Parr and Colacicco (6); Schindler, D. W., and S. E. Bayley. 1993. The biosphere as an increasing sink for atmospheric carbon: estimates from increased nitrogen deposition. *Global Biogeochemical Cycles* 7:717–733.

10. Stevenson, F. J. 1986. *Cycles of Soil*. New York: Wiley-Interscience, pp. 67, 184–185.

11. Amarasiri, S. L. 1978. In *Organic Recycling in Asia*. Rome: FAO, pp. 119–133.

12. Because of heavy reliance on relatively small-sized, and often poorly fed, oxen and horses, draft power in traditional agricultures was rather limited. In Europe and North America this changed only with the introduction of more powerful and better-fed breeds after 1700; weak animals still abound in Asia and Africa. Primitive ards, scratch plows that, unlike moldboard plows, did not overturn the furrow, also made proper incorporation of crop residues very difficult. The evolution of animal power and plows is traced in Smil (2), pp. 30–35, 40–49, 57–73, 85–86.

13. For example, extensive rural surveys showed that in China in the early 1930s between 59 and 74% of all major cereal straws and 90% of all cotton stalks were used for fuel. Most of the residues were burned by rural households, but appreciable amounts were sold by farmers in towns and cities; see Buck, J. L. 1937. *Land Utilization in China*. Nanking: Nanking University, p. 238.

14. Ruminants can digest cellulose in crop residues because the microorganisms in their rumen produce requisite enzymes; see Van Soest, P. J. 1994. *Nutritional Ecology of the Ruminant.* Ithaca: Cornell University Press. In order to maintain normal rumen activity, at least $1/7$ of their normal diet (in dry matter terms) should be in roughages; see National Research Council. 1996. *Nutrient Requirements of Beef Cattle.* Washington, D.C.: National Academy Press.

15. Wuest, P. J., D. J. Rose, and R. B. Beelman, eds. 1987. *Cultivating Edible Fungi.* Amsterdam: Elsevier; Maher, M. J., ed. 1991. *Science and Cultivation of Edible Fungi.* Rotterdam: A. A. Balkema.

16. Variability of manure production is described in Smil, V. 1983. *Biomass Energies.* New York: Plenum Press, pp. 324–340.

17. Powers, W. L., et al. 1975. Formulas for applying organic wastes to land. *Journal of Soil and Water Conservation* 30:286–289. Misra, R. V., and P. R. Hesse. 1983. *Comparative Analyses of Organic Manures.* Rome: FAO; Choudhary, M., L. D. Bailey, and C. A. Grant. 1996. Review of the use of swine manure in crop production: effects on yield and composition and on soil and water quality. *Waste Management Research* 14:581–595.

18. Not surprisingly, accurate labor accounts show that at least 10% of all work in China's traditional farming was connected with fertilizers. In parts of North China fertilization of grain crops was the single most time-consuming agricultural task, claiming close to $1/5$ of all human, and about $1/3$ of all animal, labor; see Buck, J. L. 1930. *Chinese Farm Economy.* Nanking: University of Nanking, pp. 305–309. But this investment was very rewarding; the ratio between labor (food energy) invested in manuring and additional food energy in higher yields was more than fiftyfold; see Smil (2), pp. 87–88.

19. Homer. 1944. *The Odyssey.* Translated by S. Butler. Roslyn, N.Y.: Walter J. Black, p. 216.

20. White, K. D. 1970. *Roman Farming.* Ithaca: Cornell University Press, pp. 125–135. Pigeon waste generally ranked first, but Varro regarded the droppings of thrushes and blackbirds as superior; see Varro, M. T. 1934. *Res Rusticae (On Agriculture).* Translated by W. D. Hooper. Cambridge, Mass.: Harvard University Press, p. 265. Poultry waste has commonly at least twice as high a nitrogen content as do cattle manures, which Romans considered to be an inferior fertilizer.

21. Columella, L. J. 1941. *Res rustica (On Agriculture).* Translated by H. B. Ash. Cambridge, Mass.: Harvard University Press, p. 137.

22. Buck (13), p. 258. These totals also include recycled human wastes.

23. Chorley, p. 80; Salter, R. M., and C. J. Schollenberger. 1938. Farm manure. In *Soils and Men.* Washington, D.C.: USDA, p. 460.

24. Extended composting of manure goes back to antiquity when it was believed that manure must not be used the first year, and that the best time to apply it is after three or four years when it exhaled the disagreeable smell, that is, nearly all volatile nitrogen; see Fussell, G. E. 1972. *The Classical Tradition in West European Farming.* Rutherford, N.J.: Fairleigh Dickinson University Press, p. 67. Both Cato and Varro give instructions on how to build a compost heap.

25. Insufficient manure storage capacity on farms keeping a larger number of domestic animals often required the spreading of wastes throughout the winter, many months before the nutrients could be assimilated by plants.

26. Nutrient losses are reviewed in Subba Rao, N. S. 1988. *Biofertilizers in Agriculture.* New Delhi: Oxford University Press; Hansen, J. A., and K. Henriksen, eds. 1989. *Nitrogen in Organic Waste Applied to Soils.* London: Academic Press; Eghball, B., et al. 1997. Nutrient, carbon, and mass loss during composting of beef cattle feedlot manure. *Journal of Environmental Quality* 26:189–193.

27. Smil (16), pp. 343–347. Published values range from 3 to 5 kg N/day (2.5–4.3 kg for urine and 0.5–0.7 kg N for feces); see Kirchamnn, H., and S. Petersson. 1995. Human urine: Chemical composition and fertilizer use efficiency. *Fertilizer Research* 40:149–154.

28. During the 1920s Buck (13), p. 258, assumed that every adult produced annually about 450 kg of night soil. This would have been an equivalent of about 3 kg N/capita, and around 2 kg N/capita might have been collected for recycling. For comparison, Zhu estimated that recently no more than 0.44 kg N/capita are recycled in night soil in China's cities, and that the rural mean is about 0.7 kg N/capita: Zhu Z. 1997. Nitrogen balance and cycling in agro-ecosystems of China. In Zhu Z., et al., eds. *Nitrogen in Soils of China.* Dordrecht: Kluwer Academic, p. 328.

29. In order to encourage complete recycling Edo authorities ordered payments for the removed night soil; see Tanaka, Y. 1998. The cyclical sensibility of Edo-period Japan. *Japan Echo* 25(2):12–16. This arrangement had persisted for more than three centuries; most of the inhabitants in even the largest Japanese cities received sewer connections only after 1970.

30. For China see King, F. H. 1927. *Farmers of Forty Centuries.* New York: Harcourt, Brace & Co. In Japan Chamberlain noted that "various manures are employed. The commonest is night soil, whose daily conveyance all about the country causes no distress to native noses." Chamberlain, B. H. 1890. *Japanese Things.* Tokyo: Hakubunsha, p. 20.

31. When visiting countryside near Paris, Claude, a young painter in Zola's *The Belly of Paris,* volunteers to toss the leafy refuse brought from the city market into the manure pit:

Claude had quite a liking for manure, since it symbolizes the world and its life. The strippings and parings of the vegetables, the scourings of markets, the refuse that fell from that colossal table, remained full of life, and returned to the spot where the vegetables had previously sprouted. . . . They rose again in fertile crops, and once more went to spread themselves out upon the market square. Paris rotted everything and returned everything to the soil, which never wearied of repairing the ravages of death.

E. A. Vizetelly's translation. Los Angeles: Sun & Moon Press (1996), pp. 271–272.

32. Kislev, M. E., and O. Bar-Yosef. 1988. The legumes: the earliest domesticated plants in the Near East? *Current Anthropology* 29:175–178.

33. Ibid.

34. *Genesis* 25:34: "Then Jacob gave Esau bread and pottage of lentils: and he did eat and drink. . . . [T]hus Esau despised his birthright." *Samuel* 17:28 lists "wheat, and barley, and flour, and parched corn, and beans, and lentils, and parched pulse."

35. Bray, F. 1984. *Science and Civilisation in China. Volume 6, Part II: Agriculture.* Cambridge: Cambridge University Press, pp. 510–518.

36. Sanjappa, M. 1992. *Legumes of India.* Dehra Dun: Bishen Singh Mahendra Pal Singh.

37. Delwiche, C. C. 1978. Legumes: past, present, and future. *BioScience* 28:565–570; National Academy of Sciences. 1979. *Tropical Legumes.* Washington, D.C.: NAS.

38. Bray (35), p. 293; Pieters, A. J. 1927. *Green Manuring*. New York: John Wiley, pp. 10, 292–295.

39. Theophrastus. 1949. *Historia Plantarum (Enquiry into Plants)* 8.9.1. Translated by A. Hort. London: W. Heinemann, vol. II, p. 199. But it must have been difficult to plow under entire plants with ards, primitive plows that did not overturn the soil; see Isager, S., and J. E. Skydsgaard. 1992. *Ancient Greek Agriculture*. London: Routledge, p. 110.

40. White (20), pp. 135–137. Columella's main reasons for extolling alfalfa were the durability of the crop, its high yield, and its high feeding value: "one seeding affords, for all of ten years thereafter, four harvestings regularly and sometimes six; it improves the soil; lean cattle of every kind grow fat on it" (Columella (21), p. 173).

41. Pieters (38) offers detailed descriptions of all the important green manure species and the actual practices involved in their cultivation in North America, Europe and Asia. See also Subba Rao, N. S. 1993. *Biofertilizers in Agriculture and Forestry*. New Delhi: Science Publishers, pp. 121–128; Porpavai, S., et al. 1996. Evaluation of green manures and grain legumes. *International Rice Research Newsletter* 21:72.

42. Lumpkin, Thomas A. 1982. *Azolla as a Green Manure: Use and Management in Crop Production*. Boulder: Westview Press; Khan, M. Manzoor. 1983. *A Primer on Azolla Production and Utilization in Agriculture*. Los Baños: University of the Philippines at Los Baños; Subba Rao (41), pp. 111–120; Meelu, O. P., et al. 1994. *Green Manuring for Soil Productivity Improvement*. Rome: FAO, pp. 14–17.

43. Ayanaba, A., and P. J. Dart, eds. 1977. *Biological Nitrogen Fixation in Farming Systems of the Tropics*. London: John Wiley; Broughton, W. J., and S. Puhler, eds. 1986. *Nitrogen Fixation*. Oxford: Clarendon Press; Eady, R. R. 1995. The enzymology of biological nitrogen fixation. *Science Progress* 78:1–17; Schubert, S. 1995. Nitrogen assimilation by legumes: processes and ecological limitations. *Fertilizer Research* 42:99–107.

44. LaRue, T. A., and T. G. Patterson. 1981. How much nitrogen do legumes fix? *Advances in Agronomy* 34:15–38; Smil, V. 1997. Some unorthodox perspectives on agricultural biodiversity: the case of legume cultivation. *Agricultural Ecosystems and the Environment* 62:135–144.

45. These estimates are based on many modern measurements summarized in Frame, J., and O. Newbould. 1986. Agronomy of white clover. *Advances in Agronomy* 40:1–88; Dovrat, A. 1993. *Irrigated Forage Production*. Amsterdam: Elsevier; and Meelu et al. (42).

46. Henson, R. A. 1993. Measurements of N_2 fixation by common bean in Central Brazil as affected by different reference crops. *Plant Soil* 152:53–58; Tsai, S. M., et al. 1993. Variability in nitrogen fixation in common bean (*Phaseolus vulgaris* L.) intercropped with maize. *Plant Soil* 152:93–101.

47. Detailed nitrogen budgets for modern American soybean crops showed that the yield of about 2.4 t/ha deriving 40% of plant nitrogen from symbiosis will cause, even with complete residue recycling, a net loss of more than 80 kg N/ha. Only when the soybeans could derive more than about 80% of all nitrogen from symbiosis, or when they were planted as green manures, did they add up to 90 kg N/ha, the enrichment surpassed only by plowing in good stands of clover or alfalfa; see Heichel, G. H. 1987. Legume nitrogen: symbiotic fixation and recovery by subsequent crops. In Z. Helsel, ed., *Energy in Plant Nutrition and Pest Control*. New York: Elsevier, pp. 62–80.

48. Raffles, T. S. 1817. *The History of Java.* London: Black, Parbury & Allen; Buck (13, 18).

49. Butzer, K. W. 1976. *Early Hydraulic Civilization in Egypt.* Chicago: University of Chicago Press.

50. This calculation is based on arable land estimates in Perkins, D. H. 1969. *Agricultural Development in China 1368–1968.* Chicago: Aldine, p. 16. Chao, J. 1986. *Man and Land in Chinese History: An Economic Analysis.* Stanford: Stanford University Press, p. 89.

51. According to Chorley's estimates, even as late as 1770 the recycled organic matter supplied no more than 1/3 of all nitrogen inputs in Northwest Europe's farming; see Chorley, G. P. H. 1981. The agricultural revolution in Northern Europe, 1750–1880: Nitrogen, legumes, and crop productivity. *Economic History* 34:71–93.

52. Ibid., p. 85.

53. Campbell, M. S., and M. Overton. 1993. A new perspective on medieval and early modern agriculture: six centuries of Norfolk farming c. 1250–c. 1850. *Past and Present* 141:38–105.

54. Chorley (51), p. 92.

55. Bennett, M. K. 1935. British wheat yield per acre for seven centuries. *Economic History* 3(10):12–29; Stanhill, G. 1976. Trends and deviations in the yield of the English wheat crop during the last 750 years. *Agro-Ecosystems* 3:1–10; Clark, G. 1991. Yields per acre in English agriculture, 1250–1850: evidence from labour inputs. *Economic History Review* 44:445–460.

56. Damen, J. 1978. Agro-ecosystems in the Netherlands, Part I. In M. J. Frissel, ed., *Cycling of Mineral Nutrients in Agricultural Ecosystems.* Amsterdam: Elsevier, pp. 70–78.

57. Principal staple crops rotated at these farms were rice, wheat, and corn; other crops included rapeseed, broad beans, peas, soybeans, peanuts, sweet potatoes, and green manures, with a multicropping index between 1.7 and 2.

58. Ruddle, K., and G. Zhong. 1988. *Integrated Agriculture-Aquaculture in South China: The Dike-Pond System of the Zhujiang Delta.* Cambridge: Cambridge University Press; Korn, M. 1996. The dike-pond concept: sustainable agriculture and nutrient recycling in China. *Ambio* 25:6–13.

59. The typical fish harvest was 10 t/ha (maxima 15–20 t/ha); cane yielded 75 t/ha. Most of the food energy from a typical farm would have come from carp (as much as 2/3), with sugarcane always being the second most important contributor (some farms grew no rice); protein output would be overwhelmingly (more than 9/10) from carp.

60. Stanhill, G. 1977. An urban agro-ecosystem: the example of nineteenth-century Paris. *Agro-Ecosystems* 3:269–284.

61. Traditionally, animal foods provided rarely more than 5% of all dietary energy in China's rural areas. For details see Buck (13), pp. 400–421.

62. Smil (2), p. 63.

63. But by that time virtually all of Egypt's farmland was reliably irrigated; *birsim* clover, the country's dominant green manure, was planted as a winter crop on some 20% of farmland; and the country was importing annually nearly 300,000 t of nitrogenous fertilizers; see Richards, A. 1982. *Egypt's Agricultural Development, 1800–1980.* Boulder: Westview.

64. For example, during the 1950s oilseeds were estimated to provide no more than 7% of all nitrogen in China's recycled wastes; see Buck, J. L., O. L. Dawson, and Y. Wu. 1966. *Food and Agriculture in Communist China*. New York: F. A. Praeger, p. 144.

65. Largely roughage-fed cattle in traditional agriculture would thus produce manure with as little as 0.3% N, while the manure of modern beef cattle fed high-protein mixtures may contain more than 4% N; see Powers et al. (17).

66. For example, on an early-nineteenth-century Dutch farm whose land was equally split between fields and pastures only about 45% of all cattle and horse manure was recycled to crops; see Damen (56), p. 76.

67. Ling, B., et al. 1993. Use of night soil in agriculture and fish farming. *World Health Forum* 14:67–70.

68. Maintaining a strictly anaerobic environment has not been easy even in modern mass-produced biogas digesters; see Smil, V. 1993. *China's Environmental Crisis*. Armonk, N.Y.: M. E. Sharpe, pp. 104–105.

69. Chorley (51), p. 74.

70. Buck (13), pp. 224–225.

71. FAO. 1998. *Production Yearbook*. Rome: FAO.

72. Smil (2), p. 94.

73. Unless presoaked for several hours, the minimum cooking times for mature pulses are much longer than those for cereal seeds and flours. Although the food energy content of legumes is nearly identical to that of cereal grains, their different composition (above all the absence of gluten complex) makes it impossible to use them for bread. Increased flatulence following higher intakes of legumes is due to the presence of indigestible oligosaccharides. Because the human digestive tract lacks the requisite enzyme (α-galactosidase), these sugars can become subject to anaerobic fermentation resulting in excessive intestinal gas production; see Augustin, J., and B. P. Klein. 1989. Nutrient composition of raw, cooked, canned, and sprouted legumes. In R. H. Matthews, ed., *Legumes*. New York: Marcel Dekker, pp. 197–217.

74. Ferrando, R. 1981. *Traditional and Non-traditional Foods*. Rome: FAO; Matthews, R. H., ed. 1989. *Legumes: Chemistry, Technology, and Human Nutrition*. New York: Marcel Dekker; Jones, J. M. 1992. *Food Safety*. St. Paul, Minn.: Eagan Press; Deshpande, S. S. 1992. Food legumes in human nutrition: a personal perspective. *Critical Reviews in Food Science and Nutrition* 32:333–363. Common enzyme inhibitors in legumes make it more difficult to digest proteins present in the pulses that may be eaten specifically for their high amino acid content. Fabva beans contain vicine and covicine, which may cause hemolytic anemia and hematuria in genetically sensitive individuals, and monoamine oxidase inhibitors, which engender headaches and palpitations. Allergies to peanuts are well documented, as is the risk from aflatoxins not only in unprocessed whole seeds but also in peanut butter.

75. Smil (44).

76. Buck (13), pp. 208–213.

77. Meelu et al. (42), pp. 51–55. Other Indian studies show that net returns of integrating fertilizer applications with green manures are generally lower than the use of urea alone; see

Chavan, L. S., et al. 1996. Integrated N management with *Gliricidia maculata* and *Sesbania aculeata* for transplanted rice. *International Rice Research Newsletter* 21:70.

Chapter 3

1. The three Islas Chincha (Norte, Centro, and Sur) are located at about 14° S and about 20 km off the port of Pisco in south-central Peru. Incredibly large flocks of birds, ranging from seagulls to penguins—roosting on the predator-free islands and feeding on what used to be the world's largest biomass of anchovies—deposited guano layers that were up to 32 m thick at Chincha Norte.

2. Church, A. H. 1890. Guano. In *Encyclopedia Britannica*. 9th edition. Chicago: R. S. Peale & Co., vol. XI, pp. 233–235.

3. Murra, J. V. 1980. *The Economic Organization of the Inka State*. Greenwich, Conn.: JAI Press, p. 20. The Inca name was actually *huano*. According to an Inca narrative, when an Inca queen, wife of Huayna Cápac, noticed during one of her inspection tours that no guano was used on a local corn crop, she "took one of the principal women by the hand . . . and taking a sackful of it, they planted." Murua, M. de. 1946 (1590). *Historia del origen y genealogia real de los reyes incas*. Madrid: Bibiliotheca Missionalia Hispánica, book IV, p. 378.

4. But Humboldt did not visit any of the guano islands, and he makes no mention of guano in the massive account of his South American travels; see von Humboldt, A. 1814–1829. *Personal Narrative of Travels to the Equinoctial Regions of the New Continent during the Years 1799–1804*. 7 vols. London: Longmen, Hurst, Orne, and Brown. On the history of the U.S. guano trade see Skaggs, J. M. 1994. *The Great Guano Rush: Entrepreneurs and American Overseas Expansion*. New York: St. Martin's Press.

5. Guano mania—including the scheming by the U.S. Congress, the personal involvement of two presidents, and the U.S. Guano Island Act of 1856, which made it possible to claim any uncharted island containing guano as the property of the U.S. citizen who discovered it—is described vividly in Skaggs (4) and Wines, R. A. 1985. *Fertilizer in America*. Philadelphia: Temple University Press, pp. 54–70. Between 1856 and 1903 U.S. entrepreneurs laid claim to ninety-four islands, rocks, and keys under the authority of the Guano Islands Act, passed in August 1856; phosphatic guanos were actually mined on some twenty of these islands, and nine of them remain in U.S. possession.

6. The total of 12.5 Mt is from Markham, C. R. 1892. *A History of Peru*. Chicago: C. H. Sergel & Co., p. 343; other published totals range between 8 and 10 Mt. Terrible work conditions were the norm during the frenzied decades of the Guano Age between 1850 and 1870. Dry, gypsum-like material was dug up by spades, moved by wheelbarrows and carts to sea cliffs, and dumped through canvas chutes first to shallow-draft boats and then by basket winches to ships, or it was loaded directly into waiting vessels. A thick ammonia smell enveloped the place as scores of ships were loading simultaneously. Annual shipments during the 1860s ranged between 300,000 and 400,000 t, and the guano monopoly earned the Peruvian treasury most of its income until the early 1870s. On Peru in the guano era see Marett, R. 1969. *Peru*. New York: Praeger, pp. 94–115; Werlich, D. P. 1978. *Peru*. Carbondale: Southern Illinois University Press, pp. 80–103.

7. Comparisons on the nutrient basis are uncertain. Unlike the Chilean nitrate, whose N content clustered tightly around 15%, the best guanos had ten times as much N as the poorest exported material. Moreover, with the single source of supply, Chilean records of nitrate exports are also fairly accurate, while we have no comparably reliable records of total guano shipments from more than a dozen other different locations around the world.

8. Church (2); J. A. Voelcker. 1911. Manures and manuring. In *Encyclopedia Britannica*. 11th edition. Cambridge: Cambridge University Press, vol. XVII, p. 617.

9. Schneider, D., and D. C. Duffy. 1988. Historical variation in guano production from the Peruvian and Benguela upwelling ecosystems. *Climatic Change* 13:309–316.

10. Flagg, J. W. 1874. Nitrate of soda, its locality, mode of occurrence and methods of extraction. *American Chemist* 4:403–408; Penrose, R. A. F. 1910. The nitrate deposits of Chile. *Journal of Geology* 18(1):1–32; Nitro de Chile. 1964. *Enciclopedia Universal Ilustrada*. Madrid: Espasa-Calpe, pp. 825–833.

11. Ericksen, G. E. 1981. *Geology and Origin of the Chilean Nitrate Deposits*. Geological Survey Professional Paper 1,188. Washington, D.C.: GPO.

12. Ibid., pp. 14–19.

13. Penrose (10), p. 24.

14. Nitro de Chile (10), p. 832.

15. Hancock, A. U. 1971 (1893). *A History of Chile*. New York: AMS Press, pp. 272–324.

16. Ibid., p. 268.

17. Penrose (10), pp. 22–28.

18. Partington, J. R., and L. H. Parker. 1922. *The Nitrogen Industry*. London: Constable & Company, pp. 37–84; Nitro de Chile (10), p. 832; Collier, S., and W. F. Salter. 1996. *A History of Chile, 1808–1994*. Cambridge: Cambridge University Press, pp. 162–166 give slightly different accounts.

19. Collier and Salter (18), p. 162.

20. Saltpeter plantations were common in eighteenth-century Europe. The French called them *nitrières,* and they became particularly productive under Lavoisier's supervision. The same method was also used to produce nitrates in the Southern secessionist states during the U.S. Civil War. All that was required was to let nitrifying bacteria act upon heaps of decomposing, well-aerated, moderately moist N-rich organic wastes mixed with lime, mortar, and wood ashes and protected from rain but dampened periodically with animal urine or the drainage from dung heaps.

The process produced a mixture of calcium and potassium nitrates, which were scraped off the piles after about two years and extracted with water. K_2CO_3 was added to the extracted nitrates, calcium and magnesium salts were separated from the solution, and evaporation and crystallization yielded nearly pure KNO_3. The yields for properly managed *nitrières* were around 5 kg KNO_3/m^3 in two years; see Brunborg, I., and P. B. Holmesland. 1985. The history of nitric acid. In C. Keleti, ed., *Nitric Acid and Fertilizer Nitrates*. New York: Marcel Dekker, pp. 4–5.

21. As the town's salt production from natural brines became less competitive, boring for NaCl began in 1839; in 1857 a shaft begun five years earlier passed through a layer that was

initially considered to contain just waste minerals, namely, carnallite ($KCl \cdot MgCl \cdot 6H_2O$), sylvite (KCl), and kainite ($KCl \cdot MgSO_4 \cdot 3H_2O$).

22. Discoveries of new explosives during the second half of the nineteenth century are described in McGrath, J. 1958. Explosives. In C. Singer et al., eds., *A History of Technology*. Oxford: Oxford University Press, vol. V, pp. 284–298. These explosives contain more nitrogen than the about 10% present in gunpowder; nitroglycerin, the principal ingredient of dynamite, contains 18.5% N.

23. Munroe, C. E. 1910. The nitrogen question from the military standpoint. *Annual Report of the Board of Regents of the Smithsonian Institution*. Washington, D.C.: Smithsonian Institution, p. 231.

24. Ibid., p. 233.

25. Nitro de Chile (10), p. 832.

26. These concerns were justified: the British naval blockade cut off the German imports soon after the beginning of World War I.

27. Coking is a dry, destructive distillation proceeding at temperatures between 1,100 and 1,150 °C (for metallurgical coke). Its product is a high-carbon (85–90% C), porous solid fuel used primarily as a reductant in smelting iron in blast furnaces. By-product gas (also called town gas or luminous gas) contains CH_4, H_2, CO, and CO_2, and its energy content is equal to about half of the energy density of natural gas.

Most of the nitrogen initially present in coal is released as N_2, and a part is retained in coke; the extremes of nitrogen released as NH_3 range from about 8 to almost 25%. No NH_3 is liberated when temperatures stay below 500 °C, and most of the NH_3 discharge occurs between 500 and 700 °C. For details on nineteenth and early twentieth century coking see Ost, H. 1919. *Lehrbuch der chemischen Technologie*. Leipzig: Max Jänecke, pp. 171–173; Porter, H. 1924. *Coal Carbonization*. New York: Chemical Catalog Company.

28. Abraham Darby produced pig iron with coke in 1709, but widespread replacement of charcoal came only after 1750; see Harris, J. R. 1988. *The British Iron Industry 1700–1850*. London: Macmillan. Beehive ovens are described in detail in Porter (27), pp. 106–110. By-product recovery ovens are rectangular chambers built in rows and interspersed with heating flues (coke-oven gas is the most convenient heat source). Coal is charged through their tops, ovens are sealed, and the processed fuel is pushed out after sixteen to twenty hours of coking. Tar and phenols are other usable by-products.

29. King, C. D. 1948. *Seventy-five Years of Progress in Iron and Steel*. New York: American Institute of Mining and Metallurgical Engineers, pp. 24–25.

30. Munroe (23), p. 235. The abundance of excellent coking coals and cheapness of the beehive ovens are the best explanations of the late introduction of by-product coking ovens in the United States.

31. Partington and Parker (18), pp. 85–146; Carbone, W. E., and O. F. Fissore. 1963. By-product ammonia. In A. Standen, ed., *Kirk-Othmer Encyclopedia of Chemical Technology*. 2nd edition. New York: Wiley, vol. 1, pp. 299–312.

32. Carbone and Fissore (31), p. 299.

33. Smil, V. 1994. *Energy in World History*. Boulder: Westview Press, p. 180.

34. In reality, besides the already noted persistence of beehive coke ovens in the United States, some pig iron smelting in North America, Russia, and Asia was still done with charcoal rather than with coke.

35. Coal use for coking and ammonia sulfate output according to Munroe (23), p. 235.

36. Ost (27), p. 173.

37. Ibid.

38. Gmelins Handbuch der Anorganischen Chemie. 1932. Bariumcyanid Ba(CN)$_2$. In *Barium*. Berlin: Verlag Chemie, p. 328.

39. History of cyanamide synthesis and of its rapid commercialization is described in Ost (27), pp. 181–182; Partington and Parker (18), pp. 188–232; Schmidt, A. 1934. *Die industrielle Chemie*. Berlin: Verlag Walter de Gruyter & Co., pp. 337–339; Voigtländer-Tetzner, W. 1956. *Die Bindung des Luftstickstoffes*. Unpublished typescript, BASF Unternehmensarchiv, vol. I, 1895–1918, pp. 49–56; Gmelins Handbuch der Anorganischen Chemie. 1956. *Calcium*. Vol. 28, pp. 179–187; Nelson, L. B. 1990. *History of the U.S. Fertilizer Industry*. Muscle Shoals, Ala.: TVA, pp. 196–197.

40. In 1895 Carl von Linde (1842–1934) was the first refrigeration engineer to liquefy air by using the Joule-Thomson effect (irreversible adiabatic expansion of a gas passing through a throttling valve). He produced pure nitrogen in 1903 and hydrogen from water gas in 1909; see Linde, C. P. 1916. *Aus meinem Leben und von meiner Arbeit*. Munich: R. Oldenbourg.

41. The total output in 1913—around 20,000 t of fixed N—was less than 5% of the nutrient shipped that year in Chilean nitrates. Soon after the beginning of World War I the two German plants were much enlarged (Trostberg from 30,000 to 80,000 t/year, Knapsack from 20,000 to 110,000 t), and two new plants were built in Piesteritz, the world's largest with an annual capacity of 150,000 t, and in Chorzow (110,000 t); see Voigtländer-Tetzner (34), p. 56. The U.S. government began building a large cyanamide plant in 1917 at Muscle Shoals in Alabama; although it was completed, it never produced anything; see Nelson (39), p. 197.

42. Partington, J. R. 1962. *A History of Chemistry*. London: Macmillan, vol. 3, p. 286.

43. Ibid., pp. 340–341.

44. Efficient hydroturbines had been available for some time (the Francis turbine since 1847), but hydroelectric generation took off only during the 1890s with the maturation of transmission systems, including the victory of alternating current over the direct current and the introduction of transformers (1885) and with the diffusion of electric motors (1888). The history of the nineteenth-century experiments with NO synthesis through electric discharge is reviewed in Gmelins Handbuch der Anorganischen Chemie. 1936. *Stickstoff*. Berlin: Verlag Chemie, pp. 344–347.

45. Curtis, H. A. 1932. A history of nitrogen fertilizer processes. In H. A. Curtis, ed., *Fixed Nitrogen*. New York: Chemical Catalog Company, pp. 77–89. During his study trip to the United States Fritz Haber visited the enterprise shortly before it closed down; it was his first direct encounter with nitrogen fixation: "I remember my vivid impression, when visiting (for the first time) the United States in 1902, on seeing the first industrial experimental station to be established, that of the Atmospheric Products Company. The idea of copying the process of the combination of nitrogen and oxygen by lightning in the atmosphere . . . fascinated

the world." Haber, F. 1924. *Practical Results of the Theoretical Development of Chemistry.* Philadelphia: Franklin Institute, p. 7.

46. For details on Birkeland Eyde process see Ost (27), pp. 167–169; Partington and Parker (18), pp. 233–268; Schmidt (39), pp. 334–337; Pauling, H. 1929. *Elektrische Luftverbrennung.* Halle: W. Knapp; Brunborg and Holmesland (20), pp. 8–12.

47. Partington and Parker (18), p. 264; Gmelin (44), p. 345; Haber (45), p. 9; Appl, M. 1976. A brief history of ammonia production from the early days to the present. *Nitrogen* 100:47–58.

48. $NaNO_3$ exports from Nitro de Chile (10), p. 832; $(NH_4)_2SO_4$ output from Ost (27), p. 173.

49. Historical data on population growth by continents and major countries are available in McEvedy, C., and R. Jones. 1978. *Atlas of World Population History.* London: Allen Lane.

50. Only England was highly urbanized (more than 50% of the total population living in cities) by 1850; the second half of the nineteenth century saw the urban population shares more than doubling in both the United States and Germany; see Smil (33), p. 209. On dietary transitions see Popkin, B. M. 1993. Nutritional patterns and transitions. *Population and Development Review* 19:138–157.

51. Reconstructions of European food energy intakes during the early phases of nineteenth-century industrialization show per capita rates just above 2,000 kcal/day. For example, the Belgian mean during the 1840s was just over 2,200 kcal, and the French average for the same decade was virtually identical (2,250 kcal); see Bekaert, G. 1991. Caloric consumption in industrializing Belgium. *Journal of Economic History* 51:633–655; Toutain, J.-C. 1971. La consommation alimentaire en France de 1789 à 1964. *Economies et Societes* 5:1909–2049. Not surprisingly, malnutrition was still relatively common in the mid-nineteenth-century Europe. A half century later Western European per capita daily means were close to 3,000 kcal/day.

52. The transformation of the French diet is an excellent example of these shifts; see Dupin, H., et al. 1984. Evolution of the French diet: nutritional aspects. *World Review of Nutrition and Dietetics* 44:57–84.

53. We have two sets of figures: Grigg's data for the years 1870 and 1910 imply an addition of 105 Mha in North America, Argentina, and Australia; Richards's totals for North and South America, Australasia, and Russia add up to a gain of 253 Mha between 1850 and 1920: Grigg, D. B. 1974. *The Agricultural Systems of the World.* Cambridge: Cambridge University Press, p. 45; Richards, J. F. 1990. Land transformation. In B. L. Turner II, et al., eds., *The Earth as Transformed by Human Action.* Cambridge: Cambridge University Press, p. 164.

54. European figures were calculated from data in Mitchell, B. R. 1981. *European Historical Statistics.* New York: Facts on File, pp. 207–281; U.S. data are from U.S. Bureau of the Census. 1975. *Historical Statistics of the United States: Colonial Times to 1970.* Washington, D.C.: GPO, pp. 510–512.

55. Brock, W. H. 1992. *The Norton History of Chemistry.* New York: W. W. Norton, p. 285.

56. Brock, W. H. 1997. *Justus von Liebig.* Cambridge: Cambridge University Press, p. 121.

57. Ibid.

58. Young, R. D., D. G. Westfall, and G. W. Colliver. 1985. Production, marketing, and use of phosphorus fertilizers. In Engelstad, O. P., ed., *Fertilizer Technology and Use.* Madison, Wisc.: Soil Science Society of America, pp. 323–376.

59. It has remained so ever since: strip mining of Florida phosphates now extracts about $3/4$ of the total U.S. production of phosphate rock, which now accounts for almost $1/3$ of the global output; see UNIDO and IFDC. 1998. *Fertilizer Manual.* Dordrecht: Kluwer Academic, pp. 111–121.

60. Organic wastes have less P than N: cereal straws have just between 0.05 and 0.1% N, and cattle and pig manures (in dry solids) 0.3–2.04% N; poultry manures are the richest source of organic P, containing 0.5–4.7% P; see Parr, J. F., and D. Colacicco. 1987. Organic materials as alternative nutrient sources. In Z. Helsel, ed., *Energy in Plant Nutrition and Pest Control.* Amsterdam: Elsevier, pp. 87–88.

61. Major discoveries of phosphates were made during the twentieth century. Huge Moroccan deposits were identified in 1914, and their extraction commenced in 1921. The Soviets opened their high-grade apatite mines in the Khibini tundra of the Kola Peninsula in 1930 and Kazakhstan phosphate beds in 1937. The largest post–World War II finds were made in China (now the world's second largest producer of phosphate rock) and Jordan. For details on phosphate resources see UNIDO and IFDC (59), pp. 90–126.

62. For example, a rice crop yielding 6 t/ha will remove 100 kg N, 22 kg P, and 133 kg K/ha. Bananas represent an extreme case: a harvest of 40 t/ha will remove 250 kg N and 830 kg K; see UNIDO and IFDC (59), p. 22.

63. Wheat straw has commonly twice as much K as N, rice straw up to three times as much; see Parr and Colacicco (60), p. 87. In recent worldwide applications about 4.25 kg N have been used for every kg K; see FAO. 1998. *Fertilizer Yearbook.* Rome: FAO.

64. Munson, R. D. 1992. Potassium in ecosystems, with emphasis on the U.S. Paper presented at the 23rd Colloquium of the International Potash Institute, Prague, October 14, 1992, p. 22.

65. German exports of KNO_3 were above 12,000 t a year during the second half of the 1890s, and they surpassed 15,000 t by 1913; see Ost (27), p. 173.

66. The total of 15 Mt of incorporated nitrogen is just an approximation based on incomplete historical statistics of crop production published before World War I by the Institut International d'Agriculture in *Annuaire International de Statistique Agricole* in Rome.

67. Ost (27), pp. 16, 173.

68. Import and output data are from, respectively, Munroe (23), p. 231, and Ost (27), p. 231.

69. At the end of the nineteenth century, German wheat yields were close to 2 t/ha, and the English harvests were a bit above 2 t/ha; see Mitchell (54). The United Kingdom was by far the largest importer of wheat, buying annually about 5 Mt, compared to Germany's purchases of almost 1 Mt/year.

70. His research also included work on nitrogen fixation: in 1892, at a soirée of the Royal Society, he exhibited an experiment on the flame of burning nitrogen (which does not propagate through the whole atmosphere because the ignition point of the gas is higher than the temperature of its flame) and on the continuous combustion of the gas in a strong induction current producing nitrous and nitric acids. For a biography and bibliography see Brock, W. H. 1971. Crookes, William. In C. C. Gillispie, ed., *Dictionary of Scientific Biography.* New York: Charles Scribner's Sons, vol. III, pp. 474–482.

71. Crookes, W. 1899. *The Wheat Problem.* London: Chemical News Office, p. 3. All subsequent quotations are also from this book, which contains the complete text of the Bristol

lecture; its second edition, published in 1905, contains additional supporting evidence by other researchers and rebuttals of published criticism.

72. Crookes was quite correct in discounting the potential for Russian wheat exports: after 1917 under Communist rule Russia and the Ukraine became chronic wheat-deficit areas, and, in spite of the large-scale extension of grain cultivation in Kazakhstan during the 1950s, the Soviet Union became the world's largest importer of wheat during the 1970s.

73. Crookes (70), p. 36.

74. Ibid., p. 39.

75. Ibid., pp. 45–46.

76. Ibid., p. 3.

Chapter 4

1. Haber, F. 1966 (1920). The synthesis of ammonia from its elements. In *Nobel Lectures: Chemistry 1901–1921*. Amsterdam: Elsevier, p. 339. For the German text of the lecture see Haber, F. 1924. Über die Darstellung des Ammoniaks aus Stickstoff und Wasserstoff. In F. Haber, *Fünf Vorträge aus den Jahren 1920– 1923*. Berlin: Julius Springer, pp. 1–24.

2. Kapoor, S. C. 1970. Berthollet, Claude-Louis. In C. C. Gillispie, ed., *Dictionary of Scientific Biography*. New York: Charles Scribner's Sons, vol. II, pp. 73–82. Berthollet followed Priestley's lead in decomposing ammonia in electric current and then exploding the mixture with oxygen; his conclusion was that the compound is made of 2.9 volumes of inflammable gas (H) and 1.1 volume of *moffette* (N).

3. A good review of pre-1900 attempts of ammonia synthesis from its elements is Gmelins Handbuch der anorganischen Chemie. 1936. *Stickstoff*. Berlin: Verlag Chemie, pp. 327–328.

4. For Ramsay's biography and bibliography of his work see Treen, T. J. 1975. Ramsay, William. In Gillispie (2), vol. XI, pp. 277–284.

5. Ostwald's letter to the BASF directors quoted in Stoltzenberg, D. 1994. *Fritz Haber: Chemiker, Nobelpreisträger, Deutscher, Jude*. Weinheim: VCH, pp. 140–141.

6. For an excellent biography and a detailed bibliography of Ostwald's work see Hiebert, E. W., and Kürber, H.-G. 1978. Ostwald, Friedrich Wilhelm. In C. C. Gillispie (2), vol. XV, supplement I, pp. 455–469.

7. BASF Ammoniaklaboratorium. 1939. Geschichte der Ammoniaksynthese und der Ammoniakoxydation. *BASF Unternehmensarchiv Ammoniak Geschichte* I G-6101.

8. Years later Bosch wrote down an undated confirmation preserved in the BASF archives: "The first task I was given as a newly hired chemist in the BASF under Knietsch was to test the finding which Ostwald had offered to our company at the turn of the century." *BASF Unternehmensarchiv Ammoniak Geschichte* I G-6101.

9. Quoted in Holdermann, K. 1954. *Im Banne der Chemie: Carl Bosch—Leben und Werk*. Düsseldorf: Econ-Verlag, p. 41.

10. In his patent claim (*Die Gewinnung von Ammoniak und Ammoniakverbindungen durch Vereinigung von freiem Stickstoff und Wasserstoff mittels Kontaksubstanzen*) Ostwald stated:

I have found that the synthesis . . . proceeds with measurable speed already at moderate heating to 250–300 °C. The speed rapidly increases with rising temperature. Catalysts are . . . mainly iron and copper. . . . In order to complete the synthesis ammonia must be removed from the reacting mixture. . . . To accomplish this gas mixture should be circulated. . . . As the amount of ammonia in gas mixture increases with higher pressure it is desirable to conduct the synthesis under elevated pressure.

Ostwald, W. 1933. *Lebenslinien: Eine Selbstbiographie.* Berlin: Klasing & Co., pp. 284–285.

11. Ibid., p. 285.

12. For Le Châtelier's biography and bibliography see Leicester, H. M. 1973. Le Châtelier, H. L. In Gillispie (2), vol. VIII, pp. 116–120. Le Châtelier remains best known for his eponymous principle: "Every change in one of the factors of an equilibrium occasions a rearrangement of the system in such a direction that the factor in question experiences a change in the sense opposite to the original change." As an expert in high-temperature studies and in the behavior of gas mixtures, as well as an accomplished metallurgist, he was eminently qualified to undertake the task of ammonia synthesis. In his Nobel lecture Haber (1, p. 339) recalled that Le Châtelier's findings were recorded only in "the obscurity of a French patent taken out under a foreign name"—and that he found out about this work only after he had completed his successful experiments.

13. Haber, F. 1924. *Practical Results of the Theoretical Development of Chemistry.* Philadelphia: Franklin Institute, pp. 7–8.

14. The first two comprehensive biographies, the first one in English, the second in German, were published only during the 1960s; see Goran, M. 1967. *The Story of Fritz Haber.* Norman: University of Oklahoma Press; Wille, H. H. 1969. *Der Januskopf: Leben und Wirken des Physikochemikers und Nobelpreisträger Fritz Haber.* Berlin: Verlag Neues Leben. Goran's work was based primarily on extensive interviews with Haber's colleagues and surviving members of Haber's family living in the United States. While it contains many interesting reminiscences, there are hardly any wider perspectives and virtually nothing on the invention's consequences. Wille's work is more comprehensive, but neither book is a critical, carefully annotated work.

In contrast, two new excellent German biographies—Stoltzenberg, D. 1994. *Fritz Haber: Chemiker, Nobelpreisträger, Deutscher, Jude.* Weinheim: VCH; and Szöllösi-Janze, M. 1998. *Fritz Haber: 1868–1934—Eine Biographie.* Munich: Verlag C. H. Beck—are, each in its special way, definitive accounts of Haber's life and work presented in a wider technical and historical setting. Both books are massive (669 and 928 pages, respectively) and are meticulously annotated. Stoltzenberg writes primarily from a chemist's perspective, while Szöllösi-Janze's habilitation work is a broader sociohistorical appraisal.

Shorter biographic sketches, containing a great deal of interesting material, are Günther, P. 1969. Fritz Haber: ein Mann der Jahrhundertwende. In *Fritz Haber: Ein Mann der Jahrundertwende.* Munich: R. Oldenbourg, pp. 5–29; Chmiel, G., et al. 1986. . . . *im Frieden der Menschheit, im Kriege dem Vaterlande . . . : 75 jahre Fritz-Haber-Institut der Max-Planck-Gesellschaft.* Berlin: ÖTV.

In spite of its title (*Aus Leben und Beruf*) Haber's only nonscientific book is just a collection of public lectures on topics ranging from relations between science and the state to his impressions of Japan, and it contains very little about his life; see Haber, F. 1927. *Aus Leben und Beruf.* Berlin: Julius Springer. The book published by Haber's second wife is anything but an impartial account, and it provides few insights into Haber's life before he met Charlotte Nathan in March 1917; see Haber, C. 1970. *Mein Leben mit Fritz Haber*. Düsseldorf: Econ Verlag.

15. The new German Reich was the product of success in war: "The imperial state was a thin layer of government superimposed on the states and heavily dependent on Prussia. . . . This state has to be understood in terms of its sudden and violent creation. It was not a product of a steady convergence between power and culture, state institutions and national sentiments." Breuilly, J. 1977. Revolution to Unification. In M. Fulbrook, ed., *German History since 1800*. London: Arnold, p. 139. Not surprisingly, militarism and monarchism remained the Reich's hallmarks until its demise, as a result of another war, in 1918.

16. Between 1870 and 1913 Germany's bituminous coal output increased more than sevenfold and its lignite extraction grew more than elevenfold; its steel production was nearly 140 times larger in 1913 than in 1870, and at 17.6 Mt it greatly surpassed the combined British and French total of 12.5 Mt. Its net national product (expressed in constant prices) grew 3.1 times in forty-three years; see Mitchell, B. R. 1981. *European Historical Statistics*. New York: Facts on File, pp. 383, 420–422, 817–821.

Between 1871 and 1910 the Second Reich's population increased by 58% (from 40.8 to 64.6 million), and *Landflucht* and industrialization created large cities (Berlin's total rose from 412,000 people in 1850 to 2.07 million). Urbanization and industrialization also led to better education, higher literacy, expansion of civil society marked by proliferation of organizations, associations, and the growing influence of socialist ideas. Socialists received more votes in the 1890 Reichstag elections than any other party, and by 1912 over ⅓ of the electorate chose the SPD. For details see Fullbrook (15).

17. Chemistry had a particularly important role to play in resolving a key dilemma of German modernization: how could a country with such limited resources (coal aside) and with a relatively late start of industrialization match its rivals, who began modernizing earlier and had easier access to mineral resources? Novel syntheses—inorganic and organic—invented by German chemists and based on domestically available resources were critical for closing the gap; see Johnson, J. A. 1990. *The Kaiser's Chemists: Science and Modernization in Imperial Germany*. Chapel Hill: University of North Carolina Press.

18. As a German patriot, and as a chemist who counted among his university teachers several men who laid the foundations of modern organic synthesis, Haber was very proud of this tradition: "We think with wonder about the achievements of those great men who during the past 50 years have elevated organic chemistry in our country to such a level that the world thinks of it almost as a German science." Haber, F. 1921. Das Zeitalter der Chemie, seine Aufgaben und Leistungen. In Haber (1924, 1), p. 50.

The excellence of German chemistry in general, not just of its organic branch, is objectively attested by its share of Nobel Prizes during the first two decades of the award's existence. Of the seventeen prizes awarded between 1901 and 1921 nine were to German chemists: Hermann Emil Fischer (1902), Johann Friedrich Wilhelm von Baeyer (1905), Edward Buchner (1907), Wilhelm Ostwald (1909), Otto Wallach (1910), Alfred Werner (1913), Richard Willstätter (1915), Fritz Haber (1918), and Walther Nernst (1920).

19. In arguing for close interdependence of theory and application Haber was not a typical academic of his time. Goran (14), p. 26, quotes him as saying: "It is not enough to seek and to know; we must also apply."

20. Hoffmann's wide-ranging interests included nitrogen chemistry (above all amines) and the chemistry of dyes. Bunsen's fame, besides inventing the eponymous burner with its nonluminous flame, rested on the study of gases and galvanic batteries and on pioneering spectroscopic research.

21. But the procedure used by Liebermann and Carl Graebe, based on Baeyer's demonstration that complex organic molecules could be broken into simpler compounds by heating them with zinc, was commercially impractical. The BASF synthesized the dye by using the method invented by Heinrich Caro (1834–1910), the company's leading researcher; his method was nearly identical to the process proposed by William Perkin (1838–1907). German and British patents were filed on June 25 and 26, 1869, less than twenty-four hours apart, not quite as close a contest as the patenting of telephone by Alexander Graham Bell and Elisha Gray, which was done, to Gray's great loss, just a few hours apart.

   Alizarin was the first natural dye (derived from the root of madder plant, *Rubia tinctorum*) that was prepared synthetically. Synthetic alizarin is still widely used to dye wool as well as in staining microscopic specimens. On the history of synthetic dyes see Brock, W. H. 1992. *The Norton History of Chemistry*. New York: W. W. Norton, pp. 293–310.

22. Piperonal—1,3-benzodioxole-5-carboxaldehyde—is used in a number of organic syntheses, particularly in perfumery, and in cherry and vanilla flavors.

23. Only a small number of graduates, about 5%, found permanent places in teaching and research at German universities; it was in Jena when Haber's interests shifted from organic to physical chemistry.

24. Speculations about Haber's motives for the conversion range from asserting his independence (although he was not reared in Judaism) to undergoing a spiritual awakening. Szöllösi-Janze (14), p. 59, offers Haber's own explanation, as told to a friend more than thirty years later: Haber claimed that the founding of the German Reich made his generation feel 100% German, and the era's rational philosophy separated them from the Jewish faith. Goran (14, p. 38), quoting Heinrich Heine, who called such conversions "obtaining a passport into European culture," added that "the passport was . . . looked upon askance, classified separately, and tolerated rather than appreciated."

25. Many excellent scientists held appointments at technical universities, and those schools have produced a large number of eminent physicists, chemists, and engineers. Nevertheless, in terms of scientific prestige, top universities ranked first.

26. Goran's book (14) contains a complete bibliography of Haber's scientific publications.

27. The habilitation thesis was published as Haber, F. 1896. *Experimental-Untersuchungen über Zersetzung und Verbrennung von Kohlenwasserstoffen*. Munich: R. Oldenbourg.

28. Haber, F. 1898. *Grundrisse der technischen Elektrochemie auf theoretischer Grundlage*. Munich: R. Oldenbourg.

29. Haber, F. 1905. *Thermodynamik der technischer Gasreaktionen*. Munich: R. Oldenbourg. Published in English (with additions by Haber) as Haber, F. 1908. *Thermodynamics of Technical Gas-Reactions*. London: Longmans, Green and Co.

30. Haber was most impressed by America's "self-esteem and enterprise. No nation has these virtues in a higher degree than the people of the United States." Goran, M. 1947. The present-day significance of Fritz Haber. *American Scientist* 35:401.

31. Goran (14, p. 28) claims that they met first at a dancing school in Breslau when she was fifteen and he eighteen, and that he proposed almost immediately, but their parents intervened.

32. Her doctoral thesis on solubilities of heavy metals was published in Breslau; see Immerwahr, C. 1900. *Beiträge zur Löslichkeitsbestimmung schwerlöslicher Salze des Quecksilbers, Kupfers, Bleis, Cadmiums und Zinks.* Breslau: Dr. R. Galle's Buchdruckerei.

33. Szöllösi-Janze (14), pp. 124–131, gathered many facts about the early years of the marriage. Goran (14, p. 31) concluded that "as the housewife role became her sole one, she became less spirited and spontaneous. . . . Where once similar backgrounds and similar interests had fostered deep attachment, a lethargic tolerance developed, and the marriage slowly began to disintegrate."

A similar case, which has only recently become better known, comes immediately to mind: the disintegration of the marriage between Albert Einstein and Mileva Maric. She was Einstein's fellow physics student (the only woman) at the Eidgenösische Technische Hochschule in Zurich, and their illegitimate daughter was given away for adoption (they finally married in 1903). Although no solid evidence supports such a claim, Mileva may have collaborated on some of Einstein key ideas before she transformed herself into a housewife. A great love affair ended in a bitter separation. See Renn, J., and Schulmann, R., eds. 1992. *Albert Einstein, Mileva Maric: The Love Letters.* Princeton: Princeton University Press.

34. We do not know why the brothers chose Haber; Szöllösi-Janze (14), p. 166, speculates that they may had been his father's business associates.

35. Haber (1920, 1), p. 326.

36. Ibid., p. 334.

37. Haber, F., and G. van Oordt. 1905. Über die Bildung von Ammoniaks aus den Elementen. *Zeitschrift für Anorganische Chemie* 43:111–115; 44:341–378.

38. Such low yields were far too small to be seen as a possible foundation of any commercial process.

39. Appl, M. 1986. *Ammonia Synthesis and the Development of Catalytic and High-Pressure Processes in the Chemical Industry.* H. L. Roy Memorial Lecture, Hyderabad, December 19, 1986, p. 3. Haber realized that, unlike in the case of SO_2 oxidation invented by Knietsch (see note 52) where a nearly 100% conversion of the reactants is possible in a single pass, the inherently low yield of ammonia synthesis will require repeated passes over the catalyst.

40. Haber (1920, 1), pp. 335–336.

41. E. N. Hiebert. 1978. Nernst, Hermann Walther. In Gillispie (2), vol. XV, pp. 432–453.

42. The most common definition of the third law is that a system in equilibrium at the temperature of absolute zero is in a state of perfect order and has zero entropy.

43. Nernst, W., and F. Jost. 1907. Über das Ammoniakgleichgewicht. *Zeitschrift für Elektrochemie* 13:521–524.

44. Haber, F., and R. Le Rossignol. 1907. Über das Ammoniakgleichgewicht. *Berichte der Deutschen Chemischen Gesellschaft* 40:2,144–2,154.

45. For the original text of Nernst's verdict see Mitasch, A. 1951. *Geschichte der Ammoniaksynthese.* Weinhem: Verlag Chemie, p. 68.

46. Clara Haber's letter to Richard Abegg written on July 23, 1907 is cited in Stoltzenberg (14), p. 156. Haber and Nernst were never reconciled. According to James Franck, American-

German physicist and the 1926 Nobelian, "Haber believed that he was very close to finding out the 3rd thermodynamic law and he could not forget that Nernst snatched the discovery from him. And Nernst would tell everybody that nitrogen fixation was his discovery." Quoted in Szöllösi-Janze (14), p. 169.

47. Haber, F., and R. Le Rossignol. 1907. Zur Lage des Ammoniakgleichgewichts. *Zeitschrift für Elektrochemie* 14:513–514. The following table shows percentages of NH_3 at equilibrium at different temperatures and pressures (I retain the indication of pressures in atmosphere as in the original paper; 1 atm = 101.3 kPa):

°C	1 atm	30 atm	100 atm	200 atm
200	15.3	67.6	80.6	85.8
300	2.18	31.8	42.1	62.8
500	0.129	3.62	10.4	17.6
600	0.049	1.43	4.47	8.25

48. Haber (1920, 1), p. 337.

49. But no breakthrough was achieved. As Haber noted in his Nobel lecture, the progress of synthesis in the electric arc "is limited by the fact that with a consumption of one kilowatt-hour no more than 16 grams of nitrogen are converted into nitric acid, whilst a complete conversion of electrical to chemical energy ought to yield 30 times as much. An explanation of this has been given by Muthmann and Hofer, who have demonstrated that the high-tension arc used in this process acts as a Deville's heat evaporation chamber" (Haber, 1920, 1, p. 332).

50. Haber, F., and A. König. 1907. Über die Stickoxydbildung im Hochspannungsbogen. *Zeitschrift für Elektrochemie* 13:725–743.

51. For short reviews of BASF history see BASF. 1960. *BASF Makes History.* Ludwigshafen: BASF; Wolf, G. 1970. *Die BASF.* Ludwigshafen: BASF; BASF. 1994. *BASF: Stationen ihrer Geschichte.* Ludwigshafen: BASF. The BASF catalog prepared for the Paris World Exposition of 1900 boasted "The BASF is undoubtedly the largest chemical plant in the world." By that time the company had a wide-ranging research program staffed by nearly 150 chemists and seventy-five engineers.

52. To accomplish the catalytic conversion of SO_2 to SO_3 (using platinized asbestos), the critical step in production of concentrated H_2SO_4, Knietsch invented a tube-bundle reactor in which the feed for the fixed bed inside the tubes was preheated by the countercurrent flow of hot reactants outside the tubes. Haber must have had this model in mind when he designed his preheating arrangement.

H_2SO_4, used in a wide variety of chemical syntheses, has been for decades the most important product of chemical industry by weight (more than 130 Mt a year have been produced during the 1990s). Chlorine—used above all in vinyl chloride monomer synthesis, for bleaching kraft pulp, ethylene dichloride production, and water treatment—ranks among the world's ten most important chemical products by mass.

53. Dark blue indigo, perhaps the oldest known natural cloth dye, was derived from a glucoside in several species of the *Indigofera* plant growing in India, Indonesia, and Central America. Over the years BASF sank an equivalent of the company's total capital stock in the

indigo gamble, and got a rich payoff: just three years after its introduction synthetic indigo accounted for more than 20% of the company's total earnings. Indanthren is highly wash- and light-resistant, but chlorine bleaches it easily. The global diffusion of American blue jeans since the 1960s has supported a strong demand for synthetic blue dyes.

54. Ost, H. 1919. *Lehrbuch der Chemischen Technologie.* Leipzig: Dr. Max Jänecke, p. 168; Nagel, A. von, et al. 1991. *Stickstoff.* Ludwigshafen: BASF, pp. 9–11. Schönherr's design produced somewhat higher concentrations of NO than the Birkeland-Eyde process did, thus allowing for smaller reaction and absorption units.

55. Between 1909 and 1911 larger Schönherr furnaces (700 kW) were installed in the Norsk Hydro's Notodden plant equipped originally only with Birkeland-Eyde furnaces. In 1910 the BASF withdrew from the Alz project, and in September 1911, when the design work for the first synthetic ammonia plant was well advanced, it also canceled its agreement with Norsk Hydro. The second Rjukan plant had only Birkeland-Eyde furnaces, which were cheaper to operate and easier to maintain. They worked until 1929, when the plant switched to electrolyt- ically produced hydrogen and Haber-Bosch ammonia synthesis; see Brunborg, I., and P. B. Holmesland. 1985. The history of nitric acid. In C. Keleti, ed., *Nitric Acid and Fertilizer Nitrates.* New York: Marcel Dekker, pp. 11–12.

56. Cited in Stoltzenberg (14), p. 148.

57. Ibid.

58. Le Rossignol claimed that when the basic recirculation apparatus was designed and os- mium was found to be a suitable catalyst, Geheimrat Bernthsen told Haber personally that BASF was not really interested in his experiments. Things changed only when Engler informed General Director von Brunck about Haber's progress. After von Brunck's visit to Karslruhe, the company became more supportive; see Le Rossignol, R. 1928. Zur Geschichte der Her- stellung des synthetischen Ammoniaks. *Die Naturwissenschaften* 50:1,071.

59. Haber (1920, 1), p. 332.

60. Perhaps the best summary of this work is Haber, F. 1910. Gewinnung von Salpetersäure aus Luft. *Zeitschrift für angewandte Chemie* 23:684–689.

61. Haber (1920, 1), p. 333.

62. Krassa, P. 1966. Zur Geschichte der Ammoniaksynthese. *Chemiker-Zeitung* 90:105. Le Rossignol's valve design was eventually patented, and he was promoted to Haber's private assistant at an increased salary.

63. As already noted, no catalysts were (and still are not) available to perform the reaction under otherwise much more desirable low temperatures.

64. For details see chapter 6.

65. BASF. 1908. *Patentschrift Nr 235421: Verfahren zur synthetischen Darstllung von Am- moniak aus den Elementen.* Kaiserliches Patentamt, Berlin (June 8, 1911).

66. Krassa (62), p. 105.

67. Welsbach's incandescent gas mantle patented in 1885 enabled the gas industry to compete for a few more decades with electric lights.

68. Experiments with different catalysts resulted in ammonia yields of less than 1% with CeO_2 (0.82%), 1.6–3.2% with Mn, but up to 9% with Os; see Haber, F., and R. Le Rossig-

nol. 1909. Über die technische Darstellung von Ammoniak aus den Elementen. *BASF Unternehmensarchiv* Ammoniak Geschichte I G-6101/5.

69. The letter is cited in Stoltzenberg (14), pp. 159–160.

70. Holdermann (9), p. 69; Nagel et al. (54), p. 21.

71. Josephson, M. 1959. *Edison: A Biography.* New York: McGraw-Hill; Wright, O. 1953. *How We Invented the Airplane.* New York: David McKay.

72. Haber, F. 1909. Letter to the Directors of the BASF, July 3, 1909. *BASF Unternehmensarchiv,* Fritz Haber, Allgemeine Correspondenz II, no. 92.

73. Le Rossignol (58), p. 1,071.

74. "There were difficulties with tightness and heating. We needed the pressure of 200 atm, and when it was achieved, as no ammonia was produced at 400 °C, we had to slowly raise the temperature. On the decisive day ammonia was there at 460 °C. Dr. Mittasch immediately called Dr. Bosch. He came by the next express train from Ludwigshafen and at the university entrance he called to us: 'Bravo, now we continue for another two hours, then we switch it off and proceed with it in Ludwigshafen.'" BASF Pressestelle. 1959. *50 Jahre synthetisches Ammoniak,* p. 1.

75. Haber, F. 1909. *Patentschrift Nr 238450: Verfahren zur Darstellung von Ammoniak aus den Elementen durch Katalyses unter Druck be erhöhter Temperatur.* Kaiserliche Patentamt, Berlin (September 28, 1911).

Chapter 5

1. At that time the highest routinely used pressure in industrial production was in coal-fired electricity-generating plants that produced steam at between 0.9 and 1.24 MPa; see Bannister, R. L., and G. J. Silvestri, Jr. 1989. The evolution of central station steam turbines. *Power* 111(2):70–78.

2. Bosch acknowledged the unprecedented extent of the team quest in closing sentences of his Nobel lecture: "It scarcely needs to be added that this achievement has only been made possible by a large staff of colleagues. It is probably true to assert that such numbers have never before been engaged on one single problem." He singled out Alwin Mittasch, head of the scientific laboratory, and Franz Lappe, who was in charge of technical development; see Bosch, C. 1966 (1931). The development of the chemical high pressure method during the establishment of the new ammonia industry. In *Nobel Lectures: Chemistry 1922–1941.* Amsterdam: Elsevier, p. 235.

3. Nagel, A. von, et al. 1991. *Stickstoff.* Ludwigshafen: BASF, p. 56. This represents a 2,000-fold performance jump in four years, an impressive figure even for an initial stage of commercialization of a new synthesis process in which large capacity increases are common.

4. Major high-pressure processes now include production of methanol and higher alcohols, synthesis of polyethylene, and hydrocracking of crude oils; see Furter, W. F., ed. 1980. *History of Chemical Engineering.* Washington, D.C.: American Chemical Society, pp. 255–257. Hydrocracking—the addition of hydrogen to massive hydrocarbon molecules and their subsequent catalytic cracking under pressure of 6.9–17.2 MPa to yield lighter liquid fuels—now produces most of the world's gasolines and aviation jet fuels at a rate surpassing 800 Mt and

2 Mt a year, respectively; see United Nations. 1997. *Energy Statistics Yearbook*. New York: United Nations, pp. 203, 214.

5. Somorjai, G. A. 1994. *Introduction to Surface Chemistry and Catalysis*. New York: Wiley; Moulijn, J. A., et al., eds. 1993. *Catalysis An Integrated Approach to Homogeneous, Heterogeneous, and Industrial Catalysis*. Amsterdam: Elsevier.

6. Palmaer, W. 1966 (1931). Chemistry 1931. In *Nobel Lectures: Chemistry 1922–1941*. Amsterdam: Elsevier, p. 189.

7. Ibid. Friedrich Bergius (1884–1949) solved the challenge of liquefying coal by suspending pulverized coal in oil and treating this mixture with hydrogen under pressure; suspension provided efficient means of heat distribution, preventing localized overheating; hydrogenation made it possible to convert a larger share of coal's carbon to oil than would conventional distillation; see Bergius, F. 1966 (1932). Chemical reactions under high pressure. In *Nobel Lectures: Chemistry 1922–1941*. Amsterdam: Elsevier, pp. 244–276. The first plant using catalytic hydrogenation of coal to produce gasoline was completed at Leuna near Halle, the site of at that time the world's largest ammonia synthesis plant, on April 1, 1927 (more details on this plant follow later in this chapter).

8. In contrast to the relatively rich biographical literature on Fritz Haber—including two exceptionally detailed works (see chapter 4, note 14)—there is only a single comprehensive book on Carl Bosch; see Holdermann, K. 1954. *Im Banne der Chemie: Carl Bosch—Leben und Werk*. Düsseldorf: Econ-Verlag. Although it is very informative and contains a great deal of interesting detail, it is, in Peter Hayes's accurate description, a "rather worshipful account." The book also lacks any annotations, and it has only a rudimentary bibliography (but it contains brief lists of Bosch's major publications, patents, and honors).

9. Bosch's doctoral thesis was entitled "Über die Kondensation von Dinatrium-Acetondicarbonsäurediäthylester mit Bromacetophenon."

10. For details see chapter 4.

11. Holdermann (8), p. 46. This sudden declaration in the midst of a walk should not be surprising; Bosch belonged to those scientists who always think about their work. He once confided that "My thinking comes as a continuous process that does not stop even in my sleep" (Holdermann, p. 7).

12. Ibid., pp. 47–50, 62–63.

13. Bosch's colleagues had repeatedly recalled his consuming dedication. Julius Kranz remembered an occasion when the two of them monitored an experiment for seventy-six hours, taking short turns to sleep on the same cot; see Holdermann (8), p. 78–79. Haber had no doubt where the credit was due: "Dr. Bosch has made a large-scale industry of ammonia synthesis"; see Haber, F. 1966 (1920). The synthesis of ammonia from its elements. In *Nobel Lectures: Chemistry 1901–1921*. Amsterdam: Elsevier, p. 338. The director of the American Cyanamid Company, W. S. Landis, said in 1915: "Too much honor cannot be shown the courageous chemists who have succeeded in placing this process on a commercial working basis. The difficulties seem almost unsurmountable" (Holdermann, 8, p. 141).

14. Szöllösi-Janze, M. 1998. *Fritz Haber, 1868–1934: Eine Biographie*. Munich: Verlag C. H. Beck, p. 181. Duisberg, himself a chemist, was one of Germany's leading industrialists. General Director of Farbenwerke Bayer since 1912, he was the prime mover behind the post–World War I consolidation of German chemical companies into the I. G. Farben consortium.

15. Holdermann (8), pp. 197–198.

16. Ibid., p. 200. Bosch added: "If we had charged them with osmium instead of with our new catalyst, the world's supply of this rare metal which we had already bought up would have disappeared."

17. Copper's melting point is relatively high (1,083 °C), but its creep, the continuous deformation when subject to constant stresses, already becomes pronounced at 120 °C, and the metal's tensile strength is no more than 170 MPa; see Bolton, W. 1989. *Engineering Materials Pocket Book*. Boca Raton, Fla.: CRC Press, pp. 98, 100.

18. Today's carbon steels contain 0.08–1.03% C (and usually 0.25–0.9 Mn), and their tensile strength at 650 °C is between 585 and 890 MPa; see Bolton (17), pp. 32–33, 47–48.

19. Holdermann (8), pp. 200, 202–207.

20. Ibid., p. 208

21. Johannes Fahrenhorst—cited in Nagel (3), p. 29—described in his memoirs the moment of Bosch's discovery on a Saturday morning in early February 1911. Bosch, after listening for a while to yet another futile discussion about the failing tubes, lost his patience, put on his hat, and left the room. Suddenly he was back, cracked the door open, and called: "Get a soft iron pipe." Holdermann (8), p. 95, reported that Bosch told his wife on the day of the discovery: "Today I have made either a really great breakthrough or a monumental mistake."

22. Holdermann (8), pp. 208–209.

23. Ibid., p. 209.

24. Ibid., p. 212.

25. A small amount of air was introduced into the converter through a nozzle, and some of the synthesis gas was ignited with an electrical heating element and burned: "A certain amount of water was formed but did not appear to harm the catalyst. At any rate this was the lesser evil" (Holdermann [8], p. 212). Burners designed by Lappe worked so well that the company never had any explosion associated with this heating method.

26. Eventually the cold synthesis gas was used instead of pure nitrogen. In any case, this obviated the need for making pinholes in the outer steel jacket.

27. "Although at the beginning of our tests we had counted on compensating for the heat losses by heating the convertor, even with this arrangement the utmost economies had to be made since the amount of gas passing through are enormous and the heat has invariably to be supplied to the hottest point before the catalyst" (Holdermann [8], p. 211).

28. "This represented quite a considerable advance for then we were able to dispense entirely with continuous heating during the operation" (Ibid., p. 221).

29. Ibid., pp. 221–222. Electric motors available at that time were not powerful enough to run the large compressors that were powered by coal gas produced from lignite; Oppau compressors were converted to electricity only in 1936–1937; see Nagel (3), pp. 45–46.

30. Holdermann (8), pp. 224, 227–228.

31. Ibid., pp. 222–227. Extensive and instantaneous process control had eventually received an enormous boost from the introduction of computers and inexpensive probes, helping to make the continuous chemical syntheses one of the least labor-intensive industrial undertakings.

32. BASF. 1910. Bericht über die Arbeiten de BASF auf dem Gebiete der synthetischen Darstellung von Ammoniak aus der Elementen. *BASF Unternehmensarchiv* Ammoniak Geschichte I G-6101.

33. By far the most comprehensive account of the earliest catalysis experiments for ammonia synthesis is Mittasch, A. 1951. *Geschichte der Ammoniaksynthese*. Weinheim: Verlag Chemie, pp. 93–121. Although the book was published only in 1951, the original 400-page manuscript was written by Mittasch in 1920.

34. Ibid., pp. 93–95.

35. Only during the 1980s did we get the first direct evidence of iron's catalytic action at the molecular level by using a newly developed technique of high-resolution electron energy loss. The metal lowers the energy barrier that prevents the dissociation of two nitrogen atoms in N_2 by donating its electrons to the nitrogen molecule. As iron atoms form a chemical bond with N_2, the bond between the two N atoms is reciprocally weakened; see Friend, C. M. 1993. Catalysis on surfaces. *Scientific American* 268(4):74–79.

36. Mittasch (33), pp. 97–99. The patent application listed seven examples of mixed catalysts based on iron and cobalt oxides.

37. Ibid., pp. 100–114; Gmelins Handbuch der Anorganischen Chemie. 1936. Mehrstoffkatalysatoren. In *Stickstoff*. Berlin: Verlag Chemie, pp. 335–344.

38. Other elements to be avoided are oxygen, fluorine, arsenic, selenium, bromine, iodine, tin, and lead.

39. CaO, MgO, and K_2O have been widely used as promoters in modern iron-based catalysts used for ammonia synthesis.

40. Appl, M. 1986. *Catalytic and High-Pressure Processes in the Chemical Industry*. H. L. Roy Memorial Lecture, Hyderabad, December 19, 1986, p. 5.

41. Doubly promoted iron catalysts help to produce 13–14% of the ammonia in the exit gas, compared to 8–9% in singly promoted catalysts and to just 3–5% for pure iron under identical conditions.

42. Cited in Holdermann (8), pp. 87–88; also in Stoltzenberg, D. 1994. *Fritz Haber: Chemiker, Nobelpreisträger, Deutscher, Jude*. Weinheim: VCH, p. 165.

43. Appl (40), p. 6; UNIDO and IFDC. 1998. *Fertilizer Manual*. Dordrecht: Kluwer Academic, p. 179.

44. Mittasch (33), p. 116.

45. It remains so today. The next chapter will note the modern means of lowering this cost.

46. Electrolysis of aqueous solutions of NaCl (brine) had been in commercial use since the 1880s. The overall reaction is

$2 \, NaCl + 2 \, H_2O \rightarrow 2 \, NaOH + Cl_2 + H_2$.

47. For more details on the process see Gmelins Handbuch der Anorganischen Chemie. 1936. Wasserstoffkontaktverfahren. In *Stickstoff*. Berlin: Verlag Chemie, pp. 372–377.

48. Both gases have a very low energy density (coke-oven gas just 3.3–5.0 MJ/m^3, water gas 10.9–11.7 MJ/m^3, compared to 29–39 MJ/m^3 for natural gas) and hence make better feedstocks than fuels.

49. Stoltzenberg (42), pp. 183–184; Nagel (3), p. 36. The high cost and complex chemistry of cuprous ammonia and formate (or acetate) solutions led to their replacement by nitrogen scrubbing and catalytic methanation; see the next chapter.

50. This is done because the moisture could lower the efficiency of catalysts.

51. The reaction is now done in continuous saturator-crystallizer units under a vacuum, or at atmospheric pressure; see UNIDO and IFDC (43), pp. 244–245.

52. Cited in I. G. Farbenindustrie Patent Abteilung. 1939. Geschichte der Ammoniaksynthese und der Ammoniakoxydation. *BASF Unternehmensarchiv* Ammoniak Geschichte II G-6101, p. 4.

53. What is so remarkable about this incident is that in 1839 Kuhlmann saw the importance of his work in the very same light Ostwald did more than sixty years later, under greatly changed political circumstances. Kuhlmann wrote: "One can surely hope that the knowledge of the phenomenon I had ascertained will mean that a country would not have to worry about the difficulty, indeed the impossibility of securing the needed amount of saltpetre during the time of sea war." Cited in I. G. Farbenindustrie Patent Abteilung (52), p. 4.

54. For details on modern catalytic oxidation of ammonia see Andrew, S. P. S. 1985. The oxidation of ammonia. In C. Keleti, ed., *Nitric Acid and Fertilizer Nitrates*. New York: Marcel Dekker, pp. 31–40.

55. Chronology of experimental work according to Voigtländer-Tetzner, W. 1956. *Die Bindung des Luftstickstoffes*. Unpublished typescript, *BASF Unternehmensarchiv*, pp. 78–79, 98–159, 562a, b.

56. Mittasch (33), p. 122; BASF Ammoniaklaboratorium, 1939. Geschichte der Ammoniaksynthese und der Ammoniakoxydation. *BASF Unternehmensarchiv Ammoniak Geschichte I G-6101/11.*

57. Memo by Johannes Fahrenhorst to Carl Bosch on May 4, 1917, reprinted in Voigtländer-Tetzner (55), p. 159.

58. Mittasch (33), p. 136.

59. Details on the Oppau plant are according to Partington, J. R., and L. H. Parker. 1922. *The Nitrogen Industry*. London: Constable & Company, pp. 162–165; Holdermann (8), pp. 115–135; Mittasch (33), pp. 136–140; Voigtländer-Tetzner (55), pp. 78–79, 159; Appl (40), pp. 8–9; Nagel (3), pp. 43–46.

60. Nagel (3), pp. 38–42; Stoltzenberg (42), pp. 174–178; Szöllösi-Janze (14), pp. 183–184. Nernst and Haber came together to the *Reichsgericht* patent claim hearing in Leipzig on March 4, 1912. As Bosch was also present, this was a unique litigation affair: one past Nobel Prize winner (Ostwald, 1909) was trying to wrest the claim to the invention of an enormously important synthesis away from an alliance of three future Nobel Prize laureates.

61. Holdermann (8), p. 119. BASF set up a separate nitrogen division (*Stickstoffabteilung*) in June 1912. Bosch became its head in April 1913. Mittasch's ammonia laboratory became part of it, and Mittasch remained its leader until 1933; see Mittasch (33), p. 7.

62. This achievement has lost little of its impressiveness with time: few other twentieth-century techniques have gone from a laboratory demonstration to a large-scale commercial enterprise in just four years, and few new plants were completed in such a short time.

63. Holdermann (8), pp. 130–133.

64. Ibid., pp. 134–135. As Bosch noted at that time, his demands for more productive ammonia synthesis were pushing the performance boundaries of the German steel industry. The need to forge converters from massive ingots was eliminated with the development of multilayered vessel construction initiated by A. O. Smith in 1931, but, at the same time, metallurgical advances raised the maximum limit of forged converters to well over 100 t.

65. The atmosphere of that time is vividly described by Philipp Scheidemann, the Socialist vice president of the Reichstag, and the man who proclaimed Germany a republic in 1918; see Scheidemann, P. 1929. *The Making of New Germany.* New York: Meredith Press, pp. 201–216.

66. The English text of the manifesto is in Lutz, R. H. 1932. *Fall of the German Empire, 1914–1918.* Stanford: Stanford University Press, pp. 74–75.

67. Paul Ehrlich (1854–1915; Nobel Prize for Physiology in 1908) was one of the founders of modern serology and drug therapy; Emil Fischer (1852–1919; Nobel Prize for Chemistry in 1902) and Richard Willstätter (1872–1942; Nobel Prize for Chemistry in 1915) were among the world's leaders in organic synthesis of natural compounds.

68. Holdermann (8), pp. 135–136. Walther Rathenau (1867–1922) was Germany's leading industrialist, Director of AEG (the German equivalent of General Electric), and in 1914–1915 the organizer of the raw materials division (*Kriegsrohstoffabteilung*) in the German Ministry of War.

69. Perhaps the best available English-language publication dealing with the German genesis of the complex is Johnson, J. A. 1990. *The Kaiser's Chemists: Science and Modernization in Imperial Germany.* Chapel Hill: University of North Carolina Press.

70. The BASF also found direct military uses for its chlorine and phosgene. During the course of the war BASF and Bayer delivered 11,000 t of phosgene and 27,600 t of chlorine, which were used as poisonous gases on both the Western and Eastern Fronts; see Chmiel, G., et al. 1986. *. . . im Frieden der Menschheit, im Kriege dem Vaterlande . . . 75 Jahre Fritz-Haber-Institut der Max-Planck-Gesellschaft.* Berlin: ÖTV, p. 30. Haber's key role in gas warfare will be explained in the postscript.

71. Holdermann (8), pp. 136–137. When Bosch remarked, "When the supplies of Chilean nitrate are gone, we are finished," the ministry's officers replied, "But we have the large potassium deposits at Stassfurt!" That the men responsible for German munitions supply were unaware of the difference between KNO_3 and KCl is incredible.

72. Nagel (3), p. 94.

73. Belgian chemist Ernest Solvay (1838–1922) solved the technical problem of producing Na_2CO_3 from $NaCl$ and NH_3 by reacting the rain of ammoniated salt in a countercurrent of CO_2 in a spray tower. Ammonia was recovered by reaction with lime, but $CaCl_2$ was discarded:

$NaCl + NH_3 + CO_2 + H_2O \rightarrow NH_4Cl + NaHCO_3,$

$2\ NaHCO_3 \rightarrow Na_2CO_3 + CO_2 + H_2O,$

$2\ NH_4Cl + CaO \rightarrow CaCl_2 + 2\ NH_3 + H_2O.$

HNO_3 can be then easily neutralized by Na_2CO_3 to produce $NaNO_3$:

$2\ HNO_3 + Na_2CO_3 + \rightarrow 2\ NaNO_3 + H_2CO_3.$

74. With ammonia becoming largely the feedstock for HNO_3 production, Oppau's shipments

of ammonium sulfate fell by nearly 90% in 1915 and remained much below the 1914 rate for the duration of the war; see Voigtländer-Tetzner (55), p. 283 and Anlage 1.

75. BASF. 1925. *Geschichte der BASF 1865–1925: Ammoniakwerk Merseburg GmbH. BASF Unternehmensarchiv,* A 18/22. Oppau increased its daily capacity to 70 t NH_3 by November 1915 and to 200 t NH_3 by July 1916, and it reached its peak output in December 1917 with 7133 t (230 t/day); see Mittasch (33), p. 138.

76. Bosch, C. 1916. Bericht über die Besprechung mit Herrn Leutnant Schmitz von der Kriegs-Rohstoff-Abteilung in Berlin am 4. Februar. *BASF Unternehmensarchiv: Ammoniakwerke Merseburg* A 18/32a. Bosch asked for an interest-free loan of 30 million marks from the state, specified conditions of loan repayments and profit guarantees, and promised to minimize the use of strategic metals, particularly of copper. He came to know Hermann Schmitz better during their participation in armistice negotiations in Spa and peace treaty talks in Versailles (where Bosch represented German chemical industries). He hired him as the finance director of BASF in 1921.

Schmitz eventually succeeded Bosch as Chairman of the Managing Board (*Vorstandsrat*) in 1935. In 1948 he and nineteen other top I. G. Farben executives were tried in Nuremberg for preparing and waging aggressive war but received only minor sentences (four years for Schmitz) on the plundering count. For detailed accounts of the trial see Dubois, J. E., Jr. 1953. *Generals in Grey Suits.* London: Bodley Head; Borkin, J. 1978. *The Crime and Punishment of I. G. Farben.* New York: Free Press.

77. The history of Leuna is described in BASF. 1925. Geschichte der BASF 1865–1925. Ammoniakwerk Merseburg Gmbh (Leuna Werke). *BASF Unternehmensarchiv* A 18/22; Voigtländer-Tetzner (55), p. 78; Schäfer, H.-G. 1966. Fünfzig Jahre Leunawerke: fünfzig Jahre Entwicklung der Hochdrucktechnik. *Chemiker-Zeitung* 90:427–429. The plant was known first as BASF Ammoniakwerk Merseburg, later as Ammoniak Merseburg GmbH, Leuna Werke.

78. The most obvious advantage of the Merseburg process was the ability to produce ammonium sulfate without any sulfuric acid. High energy intensity, due to the large volume of steam needed to recover solid sulfate from a relatively dilute solution, was the main drawback. Nevertheless, the process has been used in a number of European and Asian countries; see UNIDO and IFDC. 1998. *Fertilizer Manual.* Dordrecht: Kluwer Academic, pp. 246–247. Its steps are

$NH_3 + H_2O \rightarrow NH_4OH,$

$2NH_4OH + CO_2 \rightarrow (NH_4)_2CO_3 + H_2O,$

$CaSO_4 \cdot 2H_2O + (NH_4)_2CO_3 \rightarrow CaCO_3 + (NH_4)_2SO_4 + 2H_2O.$

79. According to Holdermann (8), p. 147, Bosch remarked about Krauch: "That man has guts!" Krauch became the first manager at Leuna, later the manager of I. G. Farben's entire nitrogen subsidiary.

80. Schäfer (77), p. 427.

81. Knickerbrocker, H. R. 1932. *The German Crisis.* New York: Farrar & Rinehart, p. 83.

82. According to Stoltzenberg (42), p. 188, respective annual outputs of by-product ammonia, cyanamide, and synthetic ammonia developed as follows (all figures in thousand t N): 1915/1916:90, 20, 24; 1916/1917:100, 58, 64; 1917/1918:100, 65, 105.

83. Hayes, P. 1987. Carl Bosch and Carl Krauch: chemistry and the political economy of Germany, 1925–1945. *Journal of Economic History* 47:353–363.

84. Beike, H. 1960. Zur Rolle von Fritz Haber und Carl Bosch in Politik und Gesellschaft. *Wissenschaftliche Zeitschrift der Technischen Hochschule für Chemie Leuna-Merseburg* 3:65.

Chapter 6

1. This combination of surprising continuity and impressive change is not uncommon in other fields, particularly as far as the basic energy convertors are concerned. If Gottlieb Daimler and Karl Benz could look at a modern internal combustion engine, if Charles Parsons could examine a large steam turbine, or if Nikola Tesla could take apart a new variable-speed electric motor, they would all be pleased to see how durable their inventions have been—but they would be also greatly surprised by the extent of innovation that has changed their machines.

2. Punctuated equilibria—to borrow a term from paleontological vocabulary—have also characterized the evolution of many other modern techniques. In general, the two interwar decades were a period of stagnation, with rapid growth and innovation resuming during the 1950s. Such diverse techniques as television and tankers could be used to illustrate this evolutionary pause.

3. This post-1989 slump has affected a wide range of industrial and natural products, particularly because of the sharp, and so far sustained, decline of economic performance in the successor states of the former Soviet Union: it either led the global output, or was among the world's top three to five producers, for commodities ranging from ammonia to zinc. In 1988 the country's ammonia output was the world's largest (15.6 Mt N), but in 1996 Russia and Ukraine produced less than 7 Mt N, and the output of all the other successor states was less than 1 Mt N.

4. As it was thought that the mixture of the two compounds cannot detonate, explosives were used to break up badly caked heaps of bulk fertilizer; see Shah, K. D. 1996. *Safety of Ammonium Nitrate Fertilisers*. Peterborough: Fertiliser Society. Figures for ammonium sulfate production are from Voigtländer-Tetzner, W. 1956. *Die Bindung des Luftstickstoffes*. Unpublished typescript, BASF Unternehmensarchiv, vol. I 1895–1918, pp. 30. Bosch put his protégé Carl Krauch in charge of the plant's reconstruction after the explosion.

5. The initial capacity of the urea plant was 40 t/day in the total recycle process (for the explanation see the next chapter); see Nagel, A. von, et al. 1991. *Stickstoff*. Ludwigshafen: BASF, p. 114.

6. The last syllable stands for *Kalium*, potassium in German. The first *Nitrophoska* was made in December 1927; its initial content was 17.5% N, 5.7% P, and 18% K, and the fertilizer was later reformulated several times; see Mittasch, A. 1928. Über Misch- und Volldünger. *Zeitschrift für angewandte Chemie* 41:902–916; Nagel (5), pp. 118–126.

7. Data on German fixation of nitrogen for the years 1914–1940 are from Schmidt, A. 1934. *Die industrielle Chemie*. Berlin: Verlag Walter de Gruyter & Co., pp. 330–332; Balz, O., et al. 1961. *Der Stickstoff*. Düsseldorf: Fachverband Stickstoffindustrie, pp. 432, 433; Timm, B. 1963. 50 Jahre Ammoniak-Synthese. *Chemie-Ingenieur-Technik* 35:823.

8. Voigtländer-Tetzner (4), pp. 358–359.

9. Partington, J. R., and L. H. Parker. 1922. *The Nitrogen Industry*. London: Constable & Company, pp. 175–180. After Brunner, Mond & Company joined United Alkali and a number of other British companies to form Imperial Chemical Industries (ICI) in 1926—a move similar to the great consolidation of the leading German chemical companies into the I. G. Farben in 1925—the Billingham plant became the foundation of the large ICI complex that grew on the site.

10. Jones, R. M., and R. L. Barber. 1947. Ammonia. In R. E. Kirk and D. F. Othmer, eds., *Encyclopedia of Chemical Technology*. New York: Interscience Encyclopedia, p. 784; U.S. Bureau of Commerce. 1975. *Historical Abstract of the United States*. Washington, D.C.: GPO, part 2, pp. 697; Nelson, L. B. 1990. *History of the U.S. Fertilizer Industry*. Muscle Shoals, Ala.: TVA, pp. 204–214.

An interesting report prepared by a member of the U.S. Fixed Nitrogen Commission after its 1919 visit of Germany compared the Oppau and Sheffield plants; see Tour, R. S. 1922. The German and American synthetic ammonia plants. Parts I–V. *Chemical and Metallurgical Engineering* 26(6):245–248; 26(7):307–311; 26(8):359–362; 26(9):411–415; 26(10):463–466.

11. By-product ammonia contained 426,000 t N, and Chilean nitrates provided about 510,000 t N; see Schmidt, A. 1934. *Die industrielle Chemie*. Berlin: Verlag Walter de Gruyter & Co., p. 330. The British naval blockade made it impossible to sell Chilean nitrates to Germany, the biggest European buyer, but the demand of the Allied powers for the explosives kept the industry going. Chilean exports peaked at almost 3 Mt/year between 1916 and 1918.

After the postwar slump the production recovered thanks to a new extraction technique (the Guggenheim method) that made it possible to produce the mineral from low-grade *caliche*. Exports surpassed 2.8 Mt a year in 1928 and 1929, but afterward a combination of the worldwide economic crisis of the early 1930s and steady growth of the synthetic fertilizer industry further marginalized the contribution of Chilean nitrates to global nitrogen supply. By 1932 exports were below 250,000 t, lower than a half century earlier; see Collier, S., and W. F. Sater. 1996. *A History of Chile, 1808–1994*. Cambridge: Cambridge University Press, pp. 161–166.

12. Natural gas is usually a mixture of several light alkanes—the general formula being $C_nH_{(2n+2)}$—dominated by methane with much lower amounts of ethane (C_2H_6) and propane (C_3H_8), and with trace quantities of CO_2, N_2, H_2S, and He. The presence of heavier compounds is actually an advantage for the synthesis of urea, which requires a $CO_2:NH_3$ feedstock ratio of 1:2, while the reforming of pure CH_4 produces too little CO_2 (the ratio of the two compounds is 7:16).

13. Appl, M. 1976. A brief history of ammonia production from the early days to the present. *Nitrogen* 100:51; Nagel et al. (5), p. 52.

14. Nagel et al. (5), p. 53.

15. Nelson (10), pp. 229, 329–330.

16. As the energy needed for gas compression goes up only as the logarithm of the pressure, it takes only about 1.3 times as much effort to compress the gas to 100 MPa than to just 20 MPa, but NH_3 concentration in the exit gas is roughly tripled. Because of its high conversion rates—up to 40% in one pass through the converter, and as much as 85% after passage

through the series of converters—the Claude process did not recirculate the gas; still, this meant that some 20% of unconverted synthesis gas was wasted. Other disadvantages of the process are a shorter life of converters and higher maintenance costs. For details on Claude's process see Gmelins Handbuch der Anorganischen Chemie. 1936. *Stickstoff.* Berlin: Verlag Chemie, pp. 398–401; Jones and Barber (10), pp. 798–799.

17. Gmelin (16), pp. 401–404.

18. Ibid., pp. 406–407; Jones and Barber (10), pp. 798–800. Italy's largest Fauser plant, with the capacity of 60 t NH3/day, was located in sub-Alpine Merano.

19. Gmelin (16), pp. 404–406; Jones and Barber (10), pp. 801–802; Nelson (15), pp. 204–229; Krupp Uhde. 1998. *Ammonia Technology.* Dortmund: Krupp Uhde.

20. Jones and Barber (10), p. 784. By 1962, 175 out of the world's 260 ammonia plants (67%) were based on the Haber-Bosch design and accounted for 70% of the global production capacity; see Frear, G. L., and R. L. Barber. 1963. Ammonia. In *Encyclopedia of Chemical Technology.* 2nd edition. New York: John Wiley, vol. 1, p. 268.

21. Voigtländer-Tetzner (4), Ergänzung (addition), p. 30.

22. For data sources see note 7.

23. USBC (10), pp. 697–698.

24. Jones and Barber (10), pp. 784–785; Nelson (10), pp. 325–327. As a result the U.S. government owned about 70% of the country's ammonia synthesis capacity; the largest plant (180,000 t N/year) was Du Pont's Morgantown Ordnance Works in West Virginia.

25. Food rationing in many European countries ended only during the early 1950s. Hybrid corn was introduced in the United States in 1934, and it was generally adopted during the 1950s. By the early 1960s, 100% of the country's corn fields were planted with hybrid seed.

26. USBC (10), pp. 697–698; Frear and Barber (20), pp. 266–267.

27. Nelson (10), pp. 331–332.

28. FAO. 1997. *Current World Fertilizer Situation and Outlook 1994/95–2000/2001.* Rome: FAO, p. 23; IFDC. 1998. *Worldwide Ammonia Capacity Listing by Plant.* Muscle Shoals, Ala.: IFDC.

29. Nichols, D. E., et al. 1980. Assessment of alternatives to present-day ammonia technology with emphasis on coal gasification. In W. E. Newton and W. H. Orme-Johnson, eds., *Nitrogen Fixation.* Baltimore: University Park Press, vol. I, pp. 43–60.

30. Fan, X., et al. 1998. Ammonium bicarbonate: China's bedrock. *Nitrogen & Methanol* 236:21–23.

31. Czuppon, T. A., et al. 1992. Ammonia. In M. Howe-Grant, ed., *Kirk-Othmer Encyclopedia of Chemical Technology.* 4th edition. New York: John Wiley, vol. 2, pp. 653–655.

32. While today's best natural gas-based syntheses need less than 30 GJ/t NH3, energy consumption of inherently less efficient plants based on partial oxidation of heavier hydrocarbons, or on coal, is as follows: in naphtha-based plants about 32 GJ/t, in fuel oil–based operations about 36 GJ/t, and in large coal-based synthesis 41 GJ/t; see UNIDO. 1998. *Fertilizer Manual.* Dordrecht: Kluwer Academic, p. 163.

33. Czuppon et al. (31), p. 656.

34. The Groningen field was discovered in 1959, and it began producing in 1965; the first British North Sea gas fields were discovered during the mid- to late 1960s, the first Norwegian fields in the early and mid-1970s, and the enormous Troll field in 1980. Russian gas comes from West Siberia's huge fields of Medvezhye and Urengoi.

35. For the most detailed technical descriptions of ammonia synthesis based on natural gas; see Honti, G. D., ed. 1976. *The Nitrogen Industry.* Budapest: Akadémiai Kiado; Streltzoff, S. 1981. *Technology and Manufacture of Ammonia.* New York: John Wiley; Jennings, J. R. 1991. *Catalytic Ammonia Synthesis Fundamentals and Practice.* New York: Plenum Press; Nielsen, A., ed., 1995. *Ammonia Catalysis and Manufacture.* New York: Springer Verlag. An excellent review of technical advances between the mid-1970s and the early 1990s is Appl, M. 1992–1993. Modern ammonia technology: Where have we got to, where are we going? *Nitrogen* 199 (September-October):46–75, 200 (November-December):27–38, 202 (March-April):44–53. For a good shorter account available on the Internet see European Fertilizer Manufacturers' Association. 1995. *Production of Ammonia.* Brussels: EFMA (http://www.efma.org).

36. The reaction

$$ZnO + H_2S \rightarrow ZnS + H_2O$$

removes all H_2S as well as the sulfur in mercaptans, but not any organic sulfur that might be present. Its removal requires catalytic (Co, Mo) hydrogenation producing H_2S. Activated carbon bed with added CuO has been also used for H_2S removal.

37. A revealing review of innovations in the ammonia industry can be made by comparing the four editions of the *Kirk-Othmer Encyclopedia of Chemical Technology* published between 1947 and 1992; see Jones, R. M., and R. L. Barber. 1947. Ammonia. In R. E. Kirk and D. F. Othmer, eds., *Encyclopedia of Chemical Technology.* New York: Interscience Encyclopedia, vol. 1, pp. 771–810; Frear, G. L., and R. L. Barber. 1963. Ammonia. In *Encyclopedia of Chemical Technology.* 2nd edition, vol. 1, pp. 258–298; Le Blanc, J. R., et al. 1978. Ammonia. In M. Grayson, ed., *Kirk-Othmer Encyclopedia of Chemical Technology.* 3rd edition. New York: John Wiley, vol. 2, pp. 470–511; Czuppon et al. (31), pp. 639–691.

38. These developments can be traced chronologically in the sources cited in the preceding note. They are also reviewed in Anonymous. 1992. Reforming the front end. *Nitrogen* 195: 22–36.

39. Integrated ammonia-urea plants use the recovered CO_2 in the subsequent synthesis of urea; other CO_2 uses include carbonating soft beverages and pumping the gas into oil wells to enhance crude oil recovery.

40. UNIDO (32), p. 173.

41. Bloch, H. P., and J. J. Hoefner. 1996. *Reciprocating Compressors: Operating and Maintenance.* Houston: Gulf Publishing.

42. Paul, D. A., et al. 1977. *The Changing U.S. Fertilizer Industry.* Washington, D.C.: USDA, p. 18.

43. In 1987 M. W. Kellogg was acquired by Dallas-based Dresser Industries, which specialize in the design, engineering, and construction of LNG plants. In 1998 Dresser joined with Halliburton, another Texas-based company, and as a result M. W. Kellogg was combined with Brown & Root Engineering and Construction, which was owned by Halliburton. Brown &

Root was a leading provider of a full range of services—ranging from design to process techniques (including the Braun Purifier Process™ for producing high-purity anhydrous ammonia), project management, and maintenance—to the nitrogen fertilizer industry. The new company—Kellogg Brown & Root (KBR)—thus has an even stronger position in the field of ammonia synthesis.

44. Centrifugal compressors take in gas through a central inlet, increase its kinetic energy with a high-speed impeller, and then convert its velocity into static pressure by diffusing it in a divergent outlet. They are particularly suitable for compressing large volumes of gas to moderate pressure; see Japikse, D. 1996. *Centrifugal Compressor Design and Performance.* White River Junction, Vt.: Concepts ETI.

45. The efficiency of the Rankine cycle is further enhanced by using high-pressure superheated steam (over 10 MPa at 510 °C).

46. Slack, A. V. 1977. Commercial ammonia processes. In A. V. Slack and G. R. James, eds., *Ammonia.* New York: Marcel Dekker, part III, pp. 291–369.

47. Czuppon et al. (31), pp. 659–664. KBR's largest (1850 t/day) units using ruthenium catalyst operate at just under 9 MPa; see Strait, R. 1999. Grassroots success with KAAP. *Nitrogen & Methanol* 238:37–43.

48. Nelson (10), p. 334. Because the size of the last wheel of the compressor determines the minimum flow, centrifugal machines must handle at least 400 m^3 of synthesis gas (with density corresponding to 3:1 hydrogen: nitrogen ratio) per hour; see Appl, M. 1986. *Catalytic and High-Pressure Processes in the Chemical Industry.* H. L. Roy Memorial Lecture, Hyderabad, December 19, 1986, p. 13.

49. Nelson (10), pp. 334–335.

50. China's road to becoming the world's largest producer of ammonia will be described in some detail in chapter 8.

51. As already noted (see note 9), the ICI has been in the ammonia business since its inception in 1926. Uhde GmbH (now Krupp Uhde) was founded in 1921 and completed its first ammonia plant in 1928 (19). Haldor Topsøe established his company in 1940; it supplied its first ammonia synthesis catalyst in 1951, the first Topsøe-designed ammonia converter began operating in 1956, and the first plant based completely on the company's design was finished in 1965; see Dybkjaer, I. 1989. *Large Ammonia Plants: Design and Operating Experience.* Lyngby: Haldor Topsøe A/S. Chiyoda Corporation has been a leading Japanese player in chemical and petrochemical processes; between 1963 and 1998 the company designed thirteen ammonia and thirteen urea plants in nine countries; see Chiyoda Corporation. 1998. *Major Projects as of September 1998.* Yokohama: Chiyoda Corporation. Kinetic Technology India is part of the Kinetic Technology International group established in 1963 in the Netherlands.

52. The M. W. Kellog Company. 1998. *Ammonia.* Houston: M. W. Kellogg.

53. During the same period of time the world had seen a similarly spectacular growth in the performance of other fundamental techniques, including steam turbines (from less than 500 to more than 1,000 MW) and crude oil tankers (maxima from around 100,000 t in the early 1960s to 400,000 t a decade later); see Smil, V. 1991. *General Energetics.* New York: John Wiley.

54. Nielsen, S. E., and H. Kato. 1998. *The Largest Ammonia Plant in Asia: The 2000 MTPD Topsøe-Designed Ammonia Plant for P. T. Kaltim Pasifik Ammoniak.* Paper presented at

Asia Nitrogen '98, February 22–24, Kuala Lumpur; Haldor Topsøe A/S. 1999. Topsøe's Position as Process Licensor for Ammonia Plants. Lyngby: Haldor Topsøe. Both the Kaltim and the Profertil plants were scheduled to produce ammonia before the end of 2000. Krupp Uhde believes that the largest ammonia plants of the next generation may have capacities of 2,500–3,000 t/day.

55. Other ammonia complexes with annual capacities in excess of 1 Mt N are in Sluiskil in the Netherlands (1.35 Mt N), Palembang (1.23) and Bontang (1.21) in Indonesia, Novomoskovsk and Kemerovo in Russia, Gorlovka and Cherkassy in Ukraine, and Umm Sais in Qatar (all of them about 1.1 Mt N); see International Fertilizer Development Center. 1998. *Worldwide Ammonia Capacity Listing by Plant.* Muscle Shoals, Ala.: IFDC.

56. International Fertilizer Industry Association. 1998. *Ammonia Capacity.* Paris: IFA (http://www.fertilizer.org). Almost 50% of the world's total capacity is now in the low-income countries.

57. Global synthesis of H_2SO_4 reached almost 149 Mt in 1995 and 155.6 Mt in 1997; see Kitto, M. 1998. The sulphur market in 1997. *Sulphur* 258:51–54.

58. Nelson (10), pp. 337–339.

59. Kitchen, D., et al. 1991. The ICI Leading Concept Ammonia (LCA) Process. In D. R. Waggoner and G. Hoffmeister, eds., *Environmental Impacts of Ammonia and Urea Production Units.* Muscle Shoals, Ala.: IFDC, pp. 47–54.

60. Parks, S. B., and C. M. Schillmoller. 1997. Improve alloy selection for ammonia furnaces. *Hydrocarbon Processing* 76(10):93–97.

61. Appl, M. 1992. Modern ammonia technology: where have we got to, where are we going? *Nitrogen* 200 (November-December 1992):27–38; Anonymous. 1994. Kellogg's technology quartet. *Nitrogen* 211:33–35; UNIDO (32), pp. 180–181. The catalyst is prepared by subliming ruthenium carbonyl—$Ru_3(CO)_{12}$—onto a carbon-containing support impregnated with rubidium nitrate ($RbNO_3$); it contains 5% Ru and 10% Rb by weight.

62. Le Blanc, J. R. 1996. *Ammonia 2000.* Houston: M. W. Kellogg Company; Czuppon, T. A., et al. 1996. Commercial Experience Review of KAAP and KRES. Paper presented at AIChE Safety Symposium, Boston, Mass.; Strait (47); M. W. Kellogg Company. 1998. *First of a New Generation.* Houston: M. W. Kellogg Company.

63. Information on energy use in ammonia synthesis can be found in Blouin, G. M. 1974. *Effects of Increased Energy Costs on Fertilizer Production Costs and Technology.* Muscle Shoals, Ala.: TVA; Pimentel, D., ed. 1980. *Handbook of Energy Consumption in Agriculture.* Boca Raton, Fla.: CRC Press; Mudahar, M. S., and T. P. Hignett. 1981. *Energy and Fertilizer.* Muscle Shoals, Ala.: IFDC; Appl (48), p. 18; Helsel, Z. R. 1987. Energy and alternatives for fertilizer and pesticide use. In Z. R. Helsel, ed., *Energy in Plant Nutrition and Pest Control.* Amsterdam: Elsevier, pp. 177–201; Appl, M. 1993. Modern ammonia technology: where have we got to, where are we going? *Nitrogen* 202:44–53; Radgen, P. 1996. Pinch and exergy analysis of a fertilizer complex. *Nitrogen* 224:39; UNIDO (32), p. 163; Strait (61), p. 43.

64. Le Blanc, J. R. 1984. Make ammonia with less energy. Houston: M. W. Kellogg Company.

65. Paul, D., and R. L. Kilmer. 1978. *The Manufacturing and Marketing of Nitrogen Fertilizers in the United States.* Washington, D.C.: USDA. Less efficient pre-1950 reciprocating compressors consumed 600–900 kWh/t NH_3.

66. This comes to 25 million Btu, an easy-to-remember rate that was used in an advertisement promoting these new energy-efficient plants. Kellogg's modification that reduced the total energy need below 25 million Btu was first commercialized at the Sherritt Gordon Mines ammonia plant (1,000 t NH_3/day capacity) at Fort Saskatchewan, Alberta, in 1983.

67. Dybkjaer, I. 1990. Advances in ammonia production technology. Lyngby: Haldor Topsøe; Dybkjaer, I. 1992. Large ammonia plants: design and operating experience. Paper presented at the 1992 IFA-FADINAP Regional Conference for Asia and the Pacific, Bali, November 30–December 2, 1992.

68. Heath, R., et al. 1985. *The Potential for Energy Efficiency in the Fertilizer Industry.* Washington, D.C.: World Bank.

69. The global average is pushed up by highly inefficient synthesis in China's small plants whose average energy need—according to Fan et al. (30), p. 23—was 57 GJ/t NH_3 in 1997, as well as by the relatively low efficiencies of many aging coal-based Indian plants and hydrocarbons-based European units.

70. Helsel (63), pp. 182–183. Phosphate rock can be applied directly to fields after grinding, benefication, and drying, but most of it is used to produce phosphoric acid (H_3PO_4) by the reaction with H_2SO_4. The acid is the base for synthesis of such widely used fertilizes as triple superphosphate and mono- and diammonium phosphate; see UNIDO (32), pp. 295–399.

71. Most of the world's potash comes from shaft mining, the rest from solar evaporation of water in shallow lake beds and by solution mining. As nongranular potash tends to cake, it is preferable to produce a granular fertilizer; this involves compaction, crushing, and polishing, mechanical processes that are not highly energy-intensive. As a result, K is the cheapest macronutrient to produce. For details on production of potassium fertilizers see UNIDO (32), pp. 417–431.

72. For details on synthesis of urea see UNIDO (32), pp. 257–270.

73. This calculation assumed annual output of 80 Mt N, about 15 Mt P, and 20 Mt K. Unlike in the case of nitrogen, international fertilizer statistics are not given in terms of pure nutrients, but rather as P_2O_5 (phosphoric oxide) and K_2O (potash); conversion rates to pure nutrient are 0.4364 and 0.8302, respectively. One tonne of oil equivalent (toe) contains 42 GJ; 1 m^3 of natural gas typically contains about 35 MJ.

74. Global consumption of natural gas was about 2.2 trillion m^3 (1.98 Gtoe) in 1997; the world's total consumption of coals, crude oils, and natural gases was 7.67 Gtoe; see British Petroleum-Amoco. 2000. *Statistical Review of World Energy 1999.* Also available at http://www.bpamoco.com/bpstats.

75. Such efficiency gains are trivial compared to the enormous energy conservation potential in both rich and poor societies. For example, a standard household gas furnace is about 65% efficient, while high-efficiency units (operating without any chimney) are rated at 95%. As for gasoline, in 1997 the world's 650 million vehicles consumed almost 950 Mt of the fuel, 40% of it in North America. Merely replacing the current North American private car fleet (averaging less than 29 mpg) by cars whose performance would equal the corporate average fuel economy of Honda Company (whose production is dominated by the Accord and Civic, and whose performance is above 33 mpg) would save nearly 60 Mtoe a year, enough to produce half of the world's nitrogen fertilizers.

76. The global reserve/production ratio for the natural gas was sixty-two years in 1999 (compared to forty-one years for crude oil, and 230 years for coal). So far about 180 Gtoe of natural gas have been extracted and found in proven reserves; additional natural gas resources are conservatively estimated to total another 200–300; see British Petroleum (74); Odell, P. 1999. *Fossil Fuel Resources in the 21st Century*. Vienna: IAEA.

77. Future energy demand by households (driven by urbanization and electrification) and the transportation sector (due to the increasing ownership of motorcycles and cars) in low-income countries will be particularly strong.

78. Of course, the situation is already quite different in a number of countries producing a great deal of ammonia on the basis of limited natural gas reserves. In many countries nitrogen fixation claims much higher shares of natural gas or total fossil fuel consumption than indicated by respective global means. Most notably, China, the world's largest producer of nitrogen fertilizers, already uses about $1/3$ of its limited natural gas output to synthesize ammonia. If the country were to use natural gas as both feedstock and fuel for synthesizing all of its ammonia and for converting it to urea, the resulting annual demand (at least 40 Gm^3 of gas in 1997) would be more than twice as large as China's total natural gas extraction.

Chapter 7

1. About 570,000 t, or 0.7%, of all fertilizer nitrogen applied during the latter half of the 1990s was of non-NH_3 origin; see FAO. 1997. *Current World Fertilizer Situation and Outlook 1994/95–2000/2001*. Rome: FAO, p. 23. Synthesis of urea needs 0.58 t of NH_3 to produce 1 t of the compound; the rates for the synthesis of ammonium nitrate and ammonium sulfate are 0.21 and 0.27 t NH_3, respectively.

2. Ammonium sulfate is still produced as a by-product from coke-oven gas, and small amounts also come from the waste streams of several organic syntheses, above all caprolactam and acrylonitrile.

3. As plants take up sulfur almost exclusively from soluble sulfates, sulfur deficiency is not a problem in regions receiving acid deposition originating from the combustion of S-containing coals or oils. But in areas with no or little acid deposition, sulfur deficiencies can arise from the combination of high yields (removing accumulated soil S stores) and a prolonged use of S-free fertilizers.

4. Dissolution of $(NH_4)_2SO_4$ yields a NH_4^+ cation and SO_4^{2-} anion. Subsequent nitrification (microbial oxidation) releases H^+:

$$NH_4^+ \rightarrow NO_3^- + 2\ H^+ + H_2O.$$

Applications of 1.46 kg of pure CaO are needed in order to neutralize 1 kg of ammonium sulfate, compared to just 0.59 kg CaO needed to neutralize 1 kg of ammonium nitrate; see Tisdale, S. L., and W. L. Nelson. 1975. *Soil Fertility and Fertilizers*. New York: Macmillan, pp. 178–179. On the current revival of interest in the compound see Anonymous. 1999. Sulphur is the key. *Nitrogen & Methanol* 240:12–14.

5. Calcium's protective action against the disruptive effects of a high sodium concentration has been known for decades, but its mechanism was discovered only recently; see Liu, J., and J. Zhu. 1998. A calcium sensor homolog required for plant salt tolerance. *Science* 280:

1,943–1,945. More widespread cultivation of genetically engineered salt-tolerant cultivars could also increase the future use of calcium nitrate. But the anhydride, $Ca(NO_3)_2$, contains even less N than ammonium sulfate (15.5%), and because it is extremely hygroscopic its production requires air-conditioned plants and moisture-proof packaging.

6. The Dutch are the compound's most avid users; during the mid-1990s they applied about $^4/_5$ of all fertilizer nitrogen in the form of CAN.

7. IFDC. 1998. *Worldwide Ammonium Nitrate and Calcium Ammonium Nitrate Capacity Listing by Plant*. Muscle Shoals, Ala.: IFDC. Throughout Western and Central Europe ammonium nitrate has been supplying anywhere between $^1/_5$ (in Spain) and $^3/_5$ (Austria, Hungary) of all inorganic nitrogen (the Netherlands being a notable exception).

8. UNIDO. 1998. *Fertilizer Manual*. Dordrecht: Kluwer Academic, p. 221; Eben, A., and P. Kaupas. 1998. Ammonium nitrate production and operational experience. *Nitrogen & Methanol* 235:25–30; Shah, K. D. 1996. *Safety of Ammonium Nitrate Fertilisers*. Peterborough: Fertiliser Society.

9. The two worst terrorist attacks in the United States—the bombing of the World Trade Center in New York City on February 26, 1993 (which killed six people and caused enormous material damage) and the destruction of the Alfred Murrah Federal Office Building in Oklahoma City on April 19, 1996 (which killed 169 adults and children)—involved truck bombs using an ANFO mixture. The bombings also resulted in a renewed interest in additives able to desensitize ammonium nitrate, and new tests showed that this option does not work very well with the larger volumes of explosives that are used in terrorist acts (the Oklahoma bomb contained about 1.8 t of nitrate); see Hands, R. 1996. Ammonium nitrate on trial. *Nitrogen* 219:15–18.

10. Mitsui Toatsu, Montedison, Snamprogetti, and Stamicarbon have been the leading proprietary processes of total carbamate recycling; their common design objectives are maximized heat recovery and minimized amount of carbamate recycled.

11. In 1966 Stamicarbon was the first company to decompose the unconverted carbamate by the stream of CO_2 passing through the reactor effluent solution. A few years later Snamprogetti began using NH_3 as the stripping agent.

12. For comparisons of the energy costs of different fertilizer compounds see UNIDO (8), and Helsel, Z. R., ed. 1987. *Energy in Plant Nutrition and Pest Control*. Amsterdam: Elsevier, pp. 177–201.

13. Hydrolysis of urea is catalyzed by urease, an enzyme that can survive extracellularly in organic matter-clay soil complexes; the reaction can proceed rapidly in warm, moist soils containing large amounts of the enzyme. Wider use of urease inhibitors (phosphoryl di- and triamides) may eventually reduce large ammonia volatilization losses; see Byrnes, B. H., and J. R. Freney. 1995. Recent developments on the use of urease inhibitors in the tropics. *Fertilizer Research* 42:251–259.

14. Webb, A. L., et al. 1997. Urea as a nitrogen fertilizer for cereals. *Journal of Agricultural Science* 128:263–271.

15. Cassman, K. G., et al. 1996. Long-term comparison of the agronomic efficiency and residual benefits of organic and inorganic nitrogen sources for tropical lowland rice. *Experimental Agriculture* 32:927–944.

16. Denitrification in wet fields also produces more N_2O because the water-filled pore space (WFPS) in soils is the key determinant of the relative fluxes of NO and N_2O; NO fluxes peak with 30–60% WFPS, while the N_2O flows peak when WFPS is between 80 and 90%. See Veldkamp, E., et al. 1998. Effects of pasture management on N_2O and NO emissions from soils in the humid tropics of Costa Rica. *Global Biogeochemical Cycles* 12:71–79.

17. Asian urea shares in total nitrogen applications now range from about ⅔ in the Philippines to more than 99% in Bangladesh.

18. For details on ammonia storage, transportation, and the U.S. pipelines see UNIDO (8), pp. 196–206.

19. These back- or front-swept implements are used to inject the volatile liquid 15–30 cm below the surface. Even then the volatilization losses may be significant when the liquid is injected into coarse-textured soils with low organic matter content.

20. UNIDO (8), pp. 272–273.

21. Although it has a relatively low vapor pressure, aqua ammonia must be stored in covered tanks, but it does not have to be injected as deeply into soil as the anhydrous NH_3.

22. On the other hand, fluid fertilizers lack the storage and transport flexibility of solids, and they corrode storage and handling equipment and can lead to scorching when applied directly on plants; see Isherwood, K. F. 1996. *Fertiliser Supply from Factory to Farm: A Global View.* Peterborough: Fertiliser Society. The share of nitrogen solutions in the global market is only about 5%, just slightly ahead of calcium ammonium nitrate and a bit behind direct applications of anhydrous ammonia; see International Fertilizer Industry Association. 1998. *World Nitrogen Fertilizer Consumption.* Paris: IFA (http://www.fertilizer.org).

23. In 1997 fixed nitrogen in various NP, NK, and NPK compounds added up to about 15% of the nutrient's worldwide consumption: IFA (22). But because most crops require more nitrogen than P or K, applications of compound fertilizers have to be often followed by additional amounts of nitrogen.

24. Nitro de Chile. 1964. *Enciclopedia Universal Ilustrada.* Madrid: Espasa-Calpe, p. 832; Ost, H. 1919. *Lehrbuch der chemischen Technologie.* Leipzig: Max Jänecke, p. 173.

25. For details on all of these sources of fixed nitrogen see chapter 3.

26. Timm, B. 1963. 50 Jahre Ammoniak-Synthese. *Chemie-Ingenieur-Technik* 35:823.

27. Nitrate content in the recently exploited *caliche* is below 5%. The Sociedad Química y Minera de Chile Nitratos (www.soquichim.cl), a private Chilean company, is the only enterprise in the business; it has two leaching plants and a solar evaporation plant to refine the ore. Total nitrate shipments rose from 600,000 t in 1980 to 800,000 t in 1990, with nearly ¼ now shipped as KNO_3 (Chile is now the world's second largest producer of this mineral). Less than ⅕ of annual output (not even 200,00 t/year) is now exported, and the total extraction has been equivalent to only about 0.2% of all inorganic fertilizer nitrogen. See Tejeda, H. R. 1993. Chilean nitrates: environmental issues of mining, processing, and use. In R. G. Lee, ed., *Nitric Acid–Based Fertilizers and the Environment.* Muscle Shoals, Ala.: IFDC, pp. 319–325. A few countries, most notably Japan and Italy, still also synthesize small amounts of calcium cyanamide.

28. Schmidt, A. 1934. *Die industrielle Chemie.* Berlin: Verlag Walter de Gruyter & Co., p. 324; U.S. Bureau of Commerce. 1975. *Historical Abstract of the United States.* Washington, D.C.: GPO, part 2, p. 697.

29. Increased U.S. production accounted for nearly ³/₅ of the jump during the 1950s. In 1950 the country fixed 1.565 Mt N; in 1960 the total was slightly more than three times as high at 4.818 Mt N. During the 1960s the U.S. annual output rose by more than 8 Mt, accounting for ²/₅ of the 21 Mt global increase.

30. The Green Revolution did little for yields of nonstaple cereals, legumes, and oil crops. Its diffusion has been very uneven (with Africa lacking far behind Asia), and some of its socio-economic and environmental consequences have been widely criticized in many books published since the 1960s. For a bibliography of these writings see Karim, M. B. 1986. *The Green Revolution: An International Bibliography*. New York: Greenwood Press.

31. Borlaug, N. 1970. The Green Revolution: peace and humanity. A speech on the occasion of the awarding of the 1970 Nobel Peace Prize in Oslo, Norway, December 11, 1970. Text at http://www.theatlantic.com/atlantic/issues/97jan/borlaug/speech.

32. The FAO has been publishing data on worldwide data production, trade and consumption of fertilizers annually since 1950; see FAO. 1950–. *Fertilizer Yearbook*. Rome: FAO. The International Fertilizer Industry Association in Paris publishes annually the following statistics: *Fertilizer Nutrient Consumption, by Region; Fertilizer Nutrient Consumption by Product; Fertilizer Trade*. The Internet sites are, respectively, www.fao.org and www.fertilizer. org. Consumption data are also often reported in terms of crop years, which straddle calendar years.

33. FAO. 1997. *Current World Fertilizer Situation and Outlook 1994/95–2000/2001*. Rome: FAO, p. 11.

34. The survey has been carried out every second year since 1992; see Harris, G. 1998. *An Analysis of Global Fertilizer Application Rates for Major Crops*. Paris: IFA (www.fertilizer. org).

35. All of these rates are calculated by using the FAO's farmland data, whose accuracy is satisfactory only for high-income nations. The FAO must make many in-house estimates to fill numerous data gaps for low-income countries. Moreover, some official farmland figures submitted by the FAO's member states are known to be substantial underestimates. Perhaps the most notable example of such an underestimate is the case of China's farmland: its real total is 50% larger than the official claim, at least 140 Mha rather than 95 Mha. Consequently, many nitrogen application rates calculated by using standard statistics overestimate the actual usage.

36. For comparison, even an adult man weighing 70 kg will not have more than 1.5 kg N in his body tissues.

37. The production and consumption of many other industrial commodities and the extraction of various minerals show similarly skewed distributions reflecting rapid, often even exponential, growth during the second half of the twentieth century.

38. This gap was—and in the majority of cases it still remains—the norm for almost every variable measuring both the collective performance (industrial and agricultural accomplishments) as well as personal well-being (various indicators of the quality of life, ranging from infant mortality to literacy).

39. Neither consumption total is adjusted for nitrogen fertilizer consumed in producing food for exports. Appreciable amounts of the affluent world's output of cereal grains (most notably

U.S., Canadian, and Australian wheat and U.S. corn) and meat and of the Third World's specialty crop (coffee, cocoa) and tropical fruit harvest are sold abroad. The net trade in feedstuffs alone (mostly cereals, but also roots, legumes, and oil crops) amounts to between 4 and 7 Mt N/year; see Bouwman, A. F., and H. Booij. 1998. Global use and trade of feedstuffs and consequences for the nitrogen cycle. *Nutrient Cycling in Agroecosystems* 52:261–267.

40. The actual disparity is even larger than the simple means indicate. Relatively high nitrogen applications can go into producing export cash crops in many low-income tropical and subtropical countries, while their staple food crops grown for domestic consumption receive very little inorganic nitrogen.

41. Many affluent countries have experienced pronounced declines in nitrogen applications since the period of peak consumption during the early 1980s. Leaving the sharp decline of fertilizer use in the successor states of the former Soviet Union aside, these reductions have ranged between 15 and 30% for most Western European countries and about 25% for Japan.

42. For a detailed explanation of this global account see Smil, V. 1999. Nitrogen in crop production: an account of global flows. *Global Biogeochemical Cycles* 13:647–662.

43. The best way to quantify the total output of crop residues is to use average residue/crop ratios to multiply the reported harvests. This process results only in good approximations, rather than in accurate totals, because residue/crop ratios vary among cultivars as well as for the same cultivar grown in different environments; agronomic factors (planting date, irrigation regime, N applications) can also make a substantial difference. Assumptions concerning the average straw/grain ratio of cereals will make a particularly large difference in global calculations. Using the average S/G ratio of 1.3, rather than 1.2, adds almost 200 Mt more straw, a total larger than all the residual phytomass produced by the world's tuber and root crops.

44. Cereal straws, the most important crop residues, contain between 0.4 and 1.3% N, with values around 0.6% being perhaps most common.

45. These crops include all-legume or legume-grass swards cultivated intermittently on arable land—either for direct foraging or, more often, for hay and silage—as well as a variety of green manures. The FAO does not keep track of either the areas or the average yields these crops, but there is no doubt that their worldwide extent has been declining; see Wedin, W. F., and T. J. Klopfenstein. 1995. Cropland pastures and crop residues. In R. F. Barnes et al., eds., *Forages*. Ames: Iowa State University Press, vol. II, pp. 193–206. I assumed 100–120 Mha with dry matter yields of 5 t/ha for alfalfa (planted on about 30% of cropland pastures) and 4 t/ha for other leguminous and mixed legume-grass forages.

46. Nitrogen in planted seeds and roots is one of the few inputs that are fairly easy to quantify: the standard seeding rates are simply multiplied by planted areas. Seeding rates for individual countries can be found in FAO. 1996. *Food Balance Sheets 1992–94 Average*. Rome: FAO. About 17% of the world's cropland (250 Mha) was irrigated in 1995, with nearly $2/3$ of the global total in Asia. N concentrations in irrigation water are mostly between 1 and 2 mg N/kg, and water applications are at least 9,000–10,000 m^3/ha; see Postel, S. L., G. C. Daily, and P. R. Ehrlich. 1996. Human appropriation of renewable fresh water. *Science* 271:785–788.

47. I calculated the global wet deposition flux by using separate estimates of continental averages for NO_y and NH_x deposition rates. Dry deposition is equal to at least $1/3$ of the wet NH_3 flux, and dry NO_y is equal to at least $3/4$ of the wet input; see Warneck, P. 1988. *Chemistry of the Natural Atmosphere*. London: Academic Press, pp. 476–483.

   As the total terrestrial deposition of nitrogen is now about 60 Mt N/year, farmland, accounting for only some 11% of ice-free surfaces, would receive at least ⅓ of the flux. This higher level of enrichment is due to the proximity of agricultural regions and nonagricultural sources of NO_x emissions and to a large share of NH_3 emissions originating in cropping and animal husbandry. Because of their rather brief atmospheric residence time, most ammonia compounds are deposited close to the areas of their emissions; in contrast, nitrates are carried much further before being precipitated or deposited in dry form.

48. Residue burning is an agronomically undesirable practice that also generates greenhouse, and other trace, gases; see Andreae, M. O. 1991. Biomass burning: its history, use, and distribution and its impact on environmental quality and global climate. In J. S. Levine, ed., *Global Biomass Burning: Atmospheric, Climatic, and Biospheric Implications.* Cambridge, Mass.: MIT Press, pp. 3–21.

49. The main reason for this difference is that I am assuming lower averages of body weights and poorer feeds for cattle, sheep, and goats in low-income countries (they now have about 75% of the world's bovines and almost 80% of all sheep and goats). Even so, my estimate for nitrogen in cattle manure may still be too large, as some sources credit Indian cattle manure with just 0.7% N on a dry-weight basis. The other recent estimates of the global manure output are Nevison, C. D., et al. 1996. A global model of changing N_2O emissions from natural and perturbed soils. *Climatic Change* 32:327–378; Bouwman, A. F., et al. 1997. A global high-resolution emission inventory for ammonia. *Global Biogeochemical Cycles* 11: 561–587; Mosier, A. R., et al. 1998. Assessing and mitigating N_2O emissions from agricultural soils. *Climatic Change* 40:7–38.

50. Only manure produced in confinement on mixed farms and in feeding facilities located in farming areas can be economically distributed to nearby fields. As a result, hardly any manure is returned to fields in pastoral regions, while nearly all of East Asia's pig manure is recycled. In the West the recycling rates range from only about 30% in the United States— where some ⅖ of all animal wastes are voided in confinement and about ¾ of this output is actually returned to fields—to more than 90% in several small European countries. With more than 70% of urine nitrogen voided as urea, there are large losses due to rapid hydrolysis of the compound and subsequent volatilization of NH_3 during collection, storage, composting, and handling of wastes.

51. Boddey, R. M., et al. 1995. Biological nitrogen fixation associated with sugar cane and rice: contributions and prospects for improvement. *Plant Soil* 174: 195–209.

52. No less importantly, it has the highest relative population growth, with the total fertility rate (5.8 in 1995) nearly twice Asia's steadily declining mean (3.0 in 1995).

53. Oldeman, I. R., et al. 1990. *World Map of the Status of Human-Induced Soil Degradation: An Explanatory Note.* Nairobi: UNEP.

54. Larson, B. A., and G. B. Frisvold. 1996. Fertilizers to support agricultural development in sub-Saharan Africa: what is needed and why. *Food Policy* 21:509–525.

55. Between 1985 and 1995 the continent's consumption of nitrogen fertilizers rose by about 275,000 t because of about 350,000 t of higher use in Egypt and Morocco; sub-Saharan Africa's use fell for the region as a whole, as well as for every one of its largest countries (Nigeria, Congo, Angola, Kenya, Tanzania, and even South Africa).

56. In a sharp contrast to the African situation, in 1995/96 China and India, containing 38% of the global population, applied 42% (over 33 Mt N) of all nitrogen fertilizers.

57. Reconstruction of nitrogen inputs according to Han, C., and Z. Xie. 1993. Managing nitrogen inputs and cycling. Paper presented at the Symposium on Sustainable Development, Beijing, September 10–13, 1993, p. 4.

58. My summary of China's nitrogen inputs is based on the following sources: State Statistical Bureau. 1997. *China Statistical Yearbook.* Beijing: State Statistical Bureau; Galloway, J. N., et al. 1996. Nitrogen mobilization of nitrogen in the United States of America and the People's Republic of China. *Atmospheric Environment* 30:1,551–1,561; Han and Xie (57); Zhu Z. 1997. Nitrogen balance and cycling in agroecosystems of China. In Zhu Z., Q. Wen, and J. R. Freney, eds., *Nitrogen in Soils of China.* Dordrecht: Kluwer Academic, pp. 323–338.

59. This rate is calculated by using the real extent of China's farmland—about 140 Mha— rather than the grossly underestimated official value of 95 Mha. For details on China's farmland see Smil, V. 1999. China's agricultural land. *China Quarterly* 158:130–145; MEDEA. 1997. *China Agriculture: Cultivated Land Area, Grain Projections, and Implications.* Washington, D.C.: National Intelligence Council.

60. MEDEA's figures add up to about 13.4 Mha of farmland in the four coastal provinces, compared to 9.56 Mha of officially reported agricultural land. Using the official farmland total (as most researchers studying China's farming still invariably do) would boost the average application rate of the four provinces to more than 400 kg N/ha.

61. Bleken, M. A., and L. R. Bakken. 1997. The nitrogen cost of food production: Norwegian society. *Ambio* 26:134–142.

62. Reconstructions of German nitrogen inputs for the period 1900–1985 according to Rintelen, P. 1961. Betriebswirtschaftliche Betrachtungen zum Stickstoffverbrauch. In Balz, O., et al. *Der Stickstoff.* Düsseldorf: Fachverban Stickstoffindustrie, p. 402; FAO. 1981. *Crop Production Levels and Fertilizer Use.* Rome: FAO, p. 63; Pedersen, C. A. 1990. *Practical Measures to Reduce Nutrient Losses from Arable/Field Vegetable Farm.* York: Fertiliser Society. My update for 1995 results in the following average inputs (all in kg N/ha): manures and crop residues 30–35, biofixation 10–15, atmospheric deposition 25–30, nitrogen fertilizers 145, total 210–225, fertilizer share 64–69%. For slightly different totals for the years 1991–1992 see Isermann, K., and R. Isermann. 1998. Food production and consumption in Germany: N flows and N emissions. *Nutrient Cycling in Agroecosystems* 52:289–301.

63. Royal Society Study Group. 1983. *The Nitrogen Cycle of the United Kingdom.* London: Royal Society. For the study's update see Johnston, A. E., and D. S. Jenkinson. 1989. *The Nitrogen Cycle in UK Arable Agriculture.* London: Fertiliser Society.

64. British fertilizer applications declined by 12% between 1985 and 1995, and further slow decline is likely.

65. Hood, E. M. 1982. Fertilizer trends in relation to biological productivity within the U.K. *Philosophical Transactions of the Royal Society London* B 296:317; Gooding, M. J., and W. P. Davies. 1997. *Wheat Production and Utilization.* Walsingham: CAB International, pp. 44–45.

66. Manure provided about 200 kg N/ha and atmospheric deposition 20 kg N/ha; see Smaling, E. M. A. 1993. *An Agro-ecological Framework for Integrated Nutrient Management with Special Reference to Kenya.* Ph.D. thesis, Agricultural University of Wageningen.

67. If the comparison is limited to nations, only South Korea and Taiwan now apply more nitrogen per hectare of agricultural land than does the Netherlands. But, as already noted, several Chinese provinces, whose areas and populations are considerably larger than the Dutch totals, now apply annually in excess of 250 kg N/ha, and China's nationwide mean is only about 5% below the Dutch average.

68. Dutch output of manures tripled between 1965 and 1995, and manure applications on the order of 30 t/ha have been common. In fact, the country has too much manure, and it cannot recycle all of it to its fields; see Jongbloed, A. W., and C. H. Henkens. 1996. Environmental concerns of using animal manure: the Dutch case. In Kornegay, E. T., ed., *Nutrient Management of Food Animals to Enhance and Protect Environment.* Boca Raton, Fla.: Lewis Publishers, pp. 315–332.

69. This rate was calculated, as in other cases, by using the total area of U.S. farmland. But, unlike in most countries, a significant share of this land—on the order of 10–15%, depending on economic circumstances—may not be actually farmed in any given year. Taking this into account would raise the recent average application rate on arable land and cropland pastures to between 65 and 70 kg N/ha. And unlike in many European countries or in Japan, the U.S. nitrogen applications have not declined since the early 1980s: between 1985 and 1995 they actually increased slightly, from 10.4 to 11.1 Mt N.

70. Smil, V., P. Nachman, and T. V. Long II. 1982. *Energy Analysis in Agriculture: An Application to U.S. Corn Production.* Boulder: Westview Press; Runge, C. F., et al. 1990. *Agricultural Competitiveness, Farm Fertilizer and Chemical Use, and Environmental Quality: A Descriptive Analysis.* St. Paul, Minn.: Center for International Food and Agricultural Policy, tables IV-A–IV-F; U.S. Department of Agriculture. 1998. *Agricultural Yearbook.* Washington, D.C.: USDA, table XIV-I.

71. Runge et al. (70), table VI-A; USDA (70).

72. There are about 10 Mha of alfalfa, 15 Mha of farmland pastures seeded with legume-grass mixtures, and 25 Mha of soybeans and other grain legumes, which means that symbiotic fixation proceeds on about ¼ of U.S. farmland.

73. This estimate includes both wet and dry deposition of NO_3^- and NH_4^+. For the latest map of total wet deposition see National Atmospheric Deposition Program, Colorado State University, http://nadp.nrel.colostate.edu/NADP.

74. China's cultivation of green manures reached a peak of 9.9 Mha in 1975; a subsequent steady decline brought their total plantings to only about 4 Mha. Hunan province accounts for nearly ¼ of this diminished cultivation; see Smil, V. 1993. *China's Environmental Crisis.* Armonk, N.Y.: M. E. Sharpe, pp. 184–185.

75. Biofixation, mostly by green manures, supplies 85 kg N/ha, crop residues add 15, organic wastes 40, and atmospheric deposition 10 kg N/ha. Average yields of this agroecosystem are 6.7 t/ha of rice, 3.8 t/ha of wheat, and 1.5 t/ha of rapeseed. When other managed inputs containing nitrogen—recycling of crop residues, manuring, irrigation water, and cultivation of legumes—are added to the applications of synthetic fertilizers, the share of anthropogenic nitrogen rises to more than 90% of the total supply. Managed inputs account for 90% or more of the total supply in almost every intensively cropped agroecosystem.

76. This rotation would yield 7.1 t/ha of rice and 4.3 t/ha of winter wheat.

77. Several nitrogen budgets for European wheat are presented in Frissel, M. J., ed. 1978. *Cycling of Mineral Nutrients in Agricultural Ecosystems.* Amsterdam: Elsevier. The nitrogen cycle for an English plot under continuous winter wheat is quantified in Burt, T. P., and N. E. Haycock. 1993. Controlling losses of nitrate by changing land use. In Burt, T. P., et al., eds., *Nitrate: Processes, Patterns and Management.* Chichester: John Wiley, p. 343. For national application averages see Harris (34).

78. My estimates are as follows (all rates in kg N/ha): nitrogen fertilizer 200, crop residues 65, atmospheric deposition 10, biofixation 5, seeds 3. For comparison, see a nitrogen budget for Indiana corn given in Thomas, G. W., and J. W. Gilliam. 1978. Agro-ecosystems in the U.S.A. In M. J. Frissel, ed., *Cycling of Mineral Nutrients in Agricultural Ecosystems.* Amsterdam: Elsevier, pp. 228–229. This budget lists the following inputs (all in kg N/ha): nitrogen fertilizer 112, crop residues 41, irrigation 10, atmospheric deposition 10, biofixation 5.

79. The soil enrichment by N fixed by leguminous crops depends on the degree to which legumes can satisfy their total N needs from the biofixation, and on the share of the fixed N taken up by the harvested legume seeds. The latter share ranges from as little as 30% for common beans to more than 80% for soybeans. Complete recycling of bean plant residues will thus transfer most of the fixed N for eventual use by subsequent crops, but soybeans, although they are much more prolific N fixers, may not be able to provide all of the needed nutrient, and the crop will actually claim a considerable amount of soil N.

80. A nitrogen budget for a Dutch farm growing sugar beets is in Damen, J. 1978. Agro-ecosystems in the Netherlands. Part I. In M. J. Frissel, ed., *Cycling of Mineral Nutrients in Agricultural Ecosystems.* Amsterdam: Elsevier, pp. 92–93. Inputs are (all in kg N/ha): nitrogen fertilizers 305, crop residues 71, atmospheric deposition 14.

81. A nitrogen budget for Maine potatoes is given in Thomas and Gilliam (78), pp. 234–235. Inputs are (all in kg N/ha): nitrogen fertilizer 168, crop residues 65, atmospheric deposition 6, biofixation 5.

82. For discussions of these inputs see Bleken and Bakken (61) and Damen (80).

83. I assumed typical extraction rates during milling, that is, a 15% loss for wheat and a 30% loss for rice, and I used average protein contents of 10% and 7%, respectively.

84. For more on the best performances in traditional agricultures see chapter 2.

85. Once again, this calculation was done by using 140 Mha as the most likely real extent of China's farmland (see note 59).

86. The FAO's food balance sheets (http://apps.fao.org) indicate that during the mid-1990s Japan was importing about 60% of its total food supply, and that about 6% of its food came from the ocean. Consequently, the country's crop harvests could supply the prevailing diets to no more than 9.5 people/ha. Analogical adjustments for South Korea yield the real carrying capacity of about 8 people/ha with the diets prevailing during the mid-1990s. Egypt's apparent rate of about 25 people/ha reduces to about 15 people/ha (about 40% of all food supply being imported); this high rate is linked not only to high fertilizer applications (360 kg N/ha) but also to the fact that the country is unique in having 100% of its farmland irrigated.

Chapter 8

1. Liebig, J. 1840. *Chemistry in Its Application to Agriculture and Physiology.* London: Taylor & Walton, p. 85.

2. Yields of close to 8 t/ha have actually been recent averages in the United Kingdom, Netherlands, and Denmark, and the best farms produce repeatedly more than 10 t/ha. For average yields for all major crops see http://apps.fao.org.

3. Nationwide rice yields now surpass 6 t/ha not only in Japan (6.4 t/ha in 1998) and South Korea (6.99 t/ha in 1998) but also in China (6.25 t/ha in 1998).

4. That is on a scale of at least mid-sized European countries or small Chinese provinces.

5. With lignin and cellulose making up most of this phytomass we could not digest it even if we could make it somehow palatable by grinding, mixing, and flavoring.

6. Most of the grain milling residues and oilseed cakes are also used for feeding; oilseed meals are particularly excellent feeds, containing 41–48% of crude protein.

7. This is in excellent agreement with the FAO's estimates of global per capita supply of dietary protein; see FAO. 1996. *The Sixth World Food Survey 1996.* Rome: FAO, pp. 95–97. FAO's latest (1994– 1996) data on average food energy and protein supply are available at http://apps.fao.org.

8. I am ignoring a small, but increasing, amount of phytomass feed used in aquaculture. China, with its long tradition of raising herbivorous carps, is the world's largest user of crop-based fish feed. In contrast, the Western preference for carnivorous species (most notably salmon) makes that kind of aquaculture a protein-reducing rather than protein-producing endeavor, with protein in fishmeal-based feed exceeding the output of new fish protein two- to fivefold; see Tacon, A. G. J. 1995. Aquaculture feeds and feeding in the next millennium: major challenges and issues. *FAO Aquaculture Newsletter* 1995(10):2–8.

9. For detailed regional and country breakdowns of protein supply see http://apps.fao.org.

10. For example, U.S. per capita supply of animal protein was over 50 g/day during the first decade of the twentieth century, that is, before the arrival of synthetic nitrogen fertilizers; see Gortner, W. A. 1975. Nutrition in the United States, 1900 to 1974. *Cancer Research* 35: 3,246–3,253.

11. Calculated from International Fertilizer Industry Association statistics; see http://www.fertilizer.org.

12. I am ignoring a small amount of protein coming from the hunting and collecting of wild mammals, birds, reptiles, amphibians and insects. These foods can make an appreciable difference only in some tropical countries, particularly in sub-Saharan Africa (and their provision is responsible for accelerating declines of local biodiversity).

13. Population data according to various editions of the United Nation's *Demographic Yearbook.*

14. Calculated as differences between the totals of cultivated land submitted by the member states to the FAO, or estimated in the organization's headquarters in Rome. The actual expansion of cultivated land may have been appreciably higher.

15. Detailed breakdowns of food trade by commodity and by country are available at http://apps.fao.org and in the FAO's *Trade Yearbook*.

16. The world's crop production during the first decade of the twentieth century supplied about 35 kg of protein per hectare of agricultural land, corresponding to about 50 g of dietary protein (nearly 9/10 in plant foods) a day per capita; animal products from grazing and fish catches added another 10 g/day.

17. Crop harvests of the mid-1990s produced about 100 kg of edible protein per hectare, of which about 70% was in plant foods.

18. Because of surprisingly large differences in meat consumption among affluent countries such reductions would be rather large in some countries (U.S., Canada, Denmark) and modest in others (Finland, Sweden).

19. If at least 37% of all nitrogen in today's global harvest comes from the Haber–Bosch synthesis, then a 10% reduction would lower this reliance to 33%.

20. Indeed, arguments can be made that such a shift would improve the health of Western populations; for more, see the last chapter.

21. FAO/WHO/UNU. 1985. *Energy and Protein Requirements*. Geneva: WHO, pp. 58–59. This means that adult males weighing about 70 kg lose daily between 3.5 and 5.5 g of nitrogen, or between 21 and 33 g of protein, while the losses for small women (50 kg) range from 2.5 to 3.9 g N or a mere 15–24 g of protein. Losses in urine dominate; with about 2 g of nitrogen per 100 ml, they add up to 20–25 g N/day for most adults.

22. FAO/WHO/UNU (21), p. 56.

23. Both infants and adults have relatively the highest amino acid requirement for leucine, and the lowest for tryptophan.

24. This may be difficult when the staples, such as corn or cassava meal, are not highly palatable.

25. FAO/WHO/UNU (21), p. 119.

26. Huang, P. C., and C. P. Lin. 1981. Protein requirements of young Chinese male adults. In Torun, B., et al., eds., In *Energy Requirements of Developing Countries: Evaluation of New Data*. Tokyo: United Nations University, pp. 63–70.

27. Young, V. R., et al. 1989. A theoretical basis for increasing current estimates of the amino acid requirements in adult man, with experimental support. *American Journal of Clinical Nutrition* 50:80–92; Young, V. R., and P. L. Pellett. 1990. Current concepts concerning indispensable amino acid needs in adults and their implications for international nutrition planning. *Food and Nutrition Bulletin* 12:289–300; Pellett, P. L. 1990. Protein requirements in humans. *American Journal of Clinical Nutrition* 51:723–737; FAO/WHO. 1991. *Protein Quality Evaluation*. Rome: FAO.

A critical argument supporting the conclusion that the currently recommended total protein requirements for adults are incorrect is that they are considerably lower than for any other mammals whose protein requirements have been studied in some detail, while those recommended for infancy are among the highest; see McLarney, M. J., et al. 1996. Pattern of amino acid requirements in humans: an interspecific comparison using published amino acid requirement recommendations. *Journal of Nutrition* 126:1,871–1,882.

28. P/E ratios can be calculated easily from disaggregated data on protein supply listed in FAO's food balance sheets. For the latest averages see http://apps.fao.org.

29. FAO. 1996. *The Sixth World Food Survey.* Rome: FAO, p. 69.

30. Bender, W. H. 1994. An end use analysis of global food requirements. *Food Policy* 19: 381–395.

31. FAO/WHO/UNU (21), pp. 64–66, 98–109.

32. FAO (29), pp. 68–69, 112–113.

33. Detailed breakdowns of U.S. agricultural exports are presented in both the *Statistical Abstract of the United States* published every year by the U.S. Department of Commerce and in the *Agricultural Yearbook* published by the USDA, as well as in FAO's *Trade Yearbook.*

34. Fluctuations of planted area in response to anticipated export demand have been a long-standing feature of U.S. crop farming.

35. National shares of overall grain supply fed to livestock (calculated by using USDA grain consumption and feed numbers) are listed in the World Resource Institute's biannual *World Resources: A Guide to the Global Environment.*

36. Spedding, C. R. W., et al. 1981. *Biological Efficiency in Agriculture.* London: Academic Press, pp. 250–271, and the references cited in the next note.

37. The following table presents the derivation of protein conversion efficiencies prevailing in U.S. agriculture of the 1990s:

	Milk	Eggs	Chicken	Pork	Beef
Feed (kg/kg live weight)	1.0	2.5	2.5	4.0	8.0
Edible Weight (% of (LW)	95	90	55	55	40
Feed (kg/kg EW)	1.1	2.8	4.5	7.3	20.0
Protein Content (% EW)	3.5	13	20	14	15
Protein Conversion Efficiency (%)					
Typical	30	30	20	10	5
Best	40	40	30	15	8

Principal data sources are National Research Council. 1987. *Predicting Feed Intake of Food-Producing Animals.* Washington, D.C.: National Academy Press; National Research Council. 1988. *Nutrient Requirements of Swine.* Washington, D.C.: National Academy Press; National Research Council. 1994. *Nutrient Requirements of Poultry.* Washington, D.C.: National Academy Press.

38. This assumes no changes in the typical efficiency of feed conversion to meat and dairy products.

39. The U.S. annual meat supply (retail cuts or carcasses with bones in) now averages almost 120 kg/capita, compared to about 75 kg in Switzerland and just over 60 kg in Sweden; see http://apps.fao.org.

40. The USDA's 10th Nationwide Food Consumption Survey showed that mean intake for all individuals was 1985 kcal/day; see Cleveland, L. E., et al. 1997. "What We Eat in America" Survey. *Nutrition Today* 32:37–40.

41. Keys, A., and M. Keys. 1975. *How to Eat Well and Stay Well the Mediterranean Way.* New York: Doubleday & Co.; Kushi, L. H., et al. 1995. Health implications of Mediterranean diets in light of contemporary knowledge. 1. Plant foods and dairy products. *American Journal of Clinical Nutrition* 61(supplement):1,407S– 1,415S; Nestle, M. 1995. Mediterranean diets: historical and research overview. *American Journal of Clinical Nutrition* 61(supplement): 1,313S–1,320S.

42. That amount of nitrogen could be supplied just by a combination of atmospheric deposition and biofixation by leguminous crops.

43. White-Stevens, R. 1977. Perspectives on fertilizer use, residue utilization and food production. In C. R. Loehr, ed., *Food, Fertilizer and Agricultural Residues.* Ann Arbor: Ann Arbor Science, p. 10.

44. The history of China's fertilizer production and supplies is described in CIA. 1975. *People's Republic of China: Chemical Fertilizer Supplies, 1949–74.* Washington, D.C.: CIA; Chang C. 1977. Chemical fertilizer output on the Chinese Mainland. *Issues & Studies* 13: 38–53.

45. Yang, D. 1996. *Catastrophe and Reform in China.* Stanford: Stanford University Press.

46. Lysenko's enormous, and crippling, influence on plant breeding during the Stalinist era is detailed in Medvedev, Z. A. 1969. *The Rise and Fall of T. D. Lysenko.* New York: Columbia University Press.

47. Chang, G. H., and G. J. Wen. 1997. Communal dining and the Chinese famine of 1958–1961. *Economic Development and Cultural Change* 46:1–34.

48. The official explanation still blames the natural catastrophes for the suffering, but China's own statistics (areas of farmland affected by natural disasters have been listed every year in *China Statistical Yearbook*) belie it. In 1959 the area affected by drought and flood was about 10% less than in 1957, the year of the previous record harvest. The situation worsened in 1960 and 1961, but three decades later, in 1991 and 1994, the worst drought and floods in China's modern history had only a marginal effect on the food supply (the 1994 grain harvest was less than 3% below the 1993 record).

49. Ashton, B., et al. 1984. Famine in China, 1958–61. *Population and Development Review* 10:613–645; Peng X. 1987. Demographic consequences of the Great Leap Forward in China's provinces. *Population and Development Review* 13:639–670; Banister, J. 1987. *China's Changing Population.* Stanford: Stanford University Press.

50. This record gain will not be surpassed even by India's increases during the first half of the twenty-first century; see United Nations. 1998. *World Population Prospects: The 1998 Revision.* New York: United Nations, vol. I, Comprehensive Tables, pp. 222–223.

51. Smil, V. Food energy in the PRC. 1977. *Current Scene* 15:1–11; Smil, V. 1981. China's food: availability, requirements, composition, prospects. *Food Policy* 6:67–77.

52. CIA (44), pp. 6–7; Chang (44), pp. 50–52.

53. Based on consumption and production figures for calendar years in the FAO's database at http://apps.fao.org.

54. China's annual per capita meat consumption has more than doubled since 1979, the beginning of Deng Xiaoping's reforms. The FAO's food balance sheets (http://apps.fao.org)

show that nationwide the average meat supply almost tripled between 1979 and 1996 to more than 40 kg/capita, but Chinese consumption surveys indicate a less dramatic growth, with rural per capita consumption rising from just 6 kg/year in 1978 to about 13 kg/year by 1995; urban consumption in 1995 was about 24 kg/year; see State Statistical Bureau. 1996. *China Statistical Yearbook*. Beijing: SSB, pp. 283, 309. Even when the different accounting methods (retail supply calculated from food balance sheets vs. actual food consumption) are taken into consideration, the FAO's calculations appear to be unrealistically high.

55. During the mid-1950s Chinese agriculture was able to feed about 600 million people, half of the late 1990s total, without chemical fertilizers.

56. South Korea has recently been importing ²/₃ of its cereal, and nearly ⁹/₁₀ of its oil crop, consumption; Japanese imports have supplied nearly ³/₄ of all cereals, all but a few percent of oil crops, as well as ²/₅ of all fish.

57. This is based on the medium variant of the UN's latest population forecast; see United Nations (50).

58. This outcome is predicated on the assumption that by 2050 almost all countries will have a total fertility rate of 2.1, compared to the current global mean of 2.79. Although this would require steep fertility declines in a number of populous countries with high fertilities—most notably in Nigeria and Pakistan (whose current fertilities are 5.9 and 5.0, respectively)—such reductions would not be unprecedented. China cut its fertility from around six in the mid-1960s to less than two by the mid-1990s, and a number of smaller Asian countries accomplished their demographic transition even faster. On the other hand, it is easy to argue that similar performances are much less likely throughout sub-Saharan Africa or in parts of the Middle East. Consequently, the possibility that the global population will surpass 10 billion people by the year 2050 cannot be excluded—but even the UN's latest high forecast variant of 10.7 billion means that the world's population would not double during the first half of the twenty-first century.

59. The largest decline is expected in Russia: the UN's latest projections see the country's population falling from 148.1 million people in 1998 to 121.3 million (medium variant) or to as low as 102.5 million (low variant) by the year 2050; see United Nations (50), pp. 346–347.

60. Additional populations according to United Nations (50).

61. FAO. 1981. *Agriculture: Toward 2000*. Rome: FAO, p. 66.

62. Farmland lost to nonagricultural uses often includes the best alluvial soils in the vicinity of large cities. Continued population growth and expanding industrial production make a considerable amount of such losses inevitable. For example, during the early 1990s China has been losing annually close to 100,000 ha of arable land to urban, industrial, and transportation uses. Even when yielding less than the recent average, this land could produce annually at least 0.6 Mt of grain.

The declining quality of arable soils includes a number of processes, ranging from soil compaction to acidification, and from the loss of organic matter to salinization. The actual impact of these unwelcome changes is very difficult to quantify because their effects on crop yields are masked by adopting better agronomic practices, using new cultivars, and applying more fertilizers.

63. During the 1990s Egypt's imports amounted to more than ¹/₃ of its cereal consumption and to 90% of its cooking oil. The country's food balance sheets can be seen at http://apps.fao.org.

64. This includes all other populous Asian countries, as well land- and water-scarce Middle Eastern nations with high rates of population growth.

65. The energy or protein average calculated from food balance sheets must be considerably above the actual per capita requirements in order to compensate for food losses at retail and household level and for unequal access to food (both among and within families).

66. But dietary diversification also includes higher demand for, and a wider variety of, fruits. The FAO's food balance sheets show the global per capita fruit supply increasing by 1/3 between 1970 and 1996; see http://apps.fao.org.

67. The FAO's data on protein supply for countries and regions since 1961 are available at http://apps.fao.org.

68. There are three categories of publications summarizing crop response to nitrogen fertilization: reviews of nationwide trials, long-term field experiments, and short-term field tests. Two excellent examples of the first category are FAO. 1981. *Crop Production Levels and Fertilizer Use.* Rome: FAO; FAO. 1989. *Fertilizer and Food Production Summary Review of Trial and Demonstration Results 1961–1986.* Rome: FAO. Two notable examples of the second category are Schlegel, A. J., and J. L. Havlin. 1995. Corn response to long-term nitrogen and phosphorus fertilization. *Journal of Production Agriculture* 8:181–185; Johnston, A. E. 1997. The value of long-term field experiments in agricultural, ecological, and environmental research. *Advances in Agronomy* 59:291–333. Dozens of papers are published every year in the last category, reporting on the response of crops ranging from staple grains to less common vegetables.

69. Perhaps the most worrisome has been the yield decline in most long-term experiments with continuous, irrigated rice in the Philippines; a decrease in the effective nitrogen supply from the soil has been identified as the main reason at the IRRI Research Farm at Los Baños; see Cassman, K. G., et al. 1995. Yield decline and the nitrogen economy of long-term experiments on continuous, irrigated rice systems in the tropics. In R. Lal and B. A. Stewart, eds., *Soil Management: Experimental Basis for Sustainability and Environmental Quality.* Boca Raton, Fla.: Lewis/CRC Publishers, pp. 181–222.

70. Many recent volumes have examined the likely effects of large-scale environmental change on agriculture, including Rosenzweig, C., and D. Hillel. 1997. *Climate Change and the Global Harvest: Potential Impact of the Greenhouse Effect on Agriculture.* New York: Oxford University Press; Bazzaz, F., and W. Sombroek. 1996. *Global Climate Change and Agricultural Production.* New York: John Wiley; Kimble, J., et al., eds. 1995. *Soils and Global Change.* Boca Raton, Fla.: CRC Press.

71. Their food imports aside, during the mid-1990s low-income countries needed about 12 kg N/capita to produce their food; merely maintaining the current level of average supplies would then require an additional 44 Mt N (3.7 billion × 12 kg N) in synthetic fertilizers; an additional 20% increase to eliminate malnutrition and to improve the intake of animal protein would bring the increase to about 53 Mt N, double the mid-1990s amount.

Chapter 9

1. For estimates of natural, preindustrial nitrogen fixation see Cleveland, C. C., et al. 1999. Global patterns of terrestrial biological nitrogen (N_2) fixation in natural ecosystems. *Global Biogeochemical Cycles* 13:623–645. Jaffe, D. A. 1992. The nitrogen cycle. In S. S. Butcher

et al., eds., *Global Biogeochemical Cycles*. London: Academic Press, pp. 262–284; Smil, V. 1985. *Carbon Nitrogen Sulfur*. New York: Plenum Press, pp. 126–135.

2. Different estimates of global nitrogen fixation by lightning are reviewed in Liaw, Y. P., et al. 1990. Comparison of field, laboratory, and theoretical estimates of global nitrogen fixation by lightning. *Journal of Geophysical Research* 95:22,489–22,494.

3. This comparison is based on an estimate of about 20 Mt N from fossil fuel combustion and 30–35 Mt N from agricultural biofixation.

4. Revelle, R., and H. E. Suess. 1957. Carbon dioxide exchange between atmosphere and ocean and the question of an increase of atmospheric CO_2 during the past decades. *Tellus* 9: 18–27.

5. Houghton, J. T., et al., eds. 1996. *Climate Change 1995: The Science of Climate Change*. Cambridge: Cambridge University Press, pp. 14–17.

6. For detailed reviews of the biospheric sulfur cycle see Charlson, R. J., et al. 1992. Sulfur cycle. In S. S. Butcher et al., eds., *Global Biogeochemical Cycles*, London: Academic Press, pp. 185–300; Smil, V. 1997. *Cycles of Life*. New York: Scientific American Library, pp. 141–169.

7. For the history of our concern about acid deposition see Smil (1), pp. 10–12, 303–320; major publications summarizing some two decades of North American and European research are cited in note 71.

8. For extensive reviews of possible responses see Bruce, J. P., et al., eds. 1996. *Climate Change 1995: Economic and Social Dimensions of Climate Change*. New York: Cambridge University Press, 1996; Watts, R. G., ed. 1997. *Engineering Response to Global Climate Change*. Boca Raton, Fla.: CRC Press.

9. Anthropogenic sulfur emissions have recently reached a plateau of about 80 Mt S/year. Worldwide conversion to cleaner fuels (above all to natural gas), desulfurization of more than half of the coal-based electricity generating capacity in the United States, and post-1990 declines of industrial production in the former Soviet empire were the main causes of the subsequent decrease in emissions in North America, Europe, and Russia. In contrast, China, now the world's largest consumer of coal, has been experiencing rapid increases in SO_2 emissions and in the area affected by acid rain.

10. Stevenson, F. J. 1986. *Cycles of Soil*. New York: John Wiley, pp. 199–202.

11. Paul, E. A., and F. E. Clark. 1989. *Soil Microbiology and Biochemistry*. San Diego: Academic Press, pp. 134–139. Because the immobilization—the incorporation of NH_4^+ into amino acids—depends on microbial growth, the process is limited by the supply of suitable carbon substrate and by many environmental factors restricting microbial reproduction and development.

12. Stevenson, F. J. 1994. *Humus Chemistry: Genesis, Composition, Reactions*. New York: Wiley; Chen, Y., and Y. Avnimelech. 1986. *The Role of Organic Matter in Modern Agriculture*. Dordrecht: Martinus Nijhoff.

13. Paul and Clark (11), pp. 147–158. Denitrification is also a particularly effective way of removing that fraction of fertilizer nitrogen that reaches estuary and shelf waters; see Nixon, S. W., et al. 1996. The fate of nitrogen and phosphorus at the land-sea margin of the North Atlantic Ocean. *Biogeochemistry* 35:141–180. Dissolved nitrate in ground waters with low

oxygen content can also be denitrified in the presence of organic carbon compounds or reduced iron sulfide compounds by microorganisms using the oxygen bound in the nitrate; after a long residence time such water may emerge virtually nitrate-free; see Wendland, F., et al. 1994. Potential nitrate pollution of groundwater in Germany: a supraregional differentiated model. *Environmental Geology* 24:1–6.

In contrast, a higher oxygen supply will lower the denitrification rates. During the 1980s the nitrate load carried by the Seine downstream from Paris increased not only because of a higher runoff of agricultural nitrate but also because of a sharp reduction in denitrification caused by the restoration of higher oxygen levels in the river; see Chesterikoff, A., et al. 1992. Inorganic nitrogen dynamics in the River Seine downstream from Paris (France). *Biogeochemistry* 17:147–164.

14. Smil, V. 1999. Nitrogen in crop production: an account of global flows. *Global Biogeochemical Cycles* 13:647–662.

15. Granli, T., and O. C. Bøckman. 1994. Nitrous oxide from agriculture. *Norwegian Journal of Agricultural Science* Supplement 12:7–128; Mosier, A. R. 1994. Nitrous oxide emissions from agricultural soils. *Fertilizer Research* 37:191–200; Matthews, E. 1994. Nitrogenous fertilizers: global distribution of consumption and associated emissions of nitrous oxide and ammonia. *Global Biogeochemical Cycles* 8:411–439; Potter, C. S., et al. 1996. Process modeling of controls on nitrogen trace gas emissions from soils worldwide. *Journal of Geophysical Research* 101:1,361–1,377; Bouwman, A. F. 1996. Direct emission of nitrous oxide from agricultural soils. *Nutrient Cycling in Agroecosystems* 46:53–70.

N_2O from synthetic fertilizers accounts for less than 10% of all natural and anthropogenic emissions of the gas, and N_2O is responsible for less than 10% of the global greenhouse effect; consequently, even the most liberal estimate must ascribe less than 1% of the global warming effect to N_2O released from the denitrification of nitrogen fertilizers.

16. Paul and Clark (11), p. 139–145.

17. Potter et al. (15); Shepherd, M. F., et al. 1991. The production of atmospheric NO_x and N_2O from a fertilized agricultural soil. *Atmospheric Environment* 25A:1,961–1,969; Davidson, E. A., and W. Kingerlee. 1997. A global inventory of nitric oxide emissions from soils. *Nutrient Cycling in Agroecosystems* 48:37–50; Veldkamp, E., and M. Keller. 1997. Fertilizer-induced nitric oxide emissions from agricultural soils. *Nutrient Cycling in Agroecosystems* 48:69–77.

18. Hargrove, W. L. 1988. Soil, environmental, and management factors influencing ammonia volatilization under field conditions. In B. R. Bock and D. E. Kissel, eds., *Ammonia Volatilization from Urea Fertilizers*. Muscle Shoals, Ala.: National Fertilizer Developmenter Center, pp. 17–36; Terman, G. L. 1979. Volatilization losses of nitrogen as ammonia from surface-applied fertilizers, organic amendments, and crop residues. *Advances in Agronomy* 31:189–223.

19. Jayaweera, G. R., and D. S. Mikkelsen. 1991. Assessment of ammonia volatilization from flooded soil systems. *Advances in Agronomy* 45:303–356; Gould, W. D., et al. 1986. Urea transformations and fertilizer efficiency in soil. *Advances in Agronomy* 40:208–238. The principal factors promoting the process are shallow waters, their alkaline pH, high temperature and high NH_4^+-N concentration, and higher wind speeds; dominance of urea in rice farming makes matters worse because the pH of noncalcareous soils is temporarily elevated after its application.

20. To cite just one of many similar studies, measurements done during the late 1980s showed that urea-fertilized silage corn in Rhode Island lost as much as 79 kg of NO_3-N/ha, compared to just 1.5 kg N/ha from unfertilized home lawns and forests; see Gold, A. J., et al. 1990. Nitrate-nitrogen losses to ground water from rural and suburban land use. *Journal of Soil and Water Conservation* 45:305–310. The buffering effect is largely due to increased plant protein and to increased rates of volatilization and denitrification; see Johnson, G. V., and W. R. Raun. 1995. Nitrate leaching in continuous winter wheat: use of a soil-plant buffering concept to account for fertilizer nitrogen. *Journal of Production Agriculture* 8:486–491.

21. Prunty, L., and R. Greenland. 1997. Nitrate leaching using two potato-corn N-fertilizer plans on sandy soil. *Agriculture Ecosystems Environment* 65:1–13; Powlson, D. S., et al. 1989. Leaching of nitrate from soils receiving organic or inorganic fertilizers continuously for 135 years. In J. A. Hansen and K. Henriksen, eds., *Nitrogen in Organic Wastes Applied to Soils*. London: Academic Press, p. 337.

22. Pimentel, D., ed. 1993. *World Soil Erosion and Conservation*. New York: Cambridge University Press; Pimentel, D., et al. 1995. Environmental and economic costs of soil erosion and conservation benefits. *Science* 267:1,117–1,122; Agassi, M., ed. 1995. *Soil Erosion, Conservation and Rehabilitation*. New York: Dekker. The largest international study of its kind estimated that out of the total of about 750 Mha of continental surfaces that are moderately to excessively affected by water erosion, and the 280 Mha by wind erosion, the mismanagement of arable land was responsible for excessive erosion on some 180 Mha of cropland; see Oldeman, I. R., et al. 1990. *World Map of the Status of Human-Induced Soil Degradation: An Explanatory Note*. Nairobi: UNEP.

23. Larson, W. E., et al. 1983. The threat of soil erosion to long-term crop production. *Science* 219:458–465.

24. Liu, T. S. 1990. *Loess in China*. Berlin: Springer-Verlag; Hamilton, H., and S. Luk. 1993. Nitrogen transfers in a rapidly eroding agroecosystem: Loess Plateau, China. *Journal of Environmental Quality* 22:133–140.

25. Lee, L. K. 1990. The dynamics of declining soil erosion rates. *Journal Soil and Water Conservation* 45:622–624; Bloodworth, H., and J. L. Berc. 1998. Cropland acreage, soil erosion, and installation of conservation buffer strips: preliminary estimates of the 1997 National Resources Inventory. http://www.nhq.nrcs.usda.gov/land.

26. Singh, G., et al. 1992. Soil erosion rates in India. *Journal of Soil and Water Conservation* 47:97–99; Zhu Z., Q. Wen, and J. R. Freney, eds. 1997. *Nitrogen in Soils of China*. Dordrecht: Kluwer Academic.

27. Daigger, L. A., et al. 1976. Nitrogen content of winter wheat during growth and maturation. *Agronomy Journal* 68:815–818; Harper, L. A., et al. 1987. Nitrogen cycling in a wheat crop: soil, plant and aerial nitrogen transport. *Agronomy Journal* 79:965–973; Kanampiu, F. K., et al. 1997. Effect of nitrogen rate on plant nitrogen loss in winter wheat varieties. *Journal of Plant Nutrition* 20:389–404.

28. Francis, D. D., et al. 1993. Post-anthesis nitrogen loss from corn. *Agronomy Journal* 85: 659–663; Wetselaar, R., and G. D. Farquhar. 1980. Nitrogen losses from tops of plants. *Advances in Agronomy* 33:263–302. Given the magnitude of this neglected flux it is imperative that all N balance studies should consider this neglected variable before attributing any unaccounted losses to unknown factors or to higher rates of denitrification or leaching; see

Kanampiu, F. K., et al. 1997. Effect of nitrogen rate on plant nitrogen loss in winter wheat varieties. *Journal of Plant Nutrition* 20:389–404.

29. The normal atmospheric abundance of ^{15}N is 0.366%; for uptake experiments fertilizers are enriched with 5–10% ^{15}N. Another advantage of ^{15}N-labeling is that the amount of fertilizer-derived nitrogen retained in soil, and potentially assimilable in the future, can be also measured; this makes it possible to calculate the overall retention of the nutrient in the crop-soil system and to establish the true loss from the agroecosystem.

With recoveries as high as 75–80% and with another 10–15% remaining in soil the overall retention of the fertilizer nitrogen can be higher than 90%; see Powlson, D. S. 1989. Measuring and minimising losses of fertilizer nitrogen in arable agriculture. In D. S. Jenkinson and K. A. Smith, eds. *Nitrogen Efficiency in Agricultural Soils*. London: Elsevier, pp. 234–236.

30. Frissel, M. J., and G. J. Kolenbrander. 1978. The nutrient balances: summarizing graphs and tables. In M. J. Frissel ed., *Cycling of Mineral Nutrients in Agricultural Ecosystems*. Amsterdam: Elsevier, pp. 280–281.

31. Jenkinson and Smith (29).

32. Interestingly, the reported recovery efficiencies in Greek experiments with wheat rose with higher nitrogen applications, while those for barley and corn, predictably, fell as fertilization rates increased; see Simonis, A. D. 1989. Studies on nitrogen use efficiency of cereals. In Jenkinson and Smith (29), pp. 110–124.

33. Jenkinson, D. S. 1982. The nitrogen cycle in long-term experiments. *Philosophical Transaction of the Royal Society London* B 296:563–571; Johnston, A. E. 1997. The value of long-term field experiments in agricultural, ecological, and environmental research. *Advances in Agronomy* 59:291–333.

34. Craswell, E. T., and P. L. G. Vlek. 1979. Fate of fertilizer nitrogen applied to wetland rice. In *Nitrogen and Rice,* Los Baños, Philippines: International Rice Research Institute, pp. 175–191; International Network on Soil Fertility and Fertilizer Evaluation for Rice. 1987. *Efficiency of Nitrogen Fertilizers for Rice*. Los Baños, Philippines: International Rice Research Institute; Cassman, K. G. 1993. Nitrogen use efficiency in rice reconsidered: what are the key constraints? *Plant and Soil* 155/156:359–362; Doberman, A. 1998. *Summary Report on 1997 Results of a Project of the IRRI*. Los Baños, Philippines: International Rice Research Institute. The last study found that during on-farm experiments in China N recovery by a late rice crop averaged a mere 5%. A recent survey of China's fertilizer use put the mean nationwide N utilization efficiency at 35%: Xu X., et al. 1999. Fertilizers for the future. *Fertilizer International* 369:31–31.

35. Reddy, G. B., and K. R. Reddy. 1993. Fate of N-15 enriched ammonium nitrate applied to corn. *Soil Science Society of America Journal* 57:111–115.

36. This process is, of course, absolutely necessary in order to maintain nitrogen's global biospheric cycle.

37. Frissel and Kolenbrander (30), p. 283.

38. I used the following rates: North America (17 t/ha), Europe (15 t/ha), Asia (35 t/ha, except for rice paddies where erosion is generally negligible), Africa and Latin America (30 t/ha).

39. Stevenson (10), pp. 110–112.

40. Lindert, P. H., and Wu, W. L. 1996. Trends in the soil chemistry of North China since the 1930s. *Journal of Environmental Quality* 25:1,168–1,178; Stevenson (10), p. 111.

41. Terman (18); Jayaweera and Mikkelsen (19).

42. ApSimon, H. M., et al. 1987. Ammonia emissions and their role in acid deposition. *Atmospheric Environment* 21:1,939–1,946; Bouwman, A. F., et al. 1991. A global high-resolution emission inventory for ammonia. *Global Biogeochemical Cycles* 11:561–587.

43. Hargrove, W. L. 1988. Soil, environmental, and management factors influencing ammonia volatilization under field conditions. In B. R. Bock and D. E. Kissel, eds., *Ammonia Volatilization from Urea Fertilizers,* Muscle Shoals, Ala.: National Fertilizer Development Center, pp. 17–36; Hansen, J. A., and K. Henriksen, eds. 1989. *Nitrogen in Organic Wastes Applied to Soils.* London: Academic Press. Losses from volatilization of animal wastes applied to cropland must be added to totals of inorganic-N losses in all regions with intensive organic recycling.

44. Ray, C., and Uhl, V. W. 1993. Drinking water and nitrate issues in the United States and India. *Impact Assessment* 11:391–410; General Accounting Office. 1997. *Drinking Water: Information on the Quality of Water Found at Community Water Systems and Private Wells.* Washington, D.C.: GAO.

45. Commoner, B. 1971. *Closing Circle.* New York: Knopf; Commoner, B. 1975. Threats to the integrity of the nitrogen cycle: nitrogen compounds in soil, water, atmosphere and precipitation. In J. F. Singer, ed., *The Changing Global Environment.* Dordrecht: Reidel, pp. 341–366.

46. Commoner's most effective critic was Aldrich, S. R. 1980. *Nitrogen in Relation to Food, Environment, and Energy.* Urbana-Champaign: University of Illinois Press. For more on the nitrate controversy of the 1970s see Smil, V. 1985. *Carbon Nitrogen Sulfur.* New York: Plenum Press, pp. 207–211.

47. Puckett, L. J. 1995. Identifying the major sources of nutrient water pollution. *Environmental Science & Technology* 29:408A–414A.

48. Nolan, B. T., et al. 1997. Risk of nitrate in groundwaters of the United States—a national perspective. *Environmental Science & Technology* 31:2,229–2,236. The Centers for Disease Control found that in nine Midwestern states anywhere between 5.8% (in Minnesota) and 24.3% (in Kansas) of private wells had nitrate levels above the EPA's standard; see GAO (44).

49. Van der Voet, E., et al. 1996. Nitrogen pollution in the European Union: origins and proposed solutions. *Ambio* 23:120–132.

50. In 1972 Kolenbrander reviewed forty years of Dutch records and found that during that period, when average applications of nitrogen fertilizers rose by about 150 kg N/ha, nitrate levels level rose marginally only in ⅓ of the country's waterworks and remained unchanged in the rest of them; see Kolenbrander, J. 1972. Does leaching of fertilizers affect the quality of ground water at the waterworks? *Stickstoff* 15:8–15.

But extensive surveys of groundwaters had shown that by the early 1990s the WHO's limit of 50 mg NO_3/L was surpassed on 95% of Dutch farms, with mean levels well over 100 mg/L: Jongbloed, A. W., and C. H. Henkens. 1996. Environmental concerns of using animal manure: the Dutch case. In E. T. Kornegay, ed., *Nutrient Management of Food Animals to Enhance and Protect Environment.* Boca Raton, Fla.: Lewis Publishers, pp. 315–332.

Between 1975 and 1985 nitrate concentrations in the Meuse and the Ijssel rose by 15–25%, and similar, or much higher, increases were recorded in other European rivers; nitrate levels were up almost 40% in the Rhine, and 2.5 times higher in the Po; see Johnes, P. J., and T. P. Burt. 1993. Nitrate in surface waters. In *Nitrate: Processes, Patterns and Management*, T. P. Burt et al., eds., Chichester: John Wiley, p. 298.

51. This limit—50 mg NO$_3$/L (1.3 mg NO$_3$-N/L) or 11.3 ppm—is actually the WHO recommendation, which has been accepted by nearly all Western countries. In England 100 sources of drinking water exceeded the limit during the early 1980s, but 192 sources were above it in 1992; see Redman, M. 1993. Nitrogen fertilizers: an environmental and consumer perspective. In R. G. Lee, ed., *Nitric Acid-Based Fertilizers and the Environment*. Muscle Shoals, Ala.: IFDC, p. 167.

52. Peierls, B., et al. 1991. Human influence on river nitrogen. *Nature* 350:368–387; Cole, J. J., et al. 1993. Nitrogen loading of rivers as a human-driven process. In M. J. McDonnell and S. T. A. Pickett, eds., *Humans as Components of Ecosystems*. Berlin: Springer, pp. 141–157. The highest fertilizer loadings (in kg N/ha) for whole watersheds have all been in Europe (Meuse 81, Rhine 71, Rhone 41, Vistula 39, Thames 38), with the Mississippi averaging 15 and Yangzi 12; see Caraco, N. F., and J. J. Cole. 1999. Human impact on nitrate export: an analysis using major world rivers. *Ambio* 28:167–170.

53. Turner, R. E., and N. N. Rabalais. 1991. Changes in Mississippi River water quality this century. *BioScience* 41:140–147; Zhang, J., et al. 1995. Nationwide river chemistry trends in China: Huanghe and Changjiang. *Ambio* 24:275–279.

54. American Academy of Pediatrics. 1970. Methemoglobinemia: the role of dietary nitrate. *Pediatrics* 46:475–478. The rarity of the disease is attested by the fact that the latest edition of the standard U.S. pediatric text—Finberg, L., ed. 1998. *Saunders Manual of Pediatric Practice*. Philadelphia: W. B. Saunders—does not even mention methemoglobinemia in its 1,047 pages.

55. Practically all the cases reported in the United States and the United Kingdom since 1950 have been associated with water from wells where organic wastes, rather than fertilizers, were the most likely culprit. By far the largest reported outbreak during the past generation was in Hungary, where 1,353 cases were registered between 1976 and 1982; see Addiscott, T. M., et al. 1991. *Farming, Fertilizers and the Nitrate Problem*. Wallingford: CAB International, pp. 7–8.

56. Pretty, J. N., and G. R. Conway. 1990. The Blue-Baby Syndrome and Nitrogen Fertilisers: A High Risk in the Tropics? London: IIED, pp. 4–5.

57. Nitrites are also used directly to inhibit the growth of *Clostridium botulinum* in cured red meats (hams, sausages, salamis, bacon); after their antimicrobial activity was discovered during the 1920s they gradually began replacing the traditional use of nitrates, which was eventually banned in some countries (in Canada in 1975); see Cassens, R. G., et al. 1978. The use of nitrite in meat. *BioScience* 28:633–637.

58. Zaldivar, R. 1970. Geographic pathology of oral, esophageal, gastric and intestinal cancer in Chile. *Zeitschrift für Krebsforschung* 75:1–13. For reviews of the possible chronic health risks of nitrates see Deeb, B. S., and K. W. Sloan. 1975. *Nitrates, Nitrites, and Health*. Urbana: University of Illinois at Urbana-Champaign; National Academy of Sciences. 1978. *Nitrates: An Environmental Assessment*. Washington, D.C.: National Academy Press; National Academy of Sciences. 1981. *The Health Effects of Nitrate, Nitrite, and N-Nitroso Compounds*. Washington, D.C.: National Academy Press; Hill, M. 1991. *Nitrates and Nitrites in*

Food and Water. New York: Ellis Horwood; Canter, L. W. 1997. *Nitrates in Groundwater.* Boca Raton, Fla.: CRC Lewis.

There was also a short-lived U.S. scare, begun by a single epidemiological study from Australia, claiming a connection between the drinking of nitrate-contaminated water and birth defects. For its background and critique see Black, C. A. 1989. *Reducing American Exposure to Nitrate, Nitrite, and Nitroso Compounds: The National Network to Prevent Birth Defects Proposal.* Ames, Iowa: Council for Agricultural Science and Technology. For news of nitrate effects on human health see http://www.nitrate.com.

59. National Academy of Sciences. 1972. *Accumulation of Nitrate.* Washington, D.C.: National Academy Press; Corré, W. J., and T. Breimer. 1979. *Nitrate and Nitrite in Vegetables.* Wageningen: Centre for Agricultural Publishing and Documentation.

60. Forman, D., et al. 1985. Nitrates, nitrites and gastric cancer in Great Britain. *Nature* 313:620–625.

61. Forman, D. 1987. Gastric cancer, diet, and nitrate exposure. *British Medical Journal* 294: 528–529; Al-Dabbagh, S., et al. 1986. Mortality of nitrate fertiliser workers. *British Journal of Industrial Medicine* 43:507–515; National Research Council. 1982. *Diet, Nutrition, and Cancer.* Washington, D.C.: National Academy Press.

62. Eichholzer, M., and F. Gutzwiller. 1998. Dietary nitrates, nitrites, and N-nitroso compounds and cancer risk: a review of the epidemiological evidence. *Nutrition Reviews* 56:95–105.

63. Vitousek, P. M., and R. W. Howarth. 1991. Nitrogen limitation on land and in the sea: How can it occur? *Biogeochemistry* 13:87-115. Energetic constraints on the presence of diazotrophs and their limited activity due to the shortages of other nutrients (above all P, Mo, Fe) are the leading explanation.

64. Powlson, D. S., et al. 1991. *Farming, Fertilizers and the Nitrate Problem.* Wallingford: CAB International; Bronk, D. A., et al. 1994. Nitrogen uptake, dissolved organic nitrogen release, and new production. *Science* 265:1,843–1,846; Henriksen, A., and D. O. Hessen. 1997. Whole catchment studies of nitrogen cycling. *Ambio* 26:254–257.

65. Howarth, R. W. 1998. An assessment of human influences on fluxes of nitrogen from the terrestrial landscape to the estuaries and continental shelvews of the North Atlantic Ocean. *Nutrient Cycling in Agroecosystems* 52:213–223.

66. Uunk, E. J. B. 1991. *Eutrophication of Surface Waters and the Contribution of Agriculture.* Petersborough: Fertiliser Society; Harper, D. M. 1992. *Eutrophication of Freshwaters: Principles, Problems, and Restoration.* London: Chapman & Hall; Vollenweider, R. A., et al., eds. 1992. *Marine Coastal Eutrophication.* Amsterdam: Elsevier; National Research Council. 1993. *Managing Wastewater in Coastal Urban Areas.* Washington, D.C.: National Research Council; Valiela, I., et al. 1997. Nitrogen loading from coastal watersheds to receiving estuaries: new method and application. *Ecological Applications* 7:358–380.

Substantial differences in the origin of water-borne nitrogen are illustrated by two recently prepared nutrient budgets for two polluted watersheds. According to the Valiela et al. study, nitrogen in Waquoit Bay in Massachusetts comes largely from waste water (48%) and atmospheric deposition (30%), and only 15% of it originated in fertilizers. In contrast, in a Norwegian catchment dominated by farming, atmospheric deposition accounted for just 3% of total nitrogen inputs and nitrogen fertilizers for 70% (manures supplied the rest); see Hoyas, T. R., et al. 1997. Nitrogen budget in the Auli River catchment: a catchment dominated by agriculture, in Southeastern Norway. *Ambio* 26:289–295.

67. Nixon, S. W., et al. 1996. The fate of nitrogen and phosphorus at the land-sea margin of the North Atlantic Ocean. *Biogeochemistry* 35:141–180.

68. Turner and Rabalais (53); Turner, R. E., and N. N. Rabalais. 1994. Coastal eutrophication near the Mississippi river delta. *Nature* 368:619–621; Gupta, B. K. S., et al. 1996. Seasonal oxygen depletion in continental-shelf waters of Louisiana: historical record of benthic foraminifers. *Geology* 24:227–230. Analysis of data from 374 monitoring stations the Mississippi watershed has shown that the proximity of nitrogen soure to large streams is a major determinant of the downstream transport: Alexander, R. B., et al. 2000. Effect of stream channel size on the delivery of nitrogen to the Gulf of Mexico. *Nature* 403:758–761. For news of nitrate effects on ecosystems see also: http://www.nitrate.com.

69. Bell, P. R. F., and I. Elmetri. 1995. Ecological indicators of large-scale eutrophication in the Great Barrier Reef lagoon. *Ambio* 24:208–215.

70. Larsson, U., et al. 1985. Eutrophication and the Baltic Sea: causes and consequences. *Ambio* 14:9–14. More recent studies indicate a substantial disagreement concerning the net load of nitrogen from fields to the sea, with estimates for Sweden ranging from 219,000 to 570,000 t N; see Johnsson, H., and M. Hoffman. 1998. Nitrogen leaching from agricultural land in Sweden. *Ambio* 27:481–488. In any case, the eutrophication of the Baltic Sea would be even more serious if denitrifying bacteria in sediments and hypoxic deep waters did not remove about half of all nitrogen entering the sea.

71. By far the most extensive, and expensive, U.S. effort was the National Acid Precipitation Assessment Program (NAPAP) completed in 1990; its summary report is available in The U.S. National Acid Precipitation Assessment Program. 1991. *1990 Integrated Assessment Report*. Washington, D.C.: NAPAP Office of the Director. Among many other excellent European and North American reviews of the problem are Swedish Ministry of Agriculture. 1982. *Acidification Today and Tomorrow*. Stockholm: Swedish Ministry of Agriculture; National Research Council. 1986. *Acid Deposition: Long-term Trends*. Washington, D.C.: NAS; Longhurst, J. W., ed. 1991. *Acid Deposition: Origin, Impacts and Abatement Strategies*. Berlin: Springer-Verlag; Stockholm Environment Institute. 1991. *Acid Depositions in Europe: Environmental Effects, Control Strategies, and Policy Options*. Stockholm: Stockholm Environment Institute.

72. Freer-Smith, P. H. 1998. Do pollutant-related forest declines threaten the sustainability of forests? *Ambio* 27:123–131; Ciesla, W. M., and E. Donaubauer. 1994. *Decline and Dieback of Trees and Forests*. Rome: FAO; Kandler, O. 1993. The air pollution/forest decline connection: the "Waldsterben" theory refuted. *Unasylva* 44:(174):39–49.

73. National Atmospheric Deposition Program. 1999. Colorado State University, http://nadp.nrel.colostate.edu/NADP; Erisman, J. W. 1995. *Atmospheric Deposition in Relation to Acidification and Eutrophication*. Amsterdam: Elsevier.

74. Aber, J. D., et al. 1989. Nitrogen saturation in northern forest ecosystems. *BioScience* 39:378–386; Gundersen, P., et al. 1998. Impact of nitrogen deposition on nitrogen cycling in forests: a synthesis of NITREX data. *Forest Ecology and Management* 101:37–56; Aber, J., et al. 1998. Nitrogen saturation in temperate forest ecosystems. *BioScience* 48:921–934.

75. Take, for example, chronic nitrogen deposition in the Harvard Forest: pine and mixed hardwood stands, receiving 50 and 150 kg N/ha, showed extremely high nitrogen retention (85–99%) even after the sixth year of applications, with most of the retained nutrient (50–83%) going into long-term storage in soil; see Magill, A. H., et al. 1997. Biogeochemical response of forest ecosystems to simulated chronic nitrogen deposition. *Ecological Applications* 7:402–415.

76. Fenn, M. E., et al. 1998. Nitrogen excess in North American ecosystems: predisposing factors, ecosystem responses, and management strategies. *Ecological Applications* 8:706–733.

77. Rosén, K., et al. 1992. Nitrogen enrichment of Nordic forest ecosystems. *Ambio* 21:364–368.

78. Vitousek, P. M., et al. 1997. Human alteration of the global nitrogen cycle: sources and consequences. *Ecological Applications* 73:737–750; Tilman, D. 1996. Biodiversity: population versus ecosystem stability. *Ecology* 77:350–363; Rosén (77).

79. Smil (6), pp. 182–191; McGuire, A. D., and J. M. Melillo. 1995. The role of nitrogen in the response of forest net primary production to elevated atmospheric carbon dioxide. *Annual Review of Ecology and Systematics* 26:473–503; Schindler, D. W., and S. E. Bayley. 1993. The biosphere as an increasing sink for atmospheric carbon: estimates from increased nitrogen deposition. *Global Biogeochemical Cycles* 7:717–733.

80. The puzzle of the missing carbon is easily stated. On the average, only slightly less than half of all carbon released every year through fossil fuel combustion and land-use changes has remained airborne. The rest must be stored elsewhere, and the ocean and the biota are the only two possibilities. But we are not certain how much carbon these two reservoirs can, and actually do, absorb within a relatively short time and hence the puzzle: where does the missing carbon go? For much more on the puzzle see Smil (6), pp. 179–191.

81. Biomass combustion is the largest anthropogenic source of N_2O; additional contributions come from land-use changes and fossil fuel combustion.

Chapter 10

1. Needless to say, reconstructions of historic population totals are subject to large errors. The cited figures are taken from McEvedy, C., and R. Jones. 1978. *Atlas of World Population History*. London: Allen Lane, pp. 345–349.

2. In energy terms the global food harvest rose from about 6.1 EJ in 1900 to 42.5 EJ in 2000; because of the intervening extension of agricultural land the increase in average yields was slightly less than fourfold (from 7.2 GJ/ha to 28 GJ/ha in 2000). My global harvest calculations for the beginning of the twentieth century are based on incomplete historical statistics of crop production published before World War I by the Institut International d'Agriculture in the *Annuaire International de Statistique Agricole* in Rome.

3. What is no less remarkable is that the high-pressure, high-temperature Haber–Bosch synthesis of ammonia is also less energy costly than the inconspicuous work of nitrogen-fixing bacteria. Theoretical calculations based on nitrogenase-nitrogenase reductase reactions showed that the carbon cost of N_2 reduction is about 2.57 g C/g N, and studies of whole-plant energy costs of symbiotic N_2 fixation in nodulated legumes put the costs at between

0.3–20 g C/g N, with the most common rates between 5 and 6.8 g C/g N; see Phillips, D. A. 1980. Efficiency of symbiotic nitrogen fixation in legumes. *Annual Review of Plant Physiology* 31:29–49. This would be equivalent to 150–200 MJ/kg N, the rate higher than the energy cost of the earliest commercial processes and 5–7 times as much as the best current syntheses.

4. Cited in Mittasch, A. 1951. *Geschichte der Ammoniaksynthese.* Weinheim: Verlag Chemie, p. 180.

5. Average annual prices for various nitrogen compounds in most of the world's major fertilizer users are available, beginning in 1961, from the FAO's database at http://apps.fao.org.

6. Post-1950 expansion of European forests can be followed by comparing national totals of forested land in the FAO's *Forest Yearbook;* these figures beginning in 1961 are also available at http://apps.fao.org. The land-sparing effect of nitrogenous fertilizers has been immense; without them the United States would have to cultivate about twice as much land of a similar quality as it does today; see Waggoner, P. E., et al. 1996. Lightening the tread of population on the land: American examples. *Population and Development Review* 22:531–545. Even higher, and hence practically unattainable, multiples of cultivated land would be needed in such countries as China and Indonesia. And, of course, the extensions of cultivated land would be accompanied by a widespread loss of biodiversity; see Avery, D. T. 1997. Saving nature's legacy through better farming. *Issues in Science and Technology,* fall:59–64.

7. Lanyon, L. E. 1995. Does nitrogen cycle? Changes in the spatial dynamics of nitrogen with industrial nitrogen fixation. *Journal of Production Agriculture* 8:70–78.

8. Heavy metals—mostly copper, zinc, and cadmium—come from the compounds normally added to animal diets to prevent micronutrient deficiencies. Although cadmium levels are typically two orders of magnitude lower than those of copper and zinc, the element is not (unlike Cu and Zn) an essential plant micronutrient, it is highly toxic, and its accumulation in plant tissues is a clear health hazard. For this reason a number of countries with high concentrations of domestic animals have adopted limits on Cd in waste materials, as well as in inorganic fertilizers. For more on Cd in the environment see Nordberg, G. F., et al., eds. 1992. *Cadmium in the Human Environment: Toxicity and Carcinogenicity.* Lyon: International Agency for Research in Cancer.

9. Post-1975 gains in conversion efficiency have replaced more crude oil than all new energy sources combined. The improvements—whose pace has, unfortunately, slowed down after the collapse of high oil prices in 1985—have been most impressive in industrial processes and in transportation, but considerable opportunities for demand reduction remain throughout the energy system; see Rosenfeld, A., et al. 1997. Energy demand reduction. In R. G. Watts, ed., *Engineering Response to Global Climate Change.* Boca Raton, Fla.: Lewis Publishers, pp. 137–204.

10. FAO. 1980. *Maximizing the Efficiency of Fertilizer Use by Grain Crops.* Rome: FAO; FAO. 1983. *Maximizing Fertilizer Use Efficiency.* Rome: FAO; Hargrove, W. L., ed. 1988. *Cropping Strategies for Efficient Use of Water and Nitrogen.* Madison, Wis.: American Society of Agronomy; Munson, R. D., and C. F. Runge. 1990. *Improving Fertilizer and Chemical Efficiency Through "High Precision Farming."* Center for International Food and Agricultural Policy, St. Paul, Minn.: University of Minnesota; Fragoso, M. A., ed. 1993. *Optimization of Plant Nutrition.* Amsterdam: Kluwer; Dudal, R. and R. N. Roy, eds. 1995. *Integrated Plant Nutrition Systems.* Rome: FAO; Prasad, R., and J. F. Power. 1997. *Soil Fertility Management*

for Sustainable Agriculture. Boca Raton, Fla.: Lewis Publishing; Trenkel, M. A. 1997. *Improving Fertilizer Use Efficiency.* Paris: IFA.

11. Havlin, J. L., et al., eds. 1994. *Soil Testing: Prospects for Improving Nutrient Recommendations.* Madison, Wis.: Soil Science Society of America; MacKenzie, G. H., and J.-C. Taureau. 1997. *Recommendation Systems for Nitrogen—A Review.* York: Fertiliser Society. Periodic testing for major macronutrients has been common in high-income nations for decades, but testing for micronutrient deficiencies (ranging from boron and copper in many crops to molybdenum and cobalt needed by nitrogenase in leguminous species) has been much less frequent.

12. Cassman, K. G., et al. 1993. Nitrogen use efficiency of rice reconsidered: what are the key constraints? *Plant and Soil* 155/156:359–362.

13. To cite just one impressive recent example of combined nutrient trials, apparent fertilizer nitrogen recovery in the grain in a ten-year sequence of corn response to nitrogen and phosphorus fertilization was twice as high with phosphorus as without it; see Schlegel, A. J., and J. L. Havlin. 1995. Corn response to long-term nitrogen and phosphorus fertilization. *Journal of Production Agriculture* 8:181–185.

14. Inappropriate application ratios are common in many other low-income countries. For example, El-Fouly and Fawzi concluded that proper N:P:K ratios based on soil testing and plant analysis and adjusted to the prevailing cropping sequence could raise typical Egyptian yields by 20% without using more nitrogen; see El-Fouly, M. M., and A. F. A. Fawzi. 1996. Higher and better yields with less environmental pollution in Egypt through balanced fertilizer use. *Fertilizer Research* 43:1–4.

15. Applying fertilizer at the time of seedbed preparation and planting may be convenient, and, depending on many environmental and agronomic factors, these practices may be not wasteful. But, generally, the closer to the time of the peak requirement the fertilizer applied, the higher will be the rate of recovery. A recent study of nitrogen fertilization in irrigated spring wheat in Mexico demonstrated that lower and delayed applications reduced nitrogen loss without affecting yield and grain quality while saving 12–17% of after-tax profits; see Matson, P. A., et al. 1998. Integration of environmental, agronomic, and economic aspects of fertilizer management. *Science* 280:112–115.

16. Recent studies in highly erodible loess soil of China's Shaanxi Province confirm the advantages of subsurface application. Nitrogen fertilizer uptakes were just 18% for surface application to corn and 25% to wheat, but the rates rose to between 33 and 36% for subsurface placements; see Rees, R. M., et al. 1997. The effects of fertilizer placement on nitrogen uptake and yield of wheat and maize in Chinese loess soils. *Nutrient Cycling in Agroecosystems* 47: 81–91.

17. Hefner, S. G., and P. W. Tracy. 1995. Corn production using alternate furrow nitrogen fertilization and irrigation. *Journal of Production Agriculture* 8:66–69.

18. Sawyer, J. E. 1994. Concepts of variable rate technology with considerations for fertilizer application. *Journal of Production Agriculture* 7:195–201; Franzen, D. W., and T. R. Peck. 1995. Field soil sampling density for variable rate fertilization. *Journal of Production Agriculture* 8:568–674; Lu, Y., et al. 1997. The current state of precision farming. *Food Reviews International* 13(2):141–162; Wollring, J., et al. 1998. *Variable Nitrogen Application Based on Crop Sensing.* York: International Fertiliser Society.

19. Wibawa, W. D., et al. 1993. Variable fertilizer application based on yield goal, soil fertility, and soil map unit. *Journal of Production Agriculture* 6:255–261; Lowenberg-DeBoer, J., and S. M. Swinton. 1995. *Economics of Site-Specific Management in Agronomic Crops.* East Lansing: Department of Agricultural Economics, Michigan State University; Fenton, J. P. 1998. *On-Farm Experience of Precision Farming.* York: International Fertiliser Society.

20. Prasad, R., and J. F. Power. 1995. Nitrification inhibitors for agriculture, health, and the environment. *Advances in Agronomy* 54:233–281; Scharf, P. C., and M. M. Alley. 1995. Nitrogen loss inhibitors evaluated for humid-region wheat production. *Journal of Production Agriculture* 8:269–275; Sharma, S. N., and R. Prasad. 1996. Use of nitrification inhibitors (neem and DCD) to increase N efficiency in maize-wheat cropping system. *Fertilizer Research* 44:169–175; Trenkel, M. 1997. *Controlled-Release and Stabilized Fertilizers in Agriculture.* Paris: IFA. Nitrification inhibitors include some natural products, most notably those from *Azadirachta indica,* or the neem tree, widely grown in India, and a variety of synthetic chemicals ranging from ammonium polyphosphate to KCl.

Expensive slow- or controlled-release fertilizers—including such low-solubility compounds as urea formaldehyde or products coated with sulfur or polymers—have been used almost exclusively on lawns and gardens and account for only a fraction of 1% of total nitrogen applications. This should change with the wider introduction of less expensively produced nitrification inhibitors; see Thomaschewski, D. 1998. Coming soon: smart fertilizers. *Fertilizer International* 366:76.

21. These practices are reviewed in FAO. 1980. *Maximizing the Efficiency of Fertilizer Use by Grain Crops.* Rome: FAO; Wood, C. W., and J. H. Edwards. 1992. Agroecosystem management effects on soil carbon and nitrogen. *Agriculture, Ecosystems and Environment* 39: 123–138; and Dudal and Roy (10).

22. Council Directive 91/676/EEC of 12 December 1991 concerning the protection of waters against pollution caused by nitrates agricultural sources. *Official Journal* No. L 375, 31 December 1991, pp. 1–8. Nearly a decade later most of the EU countries are not in compliance with the nitrate directive.

23. The International Fertilizer Industry Association and the European Fertilizer Manufacturers' Association. 1990. *A Code of Best Agricultural Practices to Optimize Fertilizer Use.* Paris: IFA; van der Voet et al. 1996. Nitrogen pollution in the European Union: origins and proposed solutions. *Ambio* 23:120–132; Ignazi, I. C. 1996. Code of best agricultural practices. *Fertilizer Research* 43:241; Archer, J. R., and M. J. Marks. 1997. *Control of Nutrient Losses to Water from Agriculture in Europe.* York: Fertiliser Society.

24. Data on fertilization and yields of English wheat are from Gooding, M. J., and W. P. Davies. 1997. *Wheat Production and Utilization.* Walsingham: CAB International, pp. 44–45; and from http://apps.fao.org. Data on Japanese rice are from FAO. 1981. *Crop Production Levels and Fertilizer Use.* Rome: FAO, p. 59; and from http://apps.fao.org.

25. Calculated from average corn yield and nitrogen application data in Runge, C. F., et al. 1990. *Agricultural Competitiveness, Farm Fertilizer and Chemical Use, and Environmental Quality: A Descriptive Analysis.* St. Paul, Minn.: Center for International Food and Agricultural Policy, tables I-A and IV-A; U.S. Department of Agriculture. 1999. *Agricultural Yearbook.* Washington, D.C.: USDA, tables 1–37 and 14–3. Also available at http://www. usda.gov/nass/pubs.

26. These trends can be seen particularly well in U.S. statistics by contrasting the number of mixed (crops and livestock) family farms (steadily decreasing) with the number of corporate farms, size of an average operation, and spatial concentration of specialized cropping and animal production (all increasing); see USDA (25).

27. Rising trade in feedstuffs, plant oils, and meat explains most of this change. Increasing demand for out-of-season produce in affluent countries, and the eagerness of low-income countries to participate in the globalization of the economy, are additional major reasons for growing long-distance exports of food. With modern transportation low-income countries can now engage in intercontinental exports of previously perishable fruits and vegetables.

28. The history of changing fertilities can be traced in United Nations. *Demographic Yearbook.* New York: United Nations.

29. United Nations. 1991. *World Population Prospects 1990.* New York: United Nations.

30. United Nations. 1995. *World Population Prospects: The 1994 Revision.* New York: United Nations; United Nations. 1996. *World Population Prospects: The 1996 Revision.* New York: United Nations.

31. United Nations. 1998. *World Population Prospects: The 1998 Revision.* New York: United Nations.

32. Although we can be confident about the general trend, we can specify neither the time when the global population might stabilize nor the eventual peak count. Looking even further, we do not know if the global population peak will be followed by a prolonged period of stability or by a decline to a new, lower plateau.

33. Between the early 1950s and the early 1990s the average total fertility per woman in Africa declined from 6.6 to 5.7, while in Asia it fell from 5.9 to 2.8 and in Latin America from 5.9 to 2.9; see United Nations (28), vol. I, pp. 338 and 388.

34. Thai fertility dropped from 6.4 in the late 1950s to 1.9 by 1995, and South Korean fertility declined from 6.3 to 1.7 during the same period; see United Nations (28), annex I, pp. 120–121.

35. Lutz, W., W. Sanderson, and S. Scherbov. 1997. Doubling of world population unlikely. *Nature* 387:803–805.

36. Comparison of current food supply patterns makes that quite clear. In spite of a great deal of dietary convergence throughout the modern world, there are substantial differences in the frequency of consumption and in total intake of major foodstuffs even in countries belonging to the same cultural realm and enjoying comparable levels of disposable income.

37. These mean supply rates are according to aggregate food balance sheets prepared by the FAO; see http://apps.fao.org. Actual dietary intakes are considerably lower.

38. Given the current state of extensive overfishing of large parts of the world's ocean, no other large country could replicate the Japanese consumption of ocean fish, and only further expansion of aquaculture could satisfy the demand.

39. That would be of course, an undesirable state of affairs, as it would perpetuate the current overconsumption throughout the affluent world and extend the unacceptable level of malnutrition found in today's poor countries to another 3 billion people.

40. This assumption is used merely to calculate the highest conceivable consumption value as such levels of protein intake are not at all necessary.

41. Repeated failures in forecasting span a broad range of physical and social variables ranging from exhaustion dates for mineral resources to growth rates of GDP.

42. This must be done with long payback horizons in mind. In free-market societies only a long-term commitment to health and science education and to the promotion of effective public policies can bring desirable dietary transitions.

43. Contrary to a common belief, lactose intolerance, present in some 70% of the world's population, is not a major obstacle to widespread consumption of dairy products, because the disaccharide is virtually absent in fully ripened cheeses, and it is reduced to much more tolerable levels in yogurt; see Suarez, F. L., and D. A. Savaiano. 1997. Diet, genetics, and lactose intolerance. *Food Technology* 51(3):74–76.

44. Nestle, M. 1995. Mediterranean diets: historical and research overview. *American Journal of Clinical Nutrition* 61 (supplement):1,313S–1,320S; Kushi, L. H., et al. 1995. Health implications of Mediterranean diets in light of contemporary knowledge. 1. Plant foods and dairy products. *American Journal of Clinical Nutrition* 61 (supplement):1,407S–1,415S; Trichopoulou, A., et al. 1993. The traditional Greek diet. *European Journal of Clinical Nutrition* 47 (supplement):S76–S81; Keys, A., and M. Keys. 1975. *How to Eat Well and Stay Well the Mediterranean Way*. New York: Doubleday & Co..

45. O'Brien, P. 1995. Dietary shifts and implications for U.S. agriculture. *American Journal of Clinical Nutrition* 61 (supplement):1,390S–1,396S.

46. Zhang, M., et al. 1998. Assessing groundwater nitrate contamination for resource and landscape management. *Ambio* 27:170–174.

47. Flegal, K. M. 1996. Trends in body weight and overweight in the U.S. population. *Nutrition Review* 54:S97–S100. The situation is even worse when looking at actual weights associated with the lowest mortality. For U.S. adults these values from life insurance statistics are close to the 25th percentile weights for height, putting 75% of all adults above the optimal body mass. While the actual median body weight of adult males is 79 kg, the weight associated with the lowest mortality for the median height is only around 70 kg; for females the difference is smaller, 62 vs. 59 kg.

48. Popkin, B. M., and C. M. Doak. 1998. The obesity epidemic is a worldwide phenomenon. *Nutrition Reviews* 56:106–114.

49. Cui, L. 1995. Third national nutrition survey. *Beijing Review* 38(5):31.

50. Very few jobs created in modern economy require anything but light physical exertion, and occasional short periods of exercise done by people with generally sedentary lifestyles do not add up to the energy requirements that prevailed in traditional rural societies.

51. Cassell, D. K. 1994. *Encyclopedia of Obesity and Eating Disorders*. New York: Facts on File; Belfiore, F., et al., eds. 1991. *Obesity: Basic Concepts and Clinical Aspects*. Farmington, Conn.: S. Karger; Garrow, J. S. 1988. *Obesity and Related Disorders*. New York: Churchill.

52. A much more optimistic outlook prevailed during the 1970s. Near the end of his Nobel Prize acceptance speech Norman Borlaug spoke about a dream of

green, vigorous, high-yielding fields of wheat, rice, maize, sorghum and millet which are obtaining, free of expense, 100 kilograms of nitrogen per hectare from nodule-forming, nitrogen-fixing bacteria. These mutant strains of *Rhizobium cerealis* were developed in 1990 by a massive mutation breeding program with strains of *Rhizobium* obtained from roots of legumes and other nodule-bearing plants. This scientific discovery has revolutionized agricultural production for the hundreds of millions of humble farmers throughout the world, for they now receive much of the needed fertilizer for their crops directly from these little wondrous microbes.

Borlaug, N. 1970. The Green Revolution: peace and humanity. A speech on the occasion of the awarding of the 1970 Nobel Peace Prize in Oslo, Norway, December 11, 1970; text at http://www.theatlantic.com/atlantic/issues/97jan/borlaug/speech. For other optimistic predictions see Child, J. J. 1976. New developments in nitrogen fixation research. *BioScience* 26: 614-617; Postgate, J. R. 1977. Consequences of the transfer of nitrogen fixation genes to new hosts. *Ambio* 6:178–180.

53. Freiberg, C., et al. 1997. Molecular basis of symbiosis between *Rhizobium* and legumes. *Nature* 387:394–401.

54. Dilworth, M. J. 1987. Where to now? *Philosophical Transactions of the Royal Society London* B 317:279.

55. Anonymous. 1999. Alga gene boosts yield crops. *Global Issues in Agricultural Research* 1(4):7–8.

56. Nishibayashi, Y., et al. 1998. Bimetallic system for nitrogen fixation: ruthenium-assisted protonation of coordinated N_2 on tungsten with H_2. *Science* 279:540–542; Marnellos, G., and M. Stoukides. 1998. Ammonia synthesis at atmospheric pressure. *Science* 282:98–100.

57. Leigh, G. J. 1998. Fixing nitrogen any which way. *Science* 279:506–507.

58. Johnston, A. E., and P. R. Poulston. 1977. Yields on the Exhaustion Land and changes in the NPK content of the soils due to cropping and manuring. *Reports of the Rothamsted Experimental Station 1976* Part 2:53–86.

59. In some thin and highly erodible soils such a degradation may be a matter of years, not even decades.

60. Needless to say, the average quality and diversity of diets produced by such retrograde agroecosystems would be also much below the current standard.

61. For details see chapter 7 and appendix N.

62. As shown in appendix Q, I estimated the mid-1990s rate of agricultural biofixation at 33 (25–41) Mt N/year. About 23 Mt N (70%) are managed inputs, coming from planting pulses and leguminous cover crops and from *Azolla;* the rest is fixed by nonsymbiotic and endophytic diazotrophs.

63. Smil, V. 1997. Some unorthodox perspectives on agricultural biodiversity: the case of legume cultivation, *Agricultural Ecosystems and the Environment* 62:135–144.

64. This is in spite of a great diversity of soybean-based foodstuffs and their undoubted nutritional benefits; see Wang, H. L., et al. 1979. *Soybeans as Human Food: Unprocessed and Simply Processed.* Washington, D.C.: USDA.

65. Heichel, G. H. 1987. Legume nitrogen: symbiotic fixation and recovery by subsequent crops. In Z. Helsel, ed., *Energy in Plant Nutrition and Pest Control.* New York: Elsevier, pp. 62–80.

66. In all double-cropping areas the replacement of green manure by a winter grain or oil crop will double, or nearly double, that land's carrying capacity, while inorganic fertilizers supply the necessary nitrogen and other nutrients.

67. Of course, as with any long-range agricultural prediction, bioengineered solutions and transgenic cultivars have a great potential to make today's opinions and forecasts embarrassingly incorrect.

68. Cassman, K. G. 1999. Ecological intensification of cereal production systems: yield potential, soil quality, and precision farming. *Proceedings of the National Academy of Sciences of the USA* 96:5,952–5,959. Effective extension services disseminating the information about the best methods of farming are the key to such improvements, which often require little, or no, capital investment.

69. Galloway, J. N., et al. 1998. Asian change in the context of global change: an overview. In J. N. Galloway and J. Melillo, eds., *Asian Change in the Context of Global Climate Change*. New York: Cambridge University Press, pp. 1–17.

70. Floate, M. J. S. 1978. Changes in soil pools. In M. J. Frissel, ed., *Cycling of Mineral Nutrients in Agricultural Ecosystems*. Amsterdam: Elsevier, pp. 292–295. Organic farming can produce the same results. Comparison of the long-term nitrogen balance for the Rothamsted Continuous Wheat Experiment show that the plot receiving 35 t/ha of farmyard manure more than doubled its total N soil content in 115 years, gaining the nutrient at an annual rate of about 33 kg N/ha; see Jenkinson, D. S. 1982. The nitrogen cycle in long-term field experiments. *Philosophical Transactions of the Royal Society London* B 296:563–571.

71. Van Dijk, T. A., 1993. Appropriate management practices to minimize the environmental impact of applied fertilizers. In R. G. Lee, ed., *Nitric Acid-Based Fertilizers and the Environment*. Muscle Shoals, Ala.: IFDC, pp. 349–358.

72. Larson, B. A., and G. B. Frisvold. 1996. Fertilizers to support agricultural development in sub-Saharan Africa: what is needed and why. *Food Policy* 21:509–525; Lal, R. 1995. Erosion-crop productivity relationships for soils in Africa. *Soil Science Society of America Journal* 59:661–667.

73. In 1997 both the Ecological Society of America and the Scientific Committee on Problems of the Environment stated that the "preeminent problem" of nitrogen pollution is not given enough public recognition; see Moffat, A. S. 1998. Global nitrogen overload problem grows critical. *Science* 279:988.

74. At that time Barry Commoner concluded that "oxidized forms of nitrogen, which are in nature maintained at low, steady-state concentrations in air, water, plants and animals have been elevated to levels which threaten the integrity of major processes in the ecosystem, and vital biological processes in animal and man." Commoner, B. 1975. Threats to the integrity of the nitrogen cycle: nitrogen compounds in soil, water, atmosphere and precipitation. In J. F. Singer, ed., *The Changing Global Environment*, Dordrecht: Reidel, p. 362. Clearly, the concerns and the claims of the 1990s are hardly new or original.

75. Smil, V. 1993. *Global Ecology: Environmental Change and Social Flexibility*. London: Routledge; Smil, V. 1989. Our changing environment. *Current History* 88(534):9–12, 46–48.

76. For a wide-ranging, interdisciplinary analysis of this inevitable transition see Watts (9).

77. Ekstrand, A. G. 1996. Chemistry 1918. In *Nobel Lectures: Chemistry 1901–1921*. Amsterdam: Elsevier, p. 325.

Postscript

1. Unless otherwise noted, biographical details in this postscript are drawn from books by Holdermann, Stoltzenberg, and Szöllösi-Janze; see Holdermann, K. 1954. *Im Banne der Chemie: Carl Bosch—Leben und Werk*. Düsseldorf: Econ-Verlag; Stoltzenberg, D. 1994. *Fritz Haber: Chemiker, Nobelpreisträger, Deutscher, Jude*. Weinheim: VCH; Szöllösi-Janze, M. 1998. *Fritz Haber, 1868–1934: Eine Biographie*. Munich: Verlag C. H. Beck.

2. Heine, J. U. 1990. *Verstand & Schicksal*. Weinheim: VCH, p. 15.

3. For a complete list of Bosch's honors see Holdermann (1), pp. 316–319.

4. This was not true about his successor; in March 1932 a representative of DuPont in Germany wrote that "there seems to be no doubt whatever that at least Dr. Schmitz is personally a large contributor to the Nazi Party"; see Dubois, J. E., Jr. 1953. *Generals in Grey Suits*. London: Bodley Head, p. 81.

5. Not surprisingly, Bosch wrote to Haber immediately after the imposition of the new law forbidding the employment of "non-Aryan" personnel in the government bureaucracy (for more on this see later in this postscript), offering his help. He also made sure, in contravention of the new regime's wishes, that the lecture in memory of Fritz Haber—organized by Max Planck and Max von Laue and sponsored by the Kaiser-Wilhelm-Gesellschaft a year after Haber's death, on January 21, 1935—was given to a capacity audience; he asked the BASF chemists to accompany him to Berlin, and he sent telegrams to all directors of I. G. Farben urging them to attend.

6. Holdermann (1), p. 307.

7. In 1948 I. G. Farben's top executives were indicted in Nuremberg by Brigadier General Telford Taylor "of major responsibility for visiting upon mankind the most searing and catastrophic war in modern history." Taylor accused them of "wholesale enslavement, plunder and murder. . . . These men were the master builders of the Wehrmacht. They knew (and very few others knew) every detail of the intricate, enormous engine of warfare, and they watched its growth with the pride of architects. These are the men who made the war possible and they did it because they wanted to conquer"; see Dubois (4), pp. 74–75. I. G. Farben's most infamous wartime contribution was the production of Zyclon B, the lethal gas used in the Auschwitz death chambers, in its Leverkusen plant.

8. Hayes, P. 1987. Carl Bosch and Carl Krauch: chemistry and the political economy of Germany, 1925–1945. *Journal of Economic History* 47:353–363. Even with the state guarantees it appears that the company never recovered the R&D cost of coal hydrogenation by the time the Third Reich collapsed in 1945.

9. Ibid., p. 361. Moreover, "human beings were dissipated so unmercifully that *after almost four years the construction was still unfinished*. Although methanol and gasoline were produced in quantity after 1942, buna rubber was not. At a human cost of 200,000 lives—plus a quarter-billion Reichsmarks—*not one pound of rubber was ever produced at I. G. Auschwitz!*"; see Dubois (4), p. 341.

10. In October 1912 Haber informed Koppel that the original estimate of the capital cost (700,000 marks) was exceeded by 230,000 marks, and he asked for additional reserve funds of 120,000 marks. During the opening ceremony on October 23, 1912, Koppel, "in deep thanks for your majesty's presence at today's celebration," donated another 300,000 marks, and, according to a previous arrangement, the Kaiser repeatedly singled him out for special thanks and praise; see Johnson, J. A. 1990. *The Kaiser's Chemists: Science and Modernization in Imperial Germany.* Chapel Hill: University of North Carolina Press, p. 139.

11. In 1923 Haber told to an investigative committee of the Reichstag:

I have been never preoccupied with the legal permissibility of gas weapons. . . . These matters were obviously examined personally by the Chief of the General Staff and the War Minister von Falkenhayn. Although he never asked me about my legal interpretations, he left me in no doubt that for him there were international legal limits. . . . He would have never sanctioned poisoning of foodstuffs or water wells.

Haber, F. 1923. Zur Geschichte des Gaskrieges. In F. Haber, *Fünf Vorträge aus den Jahren 1920–1923.* Berlin: Julius Springer, p. 76.

12. In his testimony to a Reichstag committee (see note 11, p. 77) Haber characterized the Ypres attack as "an indisputable military success." He believed that concentrated, large-scale chlorine attacks in 1915 could have won the war for Germany by breaking the morale of the Allied soldiers. Phosgene, chlorpicrin, yperit, and mustard gas were major later additions to gas warfare waged by both Germans and the Allies. In 1986 Haber's son Ludwig published a detailed history of World War I gas warfare; see Haber, L. F. 1986. *The Poisonous Cloud: Chemical Warfare in the First World War.* Oxford: Clarendon Press. For a broader examination of chemical weapons in history see Price, R. M. 1997. *The Chemical Weapons Taboo.* Ithaca, N.Y.: Cornell University Press.

13. In Goran, M. 1947. The present-day significance of Fritz Haber. *American Scientist* 35: 71–72, the link is made explicit. He claims that Clara

began to regard poison gas not only as a perversion of science but also as a sign of barbarism. . . . Clara Haber pleaded with her husband to forsake poison gas. . . . She brought forth all the sentiments and feelings with which women stir men. Finally she quit protesting and demanded that Fritz Haber have no part in the nefarious business. . . . Stubbornly, Haber overruled his wife's every suggestion. The argument was serious and no gentle reconciliation followed. . . . That evening Clara Haber committed suicide.

This is a plausible but speculative account; not surprisingly, neither Fritz nor Hermann Haber left behind any detailed recollections of that fateful day.

14. An extensive citation from the letter is in Stoltzenberg (1), pp. 352–353.

15. Fritz Stern, Haber's godson, had this to say on this matter in his notes to a speech he delivered during the celebration of the seventy-fifth anniversary of the Fritz Haber Wilhelm Institute for Physical Chemistry and Electrochemistry in 1986:

Haber acted at a time of chauvinistic psychosis, when academics and clergy poisoned the minds of people. It was a time when service to one's country was a universal imperative, and most scientists had not yet learned—or had not even been warned—that theirs was a moral responsibility as well. Of course, it is with sorrow that one recalls that Haber apparently never acknowledged to himself the ambiguity of his wartime work—or at least he never publicly acknowledged any misgiving, a fact that his son, not implausibly, suggests may have sprung from a reluctance to seem to be groveling.

Stern, F. 1987. Fritz Haber: the scientist in power and in exile. In *Dreams and Delusions*. New York: Knopf, p. 65. I do not find any of these explanations convincing.

16. Haber, F. (11), p. 79. This conclusion was based on a report made by the Surgeon General of the U.S. Army in 1920. Of the 275,000 American casualties, more than $1/4$ were caused by poisonous gas, but only about 2% of gassed soldiers died, compared to 25% of those who suffered injuries by other weapons. As Colonel Alden Waitt of the U.S. Army's Chemical Warfare Service concluded, "the man wounded by gas had about twelve times the chance to live, in comparison with his fellow soldier suffering from the effects of traditional weapon. Those who have opposed gas weapons on the basis of inhumanity have long since been halted by the facts." Waitt, A. H. 1942. *Gas Warfare*. New York: Duell, Sloan and Pearce, p. 5.

In stark contrast to this reassuring statistical appraisal one can point to thousands of horrific personal experiences akin to those of a British officer talking to Sir Harold Hartley, Haber's British counterpart during the World War I, long after the ordeal: "It is a hateful and terrible sensation to be *choked* and suffocated and unable to get breath: a casualty from gun fire may be dying from his wounds, but they don't give him the sensation that his life is being strangled out of him." Quoted in L. F. Haber (12), p. 292.

As far as any connection with future horrific use of gas is concerned, I can do no better than to quote Fritz Stern (15, pp. 64–65):

In 1914, Haber could not envision that thirty years later, the Germans for whom he produced poison gas would use another kind of gas for the killing of millions of his own people, not in warfare, but because of an infinitely charged racial fanaticism. It is true, of course, that once the National Socialists had resolved to exterminate all Jews, they would have found the technical means to do so. Perhaps it deserves mention as well that Hitler claimed his sense of mission as savior of Germany first came to him when he lay blinded from Allied gas in a wartime hospital; some scholars have given credence to this claim. Still, it would be morally and historically senseless to seek connections where none existed.

17. Haber, C. 1970. *Mein Leben mit Fritz Haber*. Düsseldorf: Econ Verlag.

18. See Goran (13), p. 402. Heinrich Scheüch's letter made Haber's role in prolonging the war quite explicit:

During the long duration of the war you put your broad knowledge and your energy in the service of the fatherland—beyond all measure. Thanks to the high esteem which you enjoy among your colleagues, you were able to mobilize German chemistry. It was not given to Germany to emerge victorious from this war. That it did not succumb to the supremacy of its enemies after the first few months because of lack of powder, explosives and other chemical combinations of nitrogen, is in the first instance your achievement. . . . Your splendid successes . . . will always live on in history and will remain unforgotten.

Letter of November 27, 1918, in Max Planck Archiv quoted in Stern (15), p. 67.

19. Many publications have examined the effect of the Versailles peace treaty and harsh reparations on Germany's post-World War I history. For U.S. and German perspectives see Birdsall, P. 1941. *Versailles Twenty Years After*. New York: Regnal & Hitchcock; Beumelburg, W. 1931. *Deutschland in Ketten*. Oldenburg: Stalling.

20. Haber was a leading promoter of economic, scientific, and cultural cooperation between Japan and Germany. Undoubtedly, Haber's family history played a part: his paternal uncle Ludwig was the first Prussian consul at Hakodate in Hokkaido, where he was killed by a samurai in 1874. For Haber's thoughts on Japan, see "Japanische Eindrücke" (1924), "Wirtschaftlicher Zusammenhang zwischen Deutschland und Japan" (1925), and "Ansprache bei der Eröffnung des Japaninstituts" (1926), all in Haber, F. 1927. *Aus Leben und Beruf.* Berlin: J. Springer, pp. 52–66, 70–95, 148–157.

21. Haber, F. 1927. Gold in Meerwasser. *Zeitschrift für angewandte Chemie* 40:303–314.

22. Haber, C. (17), p. 282.

23. But they were not completely estranged: a year later Charlotte accompanied Haber on a Nile cruise to celebrate his sixtieth birthday. Charlotte died in Basel in 1978. Both Eva and Ludwig became British citizens. Besides *Poisonous Cloud* (12), Ludwig, an economic historian at the University of Surrey, also published the following: Haber, L. F. 1958. *The Chemical Industry during the Nineteenth Century.* Oxford: Clarendon Press; Haber, L. F. 1971. *The Chemical Industry, 1900–1930: International Growth and Technological Change.* Oxford: Clarendon Press.

24. Hermann Haber committed suicide in November 1946, as later did Clare, the eldest of his three daughters.

25. Planck, M. 1947. Mein Besuch bei Adolf Hitler. *Physikalische Blätter* 3(1947):143.

26. Quoted in Stern (15), p. 74.

27. The theory in question was, of course, Haber's lifelong belief in a Germany where Jews could become truly equal. Einstein's letter is quoted in Stern (15), p. 73. For more on Jewish scientists and Germany, see Stern, F. 1999. *Einstein's German World.* Princeton: Princeton University Press.

28. The letter is reproduced in several publications, including Stoltzenberg (1), p. 609.

29. Stern (15), p. 65, points to some remarkable similarities between Haber and Robert Oppenheimer—the scientific leader of the Manhattan Project that built the first nuclear weapons—including a poetic strain and tragic private lives.

30. Haber, F. 1966. The synthesis of ammonia from its elements. In *Nobel Lectures: Chemistry 1901–1921.* Amsterdam: Elsevier, p. 339.

Name Index

Subject Index